CRISIS
OR CHANGE

Nils Hybel

CRISIS
OR CHANGE

The Concept of Crisis
in the Light of Agrarian Structural Reorganization
in Late Medieval England

AARHUS UNIVERSITY PRESS

Copyright: Aarhus University Press, 1989
Translation and word-processing by James Manley
Printed in Denmark by Special-Trykkeriet Viborg a-s
ISBN 87 7288 229 8

The cover shows a gold noble of Edward III,
from E. A. Gardiner, *A Student's History of England*, 1904.

Denne afhandling er af det Humanistiske Fakultet
ved Aarhus Universitet antaget til forsvar for den
filosofiske doktorgrad.

Aarhus, 9. maj 1988
Ole Høiris
decanus

Foreword

A long-standing interest in the dissolution of feudalism and the establishment of a capitalist economy in western Europe formed the starting-point for this treatise on the use of the concept of crisis in English late medieval agrarian history. The book has its origins in a time when the concept of crisis has been urgently topical and hotly disputed. There has thus been a connection between the book's historical subject and the circumstances of its genesis, in a decade when the author has been palpably aware of the financial tribulations of the younger generation of historians. Thus the situation in which the book took form has underscored the importance of the support I have received from many quarters.

The Danish Research Council for the Humanities has provided research grants in connection with the Department of History at the University of Aarhus, where I was afforded the best conceivable working conditions. The Council further bore the cost of translation into English and publication. I would therefore like to extend my heartfelt thanks here to both institutions.

Several individuals have contributed in various ways to the genesis of the book. First and foremost, Professor Erik Ulsig, *dr. phil*, has read through the manuscript and made innumerable suggestions for improvements. Valuable advice and support have come from Professor Niels Steensgaard, *dr. phil.*, and Lecturer Michael Linton, *fil. dr.*, and Dr. Edward Miller and Dr. Anthony Bridbury have given me the benefit of their inspiration. Lecturer Povl A. Hansen, *fil. kand.*, has, by patiently listening to me in the initial stages of the work, helped me to come to grips with the material. In the final stage, James Manley has been a flexible and responsive translator. Finally, my wife Annedorte Hybel, to whom I dedicate this book, has given me constant support, not least by reading proofs.

Nils Hybel   Copenhagen, May 1989

Contents

Chapter VII: Revision and Fusion (1960-   )   *217*

Chapter VIII: The Historical Turning-Point (1960-   )   *256*

Introduction

1.

It is the aim of this work to contribute to a historical clarification of the socioeconomic phenomenon of crisis. In pursuit of this aim several historical eras could conceivably be embraced by the study - not least the capitalist era and the periods when the ruptures between the various ancient cultures took place, or when the ancient mode of production based on slavery was changed to a feudal mode of production. However, the choice has fallen here on the late middle ages, the thirteenth to fifteenth centuries.

There is a very long-standing tradition which associates the history of these centuries with the related concepts of crisis and catastrophe. The plague epidemics of the middle of the fourteenth century and the mythologies that have dominated discussion of them were the original reason for this. For that reason, and because the late medieval centuries did in fact exhibit a number of far-reaching social and economic changes in the northwestern European societies, this is the epoch that has been selected.

For practical reasons the study has been restricted to a single national unit within the area mentioned. The choice fell on England for a number of different reasons. That notions of late medieval crisis have been particularly prominent in English historiography, and that they must be considered a cornerstone of European crisis research, have been crucial factors; but the very large number of primary sources and the wealth of literature on that country's social and economic history have also been important for the choice.

It is well known that there is a general consensus that feudal society was fundamentally an agrarian society. Although the fact at times seems to be neglected, even by medieval historians, agriculture then determined, in a qualitatively quite different sense from today, the course of social and economic life. Any discussion of social crisis in the feudal centuries must therefore take its point of departure in an account of the economic fluctuations in this sector. For that reason, the field of investigation has been further narrowed so that the main focus is on agriculture.

The problem area that is to concern us in the following is thus the crisis of economic reproduction in feudal England, where "economy" is to be understood as the relationship between the socioeconomic structures of the agrarian feudal relations of production and the physical exchanges between the agrarian society and nature, the relationship between feudal England's dominant social and technological structures; while by crisis is understood on the one hand the reproductive crisis of feudal society, that is, the society's inability to reproduce its members, and on the other the reproductive crisis of the feudal social system, that is the similar inability of the agrarian relations of production to reproduce themselves.

Whether one can speak of a social crisis in the late medieval centuries depends not only on whether we can establish that there was a decline in the reproduction of members of the society and/or the relations of production, but also on whether this failure to reproduce was brought on by factors endogenous in or exogenous to the society. Within the framework of the proposed definition of crisis, reproductive problems due to the influence of exogenous factors will in the following not be considered as the expression of social crisis.

The proposed a priori definition of crisis will govern the investigation of the problem area and the way of answering the main question of the study: can one speak of a general social crisis in the late middle ages in England? Did the agrarian relations of production constitute a socioeconomic structure which was in itself crisis-generating, or were any reproductive difficulties in the feudal society/social system due to factors exogenous to the determining structures of society? Did the agrarian structural changes of the late middle ages result in a breakdown of the agrarian relations of production which justifies us in speaking of a crisis? How widespread was the crisis, if crisis it was, in geographical terms? Did its impact differ within the selected area of study with regional differences in the structures and content of the relations of production and with differing natural preconditions? I further wish to clarify where, in answering these questions, we can point to areas within the scope of the problem that must be described as unresearched and of particular importance.

It will have emerged from the above that this work addresses itself to the delimitation of concepts and to the development of new ways of seeing and attacking a general problem complex, within which I also want to point out unresearched areas. This ambition grew up under the impression of the often rather casual use of economic concepts by historians in general, and of the concept of crisis in particular, and in the awareness that a lack of precision in the use of socioeconomic categories has a disturbing influence on our interpretation of history. The interdisciplinary heritage of the author has made its influence felt in this respect - nor has current controversy on the economic situation of our own society failed to have its effect.

The generalizing and concept-delimiting aim of the work has necessitated a methodology which breaks with the traditional working methods of Danish historians. For the aim of developing new ways of seeing and attacking general problem complexes cannot be met simply by fact-orientated studies of primary sources, but must necessarily be pursued mainly through a critique of existing theory formations and their theoretical and empirical presuppositions. Similarly, fully-fledged conceptual clarifications must grow out of a combination of metaconceptual and empirical studies. Nor is it conceivable that the identification of the less-researched aspects of a general problem area could emerge from the study of primary sources alone.

These considerations have meant that the study makes no use of primary source material at all. The value of primary source study would in this context not have warranted the effort. For practical reasons, any consideration of primary source material would have taken on a purely illustrative significance, perhaps lifting the veil from some small corner of a larger general complex. These deliberations have meant that the traditional hallmark of the discipline of history - source criticism - has been limited to criticism of others' use of the sources, in which respect it is a general weakness, and at some points in the presentation an obtrusive problem, that the primary material used has not been available to me. However, more consistent consideration of this material would have made the task insuperable.

Moreover, it follows from the aim of the study that one of the most important results of the present work will be the identification of certain empirical research areas, whose investigation I consider will be of importance for future interpretation of the problem area. The goal of putting research into perspective this way is precisely better-delimited studies of primary sources. Seen in terms of the longer-term research strategy of which the study is a part, this involves progress from a theoretical and empirically generalizing level of analysis towards an empirically specific one.

It should be noted that the author, in developing the methodology of the study, is indebted to Karl Marx, from whose works *Grundrisse der Kritik der Politischen Ökonomie* and *Das Kapital* he drew inspiration in his early student years. The idea of allowing the investigation of the area studied here to grow out of a critique of the development of the relevant research came from these works, even though the subject matter here is quite different and the methodology used differs considerably from Marx's mainly metatheoretical one. As will emerge from the following, much emphasis has been placed on the empirical aspects of the study, and the historiographical perspective not only comes to something like its most extreme expression, but also governs the presentation structurally.

For the method is historical in more than one sense. The problem area is not only illuminated by its primary historical background - the late medieval centuries - but also by more than a hundred years of developments in historical theory in England. Throughout the work, more than a century's investigations of the social and economic history of the late medieval centuries unfold. Thus the aim is to bring out a process of historical development in the approach to the period, by comparing various interpretations with one another, with their theoretical and empirical backgrounds and to some extent with their contemporary settings.

This approach has provided plenty of opportunities to demonstrate the relativity of the concept of crisis, and how its definition has been and is related to the development of various readings of the history of the period. Its shows that there are several true readings of this history. This does not mean that everything that has been said and written is true and nothing false, and that no distinction can be made between truth and untruth; only that a truth can be known by its consistent development from its theoretical and empirical basis. If the basis is untenable and its interpretation inconsistent, what we have is an untruth.

This result has been obtained as far as possible by allowing the development of research, in the author's interpretation of it, to speak for itself. The study takes the form, especially in the first chapters, of a confrontation between the various results and interpretations that have emerged throughout the development of research, while the author's direct participation in this constructed discussion increases towards the end of the study and the final resolution of the issues presented.

The methodological linking of the identification of the various concepts of crisis and their causality has meant that the presentation, besides fulfilling its primary aim, can be read and used as a general review of research on, or an introduction to, the study of the social and economic history of agrarian England from about 1250 to 1500. In this connection it should be mentioned that two principles of presentation have been applied: a chronological one, inasmuch as we follow the development of research through time from chapter to chapter; and a thematic one, as the content of individual chapters and sections is structured by the many facets of the problem area and their interrelations. Thus if the reader wishes to find information about one or more individual aspects, for example the development of medieval grain yield or the colonization process, he or she must look in several different chapters.

The value of this secondary result of the work is related to the fact that the study, while pursuing its primary aim, has attempted to incorporate as far as possible all the available literature within the sphere of the problem in the period from the 1860s up to the most recent contributions. It has been attempted to include not only the central items in this literature, but also the more peripheral ones, and perhaps sometimes the more curious contributions. Thus hitherto neglected bibliographical, and purely historiographical aspects, have also been given closer scrutiny.

2.

The delimitation of the period has seemed natural because in England in the middle of the nineteenth century we can see on the one hand a burgeoning interest in social and economic history, at the same time as the empirically-orientated discipline of economic history was emerging from a critique of the more speculative or metaphysical tradition. On the other hand we find in the same 1860s a discussion of the social and economic consequences of "the Black Death" in England, between two of the pioneers of the new theoretical departure, a discussion which was to take on great importance for debate in subsequent years on the social and economic history of the late medieval centuries. With this discussion began an enduring dispute over the historical significance of the Black Death.

On the basis of an attempt to calculate the demographic consequences of the first plague outbreak, Frederic Seebohm argued that the Black Death led to the breakdown of the agrarian relations of production through the emancipation of the peasants from villeinage and their expulsion from the land. In contrast, Thorold Rogers saw the plague as a stimulating factor in an ongoing process involving the dissolution of villeinage, the commutation of labour services, the development of the leasing system and the decline of the manorial system. Rogers' view of the process of agrarian transformation included both an evolutionary and a revolutionary perspective. The latter of these he associated with the plague's exogenous influence on society - its reduction of the population; the former with a "neoclassical" view that the late middle ages were typified by a free market for prices and wages.

Despite his challenge to the political economists he introduced a *"neoclassical" theory* of the *crisis of the landlords and the manorial system*, in which rising wages and stagnating corn prices before 1350, and even steeper wage increases and falling corn prices after the Black Death led to falling incomes for the estate-holders and through time to the disintegration of the manorial system.

From this pioneering discussion, after documenting how it had a long aftermath in the subsequent decades up to the close of the 1880s, we turn our gaze in the second chapter to certain general historical overviews from around 1890. Here we find a theoretical and methodological revolt against Rogers' empirically-based socioeconomic historiography with its theoretical base in the work of the political economists, a revolt which shed new light on the history of the late medieval centuries. W. J. Ashley ignored the crisis, W. Cunningham assigned it new content and W. Denton backdated its impact to the beginning of the fourteenth century.

Both Cunningham's and Denton's readings argued for a view of history in which political and constitutional factors in general were seen as determinants of the social and economic structures of society. A *"political" theory of crisis* was introduced, which in Denton's version became not only the cause of the crisis of the manorial system, but of a general *agricultural crisis*, because the small farmers and farm labourers were also considered to have been hit by the crisis, and because the breakdown of the manorial system brought with it an erosion of important agricultural techniques and a consequent over-exploitation of arable land; whereas Cunningham exclusively defined the crisis as a *political and constitutional crisis*.

Chapter III demonstrates how in the debate on the social and economic history of the late medieval centuries among historians, certain central themes were pinpointed in the pioneering phase and became the object of an intensive research effort in the period between the 1880s and 1915. In the first place, the nature and social repercussions of the late medieval epidemics were studied to an unprecedented extent in English historiography. Secondly, the origins and chronology of the commutation process and the enclosure movement were thoroughly researched with a view to more accurate dating of the end of the manorial system and the identification of its social and technological results.

The research on the epidemics led to a more sophisticated picture of the history of these outbreaks, but it enriched the field with nothing definitively new on the relationship between the epidemics and the sociological and technological changes in the agrarian world of the late middle ages - not even when Charles Creighton proposed an actual *"epidemiological" theory* of this process of transformation, which none of the scholars involved incidentally described as a direct crisis; and not even when Creighton made the not insignificant remark that the plague had a considerable long-term effect beyond the late medieval centuries.

On the other hand, with the growing empirical research on the commutation process results were achieved that were of crucial importance for the interpretation of the problem area. It was documented that this process extended over several centuries from the thirteenth century far into the fifteenth, and that there are striking geographical differences to be observed in the consideration of the issue. A similar contribution was made by the study of the history of the enclosure movement, in the awareness that the late medieval forms of enclosure were oriented towards the development of the medieval field systems in certain regions, while establishing that the form of enclosure associated with the transition from arable farming to sheep-farming, and thus also the driving of the peasants from the land, can be dated at the earliest to the very last decades of our period and the next centuries. With this, Seebohm's view of the immediate consequences of the Black Death had been refuted.

The investigation of these areas also involved the rudimentary prelude to later research's interpretation of the social significance of the epidemics, and in connection with the documentation of commutation before 1350 one can trace elements of an explanation of the early breakdown in the manorial system in the light of the population growth of the high middle ages. First and foremost, though, it provided documentation that the decline of the manorial system was an evolutionary process which, if not immediately, at least in the longer term, was hastened by the Black Death, and which in itself did not lead to the breakdown of the feudal relations of production.

Even though there was no absolute agreement in the period around 1900 on this, we see in Chapter IV that the documentation produced was at least cogent enough for scholars giving general accounts in the period between 1905 and 1915 to be convinced that they should not assign the

Black Death overwhelming historical importance. In this chapter it is further shown how studies of the Winchester estates between 1916 and 1929 did not indicate either that any serious changes occurred here that can be attributed to the plague, and how the development of research between 1900 and 1930 must in general be said to confirm the evolutionary aspects that dominated Rogers' view of the socioeconomic changes of the late middle ages.

Besides the "political" and "neoclassical" theories mentioned above and the demographic theoretical elements that can be observed in research on the commutation process in the second decade of the twentieth century, in the same decade a *soil exhaustion theory* was developed, which was thought to explain the early decline of the manorial system. This theory can be considered a technical variant of the above-mentioned demographic views, which associated the population growth of the high middle ages with problems in the agrarian organization of production, because the advocates of the theory asserted that the rising population brought with it a growing imbalance between cattle-breeding and arable farming, and that this imbalance gradually dissolved a crucial medieval system of agrarian technology - the open-field system - which had in fact been organized to allow for a certain amount of population growth. The soil exhaustion theory was however already disproved in the 1920s and, although it was taken up again later, it never gained serious importance in its original form.

The period between 1930 and 1950 was in the first place characterized by yet another attempt to extend the concept of late medieval crisis, and in the second place by empirical research in areas of importance for the development of the most recently proposed crisis theories. In Chapter V we see how M. M. Postan argued for the existence of a general crisis of society in the fifteenth century - an *agricultural crisis* with roots back in 1350 and perhaps before, and related to a corresponding crisis in the urban occupational pattern - crises which were jointly expressed as a drop in the national product. We see moreover how Postan and E. A. Kosminsky, in investigating the development of entry fines in the thirteenth and early fourteenth centuries, resumed a line of research that had been begun in the second decade of the twentieth century, and how they created with this work a crucial pillar of support for Maurice Dobb's introduction of a "*neo-Marxist*" crisis theory. Similarly, J. C. Russell's extensive demographic surveys, a number of studies of the medieval colonization process, and not least Postan's critique of the neoclassical theory, all created the preconditions for the establishment of the "*neo-Malthusian*" *theory of crisis* that was soon to follow. Finally, the "epidemiological" theory was recapitulated and strengthened by John Saltmarsh.

With Postan's establishment of the neo-Malthusian crisis theory in 1950, the five current theories in English historiography of the late medieval crisis now existed. Chapter VI deals with this, with the neo-Malthusians' attempt to root the theory in firmer demographic evidence, with a neoclassical critique of this evidence, and points out how the study of the late medieval desertion process seems to challenge the neo-Malthusian views in essential areas. Consideration is also given to the theoretical and empirical contributions of Marxist authors - R. H. Hilton and Kosminsky - between 1950 and 1960. They defined the crisis as the *crisis of feudalism*, that is the crisis of the feudal social system, and Hilton in particular saw it as deriving from growing difficulties in reproducing the population from the end of the thirteenth century on. Thus Hilton saw *the crisis of feudal society* as the cause of the crisis of feudalism.

This reading of the late medieval crisis, and criticism directed at the assumptions in Postan's definition of crisis, paved the way for the revision that followed after 1960, and which is taken up in Chapter VII. While in the great majority of cases the impact of the crisis had previously been dated to the time of the Black Death, or just before it at the earliest, whereas the next century had been considered as the crisis period *par excellence*, Anthony Bridbury and others now claimed that the fifteenth century was a period of general expansion. In opposition to Postan, who had spoken of an agrarian crisis in the fifteenth century, he asserted that agricultural productivity increased rapidly in this century, whereas it had developed negatively in the thirteenth century. The revision found support, as mentioned before, in Hilton's work; but empirical studies of estate economies after the decline of the manorial system, and to some extent the meagre accurate knowledge we have of the development of agricultural productivity before and after 1350, also seem to support it.

In Chapter VII it is further documented that in the period after 1960 one can observe the beginnings of a convergence between neo-Marxist and neo-Malthusian crisis theory. Population growth in the thirteenth century became increasingly central to the neo-Marxist *overexploitation theory*, while traditional Marxist points like the landlords' unproductive consumption and failure to invest in the same century became increasingly important elements in the neo-Malthusian *overpopulation theory*.

In the concluding Chapter VIII, the criticism in recent decades of the neo-Marxist and neo-Malthusian theories and definitions of crisis is gathered together and investigated. Assertions of the importance of falling agricultural production and population developments for the agrarian restructuring have been attacked by "neoclassical" authors, while fiscal, political and climatic factors have entered with renewed strength into the debate from other quarters. The very claim that there was overpopulation in the first decades of the fourteenth century and that a "Malthusian" situation developed, resulting in a decline in the population from about the third decade of the century, has moreover been questioned in various studies of the period between 1300 and 1348.

In these studies one finds not only fairly well-documented rejections of the neo-Malthusian overpopulation theory, and in fact also of the neo-Marxist overexploitation theory, but also - indirectly - a rehabilitation of the demographic importance of the Black Death and the subsequent epidemics, and their social consequences. This is a situation that is reflected in more general literature in recent decades on the social and economic development of the late medieval centuries.

As mentioned at the beginning of this introduction, there has been a long-standing tradition of seeing the social and economic history of the late medieval centuries as a historical turning-point, and as a difficult epoch in general or for specific population groups or productive structures - a time of crisis. Unlike the foregoing centuries, the so-called high middle ages, which have just as traditionally been considered a time of growth - a boom period - the late middle ages have been seen as a time of decline - a period of recession. It will emerge from the survey presented here that this entrenched view seems to be losing more and more of its foundations. Against the background of the comparative survey made here of the development of research, it seems meaningless to take such a simplistic analytical approach, in terms of economic cycles, to the course of medieval and late medieval history.

Primary Source Types

It seems appropriate here to provide a brief explanation of the most frequently-used and important source types.

Account rolls: Manorial accounts, where the estate administrator (the bailiff or reeve) itemized the revenues and expenditures on behalf of the landlord. It should be emphasized that these were not regular profit-and-loss accounts, but a system of bookkeeping for liabilities and their discharge. One cannot therefore assume that the balance in an account represents the factual surplus or deficit of the estate for a year.

The earliest-known account rolls come from the estates of the Winchester bishopric for the year 1208/9, but they probably existed before this. From about 1250 there are accounts preserved from other estate complexes, and the numbers increase throughout the next half-century. The most complete series of account rolls is from the Winchester estates.

Surveys: Land registers - a common term for the three known types of land registers - "rentals", "customals" and "extents".

Rentals: Registers of tenant farmers' rent liabilities.

Customals: registers of tenant farmers' land with all associated rents and liabilities.

Extents: Registers fixing rents on both peasants' land and the demesne.

The registers were constructed on the basis of information given by a sworn jury composed of "worthies" and parish farmers with local knowledge. "Surveys" are preserved from as early as the twelfth century. They become more numerous from the second half of the thirteenth century.

Court rolls: Records of cases in estate courts. These shed light on the peasants' conditions and activities, also beyond the ones in which the lord was directly involved. The earliest-preserved court roll is from 1245, but they become fairly numerous in the subsequent period.

Inquisitiones post mortem: Investigations by local juries on the death of royal vassals to evaluate their possessions, their annual worth (i.e. annual rents) and their rent liabilities, and to decide on matters of inheritance. This source type is known from as far back as the twelfth century, but is only preserved in series from the beginning of the next century. It is of great value for the study of the history of the manorial system, as "surveys" of the estates were often part of the material for these investigations.

Hundred rolls: In 1274 and 1279, because of infringements of the King's rights and unlawful actions on the part of the King's local representatives during the reign of Henry III, Edward I ordered investigations of these matters. Juries from each of the "hundreds" in the country submitted information on estate-holders, the extent and terms of their holdings, the liabilities of the tenants, defaults on payments to the Crown, the courts and prerogatives of the estates, extortion from the nobility and Royal officials, and much else that has shed light on feudal vassalage and tenancy and the manorial institution. Besides the two major surveys, scattered remnants suggest that there were similar efforts in 1251, 1255 and 1285.

Subsidy rolls: Tax lists preserved from the twelfth century on, containing accounts and assessments of a number of feudal taxes imposed on non-clerics, including:

Poll taxes: Taxes fixed at a uniform amount for almost all males above a certain age. Levied in 1377, 1379 and 1381, and

Levies: Tax on movables levied on identified individuals sixteen times between 1290 and 1332. The "levy of 1334" was fixed, in negotiations with villages, towns and other local units, as a collective tax. This collective type of tax, fixed at a tenth of the value of all movables in the towns, and a fifteenth in the countryside, became the norm thereafter.

Chapter I

The Consequences of the Black Death
(1865-1866)

I.1 The population losses

In 1865 there appeared in the journal *The Fortnightly Review* two articles by Frederic Seebohm. In these articles the author set out to present evidence of the mortality caused by the Black Death in the years 1348 - 49, and to answer some questions, in his opinion hitherto unanswered, concerning the social consequences of the plague.[1]

The first article concentrates on investigating demographic developments in England before and after the Black Death, but in dealing with this question Seebohm also has to discuss the development and decline of the towns, when he attempts to use the urbanization process as an indicator of population development. His primary indicator, however, is a group of the period's ecclesiastical records. Seebohm uses the so-called "Torr's MSS", which include lists of clerics in the parishes of the See of York, and in most cases also have information on the reasons why the benefices became vacant. The same source includes corresponding information on the See of Nottingham, which is also dealt with in the study, along with the See of Norwich - in the latter case, however, only at second hand.[2] From Clutterbuck's *History of Hertfordshire* Seebohm further obtained secondary data for this area.

His approach in interpreting the primary material is simple: he only counts a vacancy as due to a death if the source explicitly gives this as the reason a benefice became vacant in the years from 1348 to 1349. In a few places, where there is no specification of the reason, he assumes that there must be other grounds for the vacancy. Against this background Seebohm believes that his assessment of mortality among parish priests in the See of York is conservative rather than exaggerated. Nevertheless he arrives at the unsettling conclusion that more than two thirds of the priests in the West Riding died in the years in question. In the East Riding the catastrophe was apparently almost as great, while "only" a good half of the parish priests in Nottinghamshire succumbed. In the diocese

1

of Norwich about two thirds of the parish priests died, while the mortality rate for Hertfordshire corresponded to that of Nottinghamshire. There is thus a good match between the results of the study of the primary sources and the secondary data Seebohm used.

But how representative does Seebohm believe this material to be? Can it be called upon to express the immediate demographic effect of the Black Death in England in general? Seebohm finds that this is a reasonable assumption, and he concludes that between one half and two thirds of the population of England died as a direct consequence of the Black Death.[3] He justifies this very rigid interpretation of the representativeness of the material by two factors. First, he rejects the idea that the parish priests could have been a group particularly vulnerable to the contagion because of their sickbed and deathbed duties. Seebohm claims that it was mainly the mendicant monks who carried out this pastoral work.[4] Secondly, he takes the view that the parish priests belonged to a privileged class which, because of its considerably better nutritional conditions than the majority of the population, was more resistant, relatively speaking, to contagious diseases.

But besides this Seebohm, as pointed out before, also attempts to justify his general view of the demographic consequences of the Black Death with an assessment of population developments in some English cities before and after 1350. Yarmouth, Norwich and York are towns which are said to have developed considerably up through the middle ages, not least because of the Flemish immigrants and their importance for the fish trade in Yarmouth and the growing textile production in the latter two cities. Thus before 1348 Yarmouth was, according to Seebohm's calculations, a city with no less than 10,000 inhabitants, while Norwich and York had populations as high as 60,000 and 40,000 respectively. Seebohm primarily arrives at these figures through his calculations, which rely on the above-mentioned mortality rate among parish priests, the Poll Tax of 1377, and the assumption that the population did not rise between 1350 and 1377.

In the case of Yarmouth he assumes that the town had 3000 inhabitants in 1377, which means, given his mortality factor of two thirds, that there were the above-mentioned 10,000 inhabitants in the period up to 1348. In support of this estimate he cites a petition from the citizens of Yarmouth to Henry VII in the seventeenth year of his reign - that is, a good 150 years after the ravages of the plague. In this petition it is said that 7,052 people died as a result of the plague, which Seebohm sees as proof that his calculation of the population before and after the Black Death must be correct. On just as dubious grounds he thinks that the population of Norwich dwindled from about 60,000 before 1348 to 6000 in 1377. The figure for 1377 is once more derived from the Poll Tax, while the figure for the period before 1348 comes from adding the 6000 to the number of plague victims in Norwich as given by a contemporary count preserved in the City Guildhall, i.e. 57,374. As for York, no further evidence is given. His figures for population developments there are calculated purely and simply from the Poll Tax of 1377 and his mortality rate of two thirds.

Although Seebohm assesses population developments in some other English cities - Bristol, Oxford and London - in the same way, it cannot be said that he thus seriously documents his general idea of the reduction in population entailed by the Black Death.[5] Rather, he uses the mortality factor derived from his study of the ecclesiastical sources inductively in calculating the population developments in the cities - at the same time accepting without reservation the assessment of mortalities caused by the Black Death made by contemporaries and their immediate posterity.

As mentioned before, in this article Seebohm deals first and foremost with the demographic aspect of the Black Death. He does not directly look for reasons for this catastrophe, but we have seen that he connects the destructive effect of the plague indirectly with the reproduction level of the population - also stating directly that the most disadvantaged classes suffered most.[6] Seebohm's view of the development of the population after the first plague outbreak is in general closely connected with his interpretation of the ecclesiastical sources cited. Later, in connection with other authors' interpretation of demographic developments in the light of the same source type, we will come back to some of the problems raised by Seebohm's reading, specifically as regards mortality among the members of this class. At the moment we may simply note that he gets much mileage out of this material, since he allows it to form the basis of his assessment of mortalities from the Black Death in England in general as well as in the cities.

I.2 Population growth and resources before 1348

Before we turn to Seebohm's evaluation in his 1865 article of the effect of the Black Death on the reproduction of social and economic structures, it seems appropriate to mention that in an article from 1870 he persists in the general view of the demographic consequences of the Black Death that we have just seen.[7] This work is moreover interesting in our context inasmuch as Seebohm in the article directly rejects a body of theory which gained prominence much later in research on the late medieval transformation process: the theory of "Malthusian check".

He does not believe that the population growth from Domesday Book until just before 1349, which he reckons as a doubling from two to four million people,[8] can have entailed problems for the reproduction of the population. For he claims that only about half of England's potential tillage land was cultivated after the Norman Conquest, so there were plentiful resources for a gradual population growth.[9] Thus only four million of England's 23 million arable acres had been under the plough in the eleventh and twelfth centuries, while four million acres had been used as pasturage. This calculation produces a margin in favour of population growth of fifteen million acres of potential tillage land,[10] a margin that, given Seebohm's assumptions, was only reduced to eleven million acres in the period immediately before 1349.[11]

The most striking problem with this calculation is its premiss that agricultural technology did not develop for a period of three hundred years.[12] It is further a doubtful assumption that the basis for the reproduction of an individual throughout this whole period is fixed at one quarter of wheat, and that the average yield per acre is also fixed at precisely one quarter.[13] The very fact that a considerable portion of the population growth must have been absorbed by urban development, which Seebohm is aware of,[14] indicates a rise in the reproduction level of the overall population. Urban development and the development of the social division of labour precisely require, all else being equal, an increasing agricultural surplus production.

I.3 The land/manpower ratio after 1350

Seebohm found that the disastrous demographic consequences of the Black Death had been neglected by historians in favour of their descriptions of court life and the wars of the kings. For him, this was the best proof that history had been deflected from its primary task: the study of the nation's life, with all its troubles and struggles - which was fatal, all the more so as the Black Death "was followed by a great social revolution, whereby the whole course of English History was changed..."[15]

For Seebohm, the paradigm within which this social revolution took place was the changes in the relationship between land and labour that accompanied the high mortality. Feudal rents fell drastically and the wages of regular and casual agricultural labourers rose correspondingly, with the result that the value of land tended towards a permanent drop right up until the sixteenth century, because the supply of labour of all types in agricultural production suddenly fell far below the demand derived from the constant amount of cultivated and potentially cultivable land. And Seebohm saw these changes as occurring within the framework of a social class struggle triggered off by the decline in the population.

His empirical account of the development of wages reflects this view. For his documentation of the rising wages after 1350 consists of the repeated attempts at state wage regulation instituted by the so-called Statutes of Labourers from 1351. This legislation, which we will later come back to in more detail, is considered to be the main cause of the social conflicts said to be typical of the last half of the fourteenth century.

With the Statutes of Labourers the King and his Council broke with "the laws of political economy" - with the principle of the free formation of prices and wages in society.[16] The author blames the Statutes of Labourers and subsequent repetitions and extensions of this legislation not only for the rises in wages, but also for the emergence of social unrest among the agricultural labour force in England - a movement that began among the free labourers but which, at least from 1377 on, also spread among villeins and free tenants, with the goal of abolishing labour rent and the adscription also imposed by the legislation.[17] Under threat of fines, prison sentences and the branding of escaped villeins, the army of agricultural labourers continued with strikes and minor revolts until the Poll Tax became the catalyst for the culmination of the unrest in the Peasant's Revolt of 1381.

Seebohm thus believes that the social unrest arose because of the intervention of state power in a formerly free wage market, in a situation where "the laws of political economy", because of the Black Death, favoured agricultural labourers and peasants.[18] At the same time he sees the strikes, caused by the state's repeated attempts at "incomes policy" and occupational and geographical adscription of the workforce, as proof that wages really rose throughout the latter half of the century.[19] These strikes, as well as the recurrent complaints to Parliament that excessively high wages were being demanded, that villeins were fleeing, and that the peasants were not paying their legal rents, and the central authorities' attempts to follow up these complaints, are Seebohm's documentation for the general rise in wages in the latter half of the century.

Nor is the evidence he cites for the corresponding fall in rents and land value convincing. The chronicler Knighton's description of how the estate-holders were under such pressure that they waived parts of their rent or sometimes the whole rent, is accepted unquestioningly.[20] And without having to deal in detail with his use of Clutterbuck's calculations of the average value of land, we can say that in the table Seebohm presents for the reader, one observes no significant difference in land values immediately before and after 1350. In fact, we have to go all the way to 1417 before the table shows a drop.[21]

I.4 The emancipation of the peasants

The efforts of the King and Parliament to stave off the consequences of the plague's depopulation of England through legislation do not seem to have worked - among other reasons because the state, through its engagement in the Hundred Years' War, directly aggravated these consequences. At any rate, Seebohm connects the Poll Tax levied by Parliament in 1377 with both the war effort in France and the culmination of the social unrest. But the misrule of the state was in this respect turned to the advantage of the people. "Now was the time for them to assert that freedom which by law of nature they had a right to demand."[22]

The imposition of this tax was the spark that made forty years of social protest explode into open rebellion - a rebellion which, although it was quelled, was so successful that "in order to induce the insurgents to lay down their arms, the abolition of villeinage was promised, as well as a fixed rent on land, instead of those personal services, which the serfs in their condition of villeinage, were bound to give."[23] "The strike of the labourers, though long and hard fought, had ended in their victory, and the defeat of the landlords."[24]

Seebohm was thus convinced that the peasants had achieved their freedom at the close of the fourteenth century. As his only attempt to document this,[25] he cites the beginnings of a proper poor law introduced in the reign of Richard II. For "a provision for the really poor must ever immediately follow as the natural result of emancipation."[26] Freedom apparently had its price.

For the peasants did not win their freedom as a result of "a onesided bargain". The landlords took measures against the rising wages, falling rents and the peasants' demands for emancipation. They gave the peasants their freedom while converting tillage land to pasture. They began producing wool instead of corn. Large areas of the country were enclosed for sheep-farming. Villages were abandoned, villeins driven from their farms and homes. This restructuring continued until the beginning of the sixteenth century, and then stopped: the supply of manpower did not now exceed the demand because of the rise in the supply, but because of the drop in demand. The economic benefits of the restructuring for the landlords in the circumstances are obvious to Seebohm - as are the wretched social and economic consequences for the peasants. Despite the fact that from 1489 on Parliament began legislating to prevent further conversion of tillage to pasture, the parish churches decayed and the villages vanished with the result, among others, of "a terrible increase in crime".[27]

It is notable that the account given here corresponds to that of the historical beginnings of capitalist accumulation given by Karl Marx, in a formulation borrowed from Adam Smith, as "die sogenannte ursprüngliche Akkumulation". Political economy's fall from grace, as Marx calls this process, mainly consists, as we have seen, of the fact that the primary producers - the peasants - are driven from the land and proletarianized. The peasant becomes free: "...frei in dem Doppelsinn, dass er als frei Person über seine Arbeitskraft als eine Ware verfügt, dass er anderseits andre Waren nicht zu verkaufen hat, los und ledig, frei ist von allen zur Verwirklichung seiner Arbeitskraft nötigen Sachen."[28] As we have seen, this status is according to Seebohm the general result of the Black Death.

The English peasants did not escape villeinage "into a condition of peasant proprietorship (as through gradual stages of tenancy-at-will, copyhold tenure, and so forth, some of them undoubtedly did)...". "The masses of the people" were "detached from the land and made dependent upon daily wages" by the results of the Black Death. So Seebohm believes that what Adam Smith called "the previous accumulation" gathered force in earnest as early as the end of the fourteenth century.[29] Thus he does not attribute any crucial importance to the changes in the agrarian feudal relations of production, arguing rather that they were generally neutralized.

7

I.5 The revolutionary effect of the plague

Seebohm assigns the Black Death crucial and revolutionary importance for this neutralization of the agrarian feudal reproductive structures. The plague interferes as an external factor with the continued reproduction of the agrarian feudal relations of production, as it disturbs the former balance between manpower and land - a balance which for him was apparently crucial to the maintenance of the feudal reproductive structures in the agricultural sector.

The plague's drastic reduction in the population creates the basis for the emancipation of the primary producers - on the one hand because these, according to Seebohm, had a direct subjective interest in exploiting this basis in that way, and on the other because the landlords were not content to remain passive, but began taking appropriate countermeasures, with the result that the peasants lost their former ties with the land - in other words, the proletarianization of the primary producers.

The liberal ideological background of the author is clearly visible in the present context, and expresses itself in certain interesting contradictions in the theoretical content of the account.[30] He uses a classic liberal body of economic theory to investigate a society which he is in fact eminently aware was not liberal. Both the maintenance and the breakdown of the agrarian reproductive structures are seen as derived from a supply/ demand relationship, the constitution of which requires at least a free labour market. But the very institution of villeinage was an obstacle to the formation of such a market.

In fact Seebohm argues that this market existed as the basis of the social and economic process that led to the abolition of villeinage. And there is thus theoretically no difference between the cause and the effect of the process - beyond the fact, noted already, that the Black Death was the factor exogenous to society that triggered off the imbalance in the market and its grave effects.

What this translates into methodologically speaking is that Seebohm can concentrate on the documentation of plague mortalities. And it is also characteristic that after he has "verified" the drastic mortality, its consequences are rooted in theoretical assumptions rather than empirical data. The subjective interest of the peasants in throwing off the yoke of villeinage is taken to be self-evident, and empirically is only sustained by the proclamations of the ideological leaders of 1381. For Seebohm the peasants are a progressive class, who justifiably exploit the favourable conditions created by the plague to have their historically just demand for freedom granted; whereas the landlords at first make a reactionary attempt to deny them this justice with the help of state power, and then, when their reaction fails, are forced to fulfil their historical mission:[31] the expropriation of the primary producers. Nor does Seebohm present convincing evidence for this aspect of the process. Rather, he invokes the political economists' view of the so-called previous accumulation process and applies it in his description of social and economic development in the fifteenth century.

In the light of Seebohm's version of the late medieval transformation process one cannot speak of a social crisis. He does not use the concept himself. But against this it should be made clear that the serious demographic reduction in England's population that he argues for after the Black Death, and its consequence - the neutralization of the feudal reproductive structures in agriculture - are seen as derived from a factor exogenous to society: the Black Death.

I. 6 The population losses

The reaction to Seebohm's articles was not long in coming. Already the next year, in 1866, *The Fortnightly Review* published an article by James E. Thorold Rogers,[32] where the author presented certain results concerning the Black Death's demographic and production-structural consequences from the first two volumes of his major work, *A History of Agriculture and Prices in England.*[33] Rogers was already aware when these two first volumes were published that they addressed an extremely narrow public.[34] He tried to solve this communicative problem later by publishing the two-volume work *Six Centuries of Work and Wages* in 1884, three years after Volumes III and IV of his main work had appeared. In the following account of Thorold Rogers' view of the late medieval transformation process, all three of the above works by this author will be considered, irrespective of the fact that the 1884 version is primarily of value as a concentration of the four first volumes of his main work as far as "the history of labour and wages" is concerned.[35]

In the 1866 article Rogers makes no attempt to present a developed critique of Seebohm's views. He is, however, sceptical of Seebohm's selection and use of sources. He doubts the reliability of the Poll Tax figures as a source,[36] and considers Seebohm's very firm acceptance of contemporary mortality counts to be mistaken. He does not think that one can rely on Knighton or other chroniclers in this respect, because they probably greatly exaggerated the number of plague victims under the impression of the actual horror of the events.[37] On the other hand, he does not directly question Seebohm's use of the ecclesiastical annals and his inductive projection of the mortality rate derived from these on to the whole population. But he considers the calculations done to establish population density before 1350, on the basis of the number and size of churches, to be misleading.[38]

Not surprisingly, Rogers believes he can find a much surer indicator of population developments in the development of wages and productivity. In the light of his estimates of the latter he arrives at the conclusion that the population of England and Wales in the first half of the fourteenth century cannot have exceeded 2.5 million, but was probably only two million -that is, only about half of Seebohm's estimate.[39] On the other hand, Rogers will not accept a mortality due to the plague that annihilates more than half the population,[40] a mortality rate that corresponds very well with his estimates of wage developments.[41]

But it must be emphasized that Rogers is relatively cautious in his assessments of the demographic consequences of the plague. His formulations are characteristically indeterminate when he touches on the problem. And we can observe a moderate shift in his assessments in the period between the appearance of his main work and the publication of *Six Centuries of Work and Wages.*[42]

I.7 The accelerating effect of the plague

Rogers expresses himself much more categorically when he comes to the socioeconomic repercussions of the Black Death. "In consequence there ensued a total revolution in the system of tenancies, a revolution wholly unknown to historians, and wholly different from that which Mr. Seebohm has suggested", he writes in his 1866 article. He is "certain that it effected a complete revolution in the relation of labour and land..."[43] "All at once, then, and as by a stroke, the labourer, both peasant and artizan, became the master of the situation in England..."[44] The Black Death "forms an era in the history of personal and political freedom."[45]

Given such formulations and many others just as absolute in Rogers' work on the revolutionary influence of the Black Death on socioeconomic development in England, one could on the face of it consider his stance very unequivocal, as later historians have done. Rogers provides good grounds to do so. But reading his works as a whole justifies a rather less absolute reading. This will be demonstrated in the following.

It is well known that we have been left very important source material from as far back as the thirteenth century for the study of England's economic history - that is, the account rolls. In this century certain landlords, especially the great church lords, apparently began to keep regular accounts. But what was the purpose of this practice? Rogers thinks that this bookkeeping was begun for several reasons - partly fiscal, partly as an effect of the commercialization of manorial production that began with, among other things, the growing trade in estates. But he further believes that the extant account rolls can be used to prove that villeinage was in the process of dissolution and that commutation was accelerating.[46] "Anything like the extreme theory of villeinage was, I am convinced, extinct before the close of the thirteenth century", writes Rogers, since he reads the recurrent insistence on imposing this lack of legal rights on the peasants in the period's legal sources as a sign that villeinage was a dying or dead institution.[47] The villein became "tenant by copy or custom" as commutation gathered speed under Edward II (1307 - 1327), "and by the end of the first quarter of the century the rule had become almost universal" at least on the secular estates.[48]

It is thus asserted in Rogers' account that the feudal agrarian relations of production were undergoing rapid development even before the Black Death. Villeinage was on the wane and commutation was gathering speed. To this it can be added that the existing type of manorial production partly took place with the help of landless labourers, and that Rogers gives us examples of how the manors leased out land as far back as the close of the thirteenth century.[49] We can therefore conclude from Rogers' account that the period before 1350 was not only typified by a transformation of the agrarian relations of production, but also that this process included elements that looked forward to the dissolution of the feudal structures. The excess mortality that occurred in England in the wake of the plague accelerated these developmental tendencies according to Rogers. The shortage of manpower which had already created difficulties for manorial production before the Black Death, among other reasons because of the attraction of the cities, now became so serious that the last

11

remaining villeins were gradually freed with the abandonment of demesne production and as commutation grew apace.[50]

Rogers thinks that the phasing-out of the manorial system in England was a process that extended all the way up to the fifteenth century, when certain religious associations, especially the bigger monasteries, still seem to have maintained manorial production on one or two of the estates nearest the monastery.[51] The same evolutionary view appears in his approach to the dissolution of villeinage. The decline began as early as the thirteenth century, and was in full swing even before the reduction in population in the 1348 - 49 period further accelerated the process. He believes that villeinage in the sense of arbitrary rent and adscription had generally been abolished before 1350. In the sense of labour rent, "the effect of the Peasants' War of 1381 was the practical extinction of villeinage" - although money equivalents of enforced labour services were paid right up to the end of the fifteenth century.[52]

I.8 The crisis of the landlords

But what are the revolutionary consequences of the Black Death? What is the revolution this event triggers off? Rogers establishes the following theoretical pattern and supports it with his estimates of the development in prices and wages.

It is certain that the immediate consequences of the plague were the annihilation of manpower, a drastic increase in wages and serious harvesting difficulties for those who were dependent on wage labour.[53] Although corn prices also rose, profitability, even on manors that could still get the necessary manpower, was falling fast. "The landowners began to exhibit symptoms of what in modern times is called agricultural distress."[54]

By contrast, the free peasants were relatively better off. Their rents were fixed, and in particular they were not dependent on outside labour, Rogers believed. Even better off after the plague was the free labourer, as for that matter were those who were still villeins. The period after 1350 gave these groups the chance of rising on the social ladder. It was a time when economic scope expanded for these, the lowest classes of society.[55] So "if...the great mass of the labouring classes had achieved comparative freedom before the great Plague began, it is reasonable to conclude on the simplest economical grounds, that their condition was much better in the end of that calamity."[56] But they were better off at the expense of the estate-holders,[57] whose reactions were not long in coming.

"The difficulties were met by an arrangement analogous to métairie holding" where the estate-holder made livestock, seed, land etc. available to the tenant on an annual contract against a fixed sum.[58] However, this special leasing system existed according to Rogers only for about fifty years; but it was the predecessor of the capitalist leasing system that gained ground in the fifteenth and particularly in the sixteenth century.[59] Although Rogers gives examples of how this "stock and land leasing system" can be traced back to the high middle ages, it is precisely the introduction of this arrangement that he believes to be the revolutionary social consequence of the Black Death - a revolutionizing of the agrarian reproductive structures which made its impact as an economic necessity both for estate-holders and peasants, particularly the free tenants,[60] and which is considered advantageous to both parties,[61] but most advantageous to the peasants.[62] Here, as in other contexts where Rogers oversteps the limits of value-free research and ventures on an evaluation of the historical phenomena, his moral/logical imperative is unsurprisingly a liberal one.[63] The economic and legal liberalization of the agrarian relations of production is judged to be a step forward for the English people.

But the reaction of the estate-holders to what for them was an agricultural crisis did not just result in an amicable and progressive introduction of the above-mentioned leasing system; it also resulted in a direct reaction to progress. For Rogers proposes the thesis that the Peasants' Revolt of 1381 was triggered off by the estate-holders' attempt to reintroduce villeinage, labour rent and the other obligations that had burdened the peasants in the first half of the century.[64] Although he is not directly able to verify this hypothesis, at least two factors suggest that he may be right. First, one of the insurgents' main demands was precisely, as we know, the abolition of villeinage.[65] Secondly, the Statutes of Labourers which Parliament adopted and continuously made more severe, had, besides what we in modern terminology would call "incomes policy" elements, the clear intention of binding the peasants once more occupationally and personally to the lords whose subjects they traditionally were.[66]

We can sum up by saying that Rogers' view of the social and economic consequences of the Black Death provides no grounds for the notion that revolutionary changes took place in the agrarian relations of production exclusively as the result of the plague; this despite the fact that his sometimes rather rigid and absolute formulations might tend, and have tended, to give that impression. Rogers' interpretation rather outlines an evolutionary process, which, it is true, accelerated very considerably with the reduction in population caused by the plague. Neither the abolition of villeinage as such nor commutation seem to have been phenomena associated with the plague alone. The only revolutionary change connected with this seems to have been the development of the "stock and land lease". Rogers was already calling the economic situation after the plague the crisis of the landlords and of the manorial system, while taking quite the opposite view of the peasants' and wage labourers' economic conditions in this and the next period.

But he did not regard the crisis of the manorial system as a phenomenon triggered off like a bolt from the blue by the demographic results of the Black Death. Manorial production had long been a dubious form of production. The yield from this type of production was so poor that "even a slight change in the circumstances which surrounded [it] would wholly alter its method..."[67] - which was also why the revenues used unproductively by the upper class on imports of foreign luxury goods, on the maintenance of splendid followings, on military accoutrements and sometimes on public alms,[68] only partly came from production on the manors. The bulk of the landlords' incomes came to them from fines, money rents, taxes on markets, ferries etc.[69] All the same, one cannot truly speak of any crisis of the manorial system until after 1350. Here the reader rather gains the impression of a form of production soon to be obsolete, reaching the limits of its expansion,[70] gradually to be replaced by the death of villeinage and commutation - a system which receives its death blow with the population losses, first in the famine years around 1315, then in connection with the Black Death.

It is especially interesting to note how Rogers argues that the phasing out of the manorial system long before, through commutation, was caused by internal contradictions which inhibited agricultural productivity and led to economic waste for both estate-holders and peasants.[71] Against this, there is nothing in his accounts to suggest that the crisis could have been set off by technological problems. The changes in the agrarian relations of production that are described do not seem to have had repercussions in the thirteenth and fourteenth centuries in the form of changes in agricultural techniques. The description of the technological structures is static.

For the inner contradictions of the manorial system hampered the material development of the agricultural production process as far as the adoption of new cultivation methods and crops and the development of stock breeding were concerned, and froze productivity, on both the demesne and the peasants' plots, at a level which led to a constant shortage situation at the beginning of the fourteenth century.[72] Natural factors, especially the climate, are therefore considered to have placed certain absolute limits on agricultural productivity, which for Rogers is very clearly emphasized by the fact that the famine years around 1315 were caused by exceptional amounts of rainfall, since "the cause of bad harvest is always excessive rain."[73]

Yet he does not believe that the climate before and after 1350 changed in general, apart perhaps from becoming a little cooler after 1300.[74] And agricultural productivity and the reproduction of the population cannot therefore have been particularly unfavourable in the period up to the outbreak of the plague in 1348. Rather the opposite: the period from 1317 to 1350 was characterized by good harvests, rising wages (because of the mortalities caused by the famine) and low prices. Nor does Rogers think, therefore, that the plague outbreak had any direct connection with the reproduction level of the population. "The best conditions of life do not appear to have given any immunity from the plague....all classes were equally affected."[75]

I.9 Evolution or revolution?

Rogers describes an evolutionary development of the agrarian relations of production in the thirteenth and the beginning of the fourteenth century. This change takes place in accordance with the estate-holders' and peasants' joint interest in creating new social conditions for the raising of agricultural productivity. The main conditions for this are the liberalization and codification of the relations between landlord and peasant -which means the abandonment of personal relations for purely financial relations and the transformation of the enforced and customary, or arbitrary relations to legally reciprocal, clearly defined relations between landlord and peasant. The reason for this historical metamorphosis is thus not to be found in any socially antagonistic content of the relations of production, but in the social unit constituted by a common antithetical relation to the natural basis. This paradigm reflects the way Rogers explains the early commutation process and the incipient dissolution of villeinage in the light of anachronistic liberal forms of consciousness which he applies to the bearers of the agrarian relations of production.

So far, one can say that Rogers sees the manorial system's dissolution as derived from the effect of factors endogenous in society. The manorial system is broken up from within; but, as we have also seen, the process is accelerated under the influence of a factor exogenous to society - that is, the Black Death and its demographic consequences; and this to such an extent that Rogers believes he is justified in speaking of the crisis of the manorial system after 1350 - a crisis which on the one hand stimu-lated the existing process of metamorphosis of the agrarian relations of production, for example through the temporary introduction of the "stock and land lease"; and which on the other hand also meant a temporary check on this transformation until after the revolt of 1381, because of the reaction of the landlords.

Rogers' general evaluation of the late medieval agrarian transformation process thus has an evolutionary as well as a revolutionary perspective. The former is associated with the play of factors endogenous in society, while the latter perspective is associated with the influence of a single exogenous factor on the development of the agrarian reproduction process.

I.10 The "neoclassical" theory

Rogers rejected any attempt to establish general historical theories. If there was a science of history at all, he thought, it was a method of gathering and analysing so-called economic facts, so as to interpret the past on this basis, and perhaps predict the future.[76] He thus sought, typ-ically for his time,[77] to distance himself from the classical political economists, for example Mill and Ricardo, whose efforts he considered dogmatic, metaphysical and full of unverified hypotheses.[78] At the same time, in opposition to previous historians' emphasis on political history, he asserted the primacy of economics in the historical process.

His leviathanic principal work is the best expression of these views. The first two volumes, where the tabular material alone fills over six hundred pages, while the body of the account mounts up to about seven hundred pages, is said to be built up on the basis of Rogers' scrutiny of well nigh eight thousand original documents.[79] The idea for this pioneering empirical effort was fostered at an international statistical congress in 1860, where it was proposed that analyses should be made of the relationships between the prices of labour and food in former times.[80] Apart from extending his empirical apparatus to include the prices of other products of work than just food, he persisted in this structuring of the problem with exemplary consistency right up to our own period.

The tabular apparatus, the empirical foundation on which the account of socioeconomic development in the late middle ages, among other things, is raised, thus consists of a presentation of prices, primarily of agricultural and urban products, secondarily of agricultural labour. These tables are drawn up, in the case of the late medieval period, on the basis of account rolls from estates spread over all the English counties apart from Cornwall, Lancashire and Westmoreland. Norfolk is best represented geographically, but material from Oxfordshire is more plentiful and cohesive than is the case for the other counties represented. Another less important type of source is the accounts of various religious institutions.

The tabular apparatus has thus been drawn up on the basis of sales records for the surplus products of a large number of manors, and their, and various religious institutions', purchase of manufactures and outside commodities, making up about five hundred pages. Besides this, about sixty pages of tables have been drawn up for wage developments among casual workers and full-time servants of the estates.

Rogers' rejection of general historical theories is purely rhetorical. His empirical method cannot help but constitute a historical theory - one which is inadequate to the understanding of the general socioeconomic transformations he deals with. This is because its basis, his empirical foundation, regardless of its enormous volume, is constructed from the marginal aspects of the economic realities of the period, which had assumed the character of reciprocal forms of appropriation, or if one will simple forms of exchange.

It will be recalled how Rogers demonstrates that wages in agriculture rose steeply in two tempi, partly after 1315 and partly after 1350. This, taken together with the fact that the prices of manufactured products necessary to the agricultural work process rose correspondingly, and that the price of corn in particular did not rise correspondingly, makes up the main components of Rogers' argument for speaking of "agricultural distress" after 1350. It would appear that manorial production was mainly based on the use of wage labour.

17

However, labour rent is mentioned, as are the money rents that were also a part of the manorial economy; but neither of these feudal forms of appropriation are considered to be of such crucial importance that systematic documentation is presented for their actual development. We must be content with scattered illustrations and logical conclusions which are perhaps not so logical after all. Manorial production first and foremost suffered from "agricultural distress" because of the development of wages, and this "distress" is simply reinforced by the fact that the feudal rent also fell. But why did it fall?

In the absence of empirical documentation, Rogers the empiricist resorts to Seebohm's documentation that rents necessarily had to fall when the number of actual and potential tenants fell as a result of the Black Death - and to the view that rents developed in relation to agricultural yield.[81] This despite his belief that it was precisely the fixed money rent that "saved English society from a severer shock than it would have otherwise experienced",[82] and although the succour of the landowners - "quit-rents", that is, fixed periodical money rents - were "at their first creation, high rates for use of land [which] were easily borne when, in the fifteenth century, agriculture improved."[83]

When, in other words, the profitability of manorial production was on the decline because of rising wages, the rising price of working materials and the lack of a corresponding rise in the prices of agricultural products, the drop in profitability was passed on to the feudal rents, for Rogers apparently a marginal aspect of the manorial economy; but no more so than that the landlords succeeded in forcing the money rents up. And no more so again than that the free rent-paying peasants after the plague are considered to be in a better position than the landlords, among other reasons precisely because of their fixed rents.

The most general objection that can be raised against Rogers' method is the obvious one that it is a doubtful assumption that there was a free market for prices, incomes and for that matter rents, and that this influenced the structural changes in the agrarian sector of the late middle ages. Although it is acknowledged that the exchange of products continued to increase into the late medieval centuries, we must remember that both manorial production and the economy of the peasants was still first and foremost aimed at self-reproduction. The same was also partly true of the wage-dependent group.[84] And despite the progress of the commutation process and the existence of wage labourers in agriculture, the relationship between lord and peasant was still quite a different matter from an

18

impersonal financial relationship whose regulatory nexus is the free market.

In principle, Rogers' offence is the same as Seebohm's, but it is far more sophisticated. He too is caught in a number of inconsistencies which generally make it difficult to distinguish between the overall aims of the late medieval transformation process and its origins, to distinguish purpose from process. The liberal goal is achieved through the unfolding of a liberalizing process.

For example, he is perfectly aware that "there was no competition for holdings in that state of society in which the great landowner cultivated his property with his own capital, and the small tenant had a genuine fixity of tenure under traditional, customary and certain payments."[85] Nevertheless, he describes the manorial system as "the old system of capitalist cultivation",[86] and its dissolution, especially after 1350, as being due to a competitive relationship not only between lord and peasant, but also to similar internal competition among the landlords and among the peasants. Thus he vacillates between explaining the breakdown of the manorial system on the basis of the above-mentioned notion of a social coincidence of the interests of lord and peasant before 1350 and a competitive element inherent in the manorial system, aggravated after the Black Death - this although he is demonstrably well aware that the manorial system in principle did not give rise to competition over either land or rents.

I.11   The aftermath (1875 - 1885)

Rogers' works have gained great importance for the later study of England's social and economic history in the late middle ages. This can be documented for the moment by the fact that twenty years were to pass after the publication of the first volume of his principal work before serious attempts were made to question the theoretical and methodological foundation of his account. Before we turn to these attempts in the next chapter, some examples will be given of Rogers' importance for views of the late medieval transformation process in the intervening years. For this purpose two well known and three less well known accounts have been chosen, as well as one article. The works in question are William Stubbs' highly-esteemed account of England's constitutional history;[87] J. R. Green's overview of the history of the English character, constitution and society;[88] Charles H. Pearson's outline, not quite so well known, of

19

the history of England in the fourteenth century;[89] the similarly little-discussed account of occupational developments in England by the German scholar W. von Ochenkowski;[90] N. M. Hyndman's book on the foundations of socialism in England;[91] and finally R. Ernle's article on agricultural techniques from 1885.[92]

In the first of these works the problem area with which we are dealing is, for good reasons, only treated very tangentially. But in the few pages where Stubbs concerns himself with social and economic development in the last half of the 1300s, he manages to make very clear use of some of Rogers' main views. For example, when Stubbs asserts that "the variety of effects that follow [the Black Death] must be referred not to the plague simply, but to the state of things which existed when the plague came and the liability of that state of things to be modified by its influence".[93] He refuses to credit the plague with "nearly all the social changes which take place in England down to the Reformation",[94] and, like Rogers, assigns the Peasants' Revolt of 1381 great importance for the real changes that occurred in the relations of production. As for the origins of this revolt, he combines Seebohm's and Rogers' main causes - the Poll Tax and the attempts of the landlords to reintroduce villeinage and labour rent.[95]

As far as the issues of the demographic consequences of the plague and the changes in the relations of production that followed are concerned, Stubbs closely follows the Rogers model. He directly cites Rogers on mortalities, and in the light of the rising wages he recounts how "the whole system of farming was changed in consequence." "The great landlords and the monastic corporations ceased to manage their estates by farming stewards, and after a short interval, during which the lands with the stock on them were let to the cultivator on short lease, the modern system of letting was introduced, and the permanent distinction between the farmer and the labourer established."[96]

In the foreword to his *Short History of the English People* Green takes the same line as Rogers on the determining relationship between legal and constitutional history on the one hand and social and economic history on the other.[97] In his view of the reasons for the socioeconomic changes of the late middle ages he is just as much in agreement with Rogers. But his interpretation of the content of the changes is more in accordance with Seebohm's. For precisely the latter's essays on the Black Death, and Rogers' *A History of Agriculture...* along with the primary sources, "the Domesday Book of St. Pauls", Knighton's and Walsingham's chronicles and the Labour Statutes, form the background against which Green describes social and economic developments in the fourteenth century. Characteristically for Green's work, which is particularly set on connecting the development of the English mind and constitution with social and economic development, we find the section relevant to our subject under the title "The Peasant Revolt".

Green thinks like Rogers that manorial production had begun its decline as early as the period before 1350. Indeed he is even more radical than Rogers, for he asserts that he has found an example of lands being let out as far back as the twelfth century - which is in fact not improbable. But nothing in his account suggests that in the case in question there was a "lease" in the sense we know from the late middle ages. Green confuses the concept of "lease" with money and product rent, and we cannot therefore with any great confidence take this as a sign of "the first disturbance of the system of tenure".[98]

Both Green and Rogers thought that commutation not only entailed a change in the form of rents and the break-up of the manorial system. With commutation demesne production continued under semi-capitalist auspices. For this reason, "at the close of Edward's reign, in fact, the lord of the manor had been reduced over a large part of England to the position of a modern landlord, receiving a rental in money form from his tenants and dependent for the cultivation of his own demesne on hired labour..."[99] The picture Green paints of the relations of production before 1350, to the extent that commutation had taken place, is thus like that of Rogers characterized by the emancipation of the primary producers from dependence on the landlords. The former personal relationship between the parties has been replaced by purely financial relations and contractual freedom. The primary producers were apparently free to choose which landlord they would sign a contract with, whether this consisted of rights to land in exchange for money or payments in kind, or of the sale of labour for money or wages in kind.

Green sees the emergence of commutation and the free labourer on the arena of history as based on two factors: first, the population growth of the high middle ages, which led to a continuing fragmentation of the peasant plots and a corresponding subdivision of labour rent obligations which made it difficult to check that rents were being duly paid; and secondly the rising unproductive consumption of the landlords, which tempted them to release the villeins from their labour obligations for money payments. These views are neither developed logically or verified empirically. Nevertheless, Green believes that against this background the commutation process began as far back as the thirteenth century - in a process which accelerated throughout the reigns of the Edwards. The Black Death plays the same reinforcing role in this process in Green's interpretation as we saw in Rogers'. Because of the falling labour supply caused by the reduction in the population, it led to the reactionary countermeasures of the estate-holders, which in turn triggered off the Peasants' Revolt of 1381. With this insurgency the liberalization process achieves its ultimate aim - the general abolition of villeinage and the labour rent.

But where this process for Rogers leads to the decline of manorial production, first through stock and land leasing then through the spread of the leasing system, Green believes like Seebohm that the end of manorial production meant that the landlords converted their tillage land to pastures. For this reason, among others, Green takes a much gloomier view than Rogers of the conditions of the peasants and free labourers in the subsequent period. And, similarly, Green is aware that the latter half of the fourteenth century was not all cakes and ale for this population group, despite the fact that the demand for labour was increasing.[100]

I.12 "The Golden Age of the People"

In Green's historical overview work, one sees Rogers' deliberations on socioeconomic development before 1350 reproduced in summary form. He too believes that the Black Death inaugurated the crisis of the landlords. But Green also has the crisis involve the peasants and free labourers. His view of the consequences of this crisis are identical to Seebohm's, except that he emphasizes that the lower social strata of the rural population were generally impoverished in the next century.

The same bleak description of the condition of the rural labouring population in the period after 1350 can be found in Charles E. Pearson's outline of English history in the fourteenth century. The crisis started by the high mortality affected peasants and farm workers as well as the smaller landlords. On the other hand, Pearson thought that the big estate-holders managed to survive the crisis, partly by being able to retain their own villeins forcibly, partly because they could tempt others' villeins, free peasants and farm labourers with high wages and low rents. So "under these circumstances, Parliament interfered again and again to protect the small gentry" by issuing the Statutes of Labourers. But these moves neither helped the small estate-holders or the rural labourers - particularly the latter, as "the wages decreed by the Statutes of Labourers were insufficient." The lowest classes of society were in difficulties and, to make matters worse, the lords in many areas of the country began to reintroduce villeinage. "In this way a part of the population was practically brought back into serfdom, and constrained to labour by irons and imprisonment".[101]

Where Rogers spoke of the crisis of the landowners in the period after the Black Death, and where Green extended the crisis to affect more or less the whole agrarian population, Pearson took the view that the crisis was worst for the lower strata. He arrived at this opinion without considering the possible restructuring or decline of the manorial economy; and because he claimed that the agricultural sector was in a process of constant expansion in the latter half of the fourteenth century, as an effect of, among other things, the increasing social division of labour. The crisis for the lowest classes of society was thus not caused, if we are to believe Pearson, by the plague alone. It was a crisis that gathered force for these classes because of the upturn in agriculture and manufactures, which meant that a large number of the peasants' and rural labourers' customary rights were trampled on.[102]

It must be stressed that Green does not bother - and even less so Pearson - to document these views. It is as if they are sometimes carried away by their narrative urge and imaginations. They are both strongly influenced by Rogers. But although Pearson pays homage in his foreword to this valuable and original work, they consider themselves able, as we have seen, to contradict Rogers on the not insignificant issue of the condition of the working population in the century after the Black Death.

Interestingly enough, Rogers' opinion of the situation of the rural population after the Black Death and the Peasants' Revolt is a major theme in H. M. Hyndman's book *The Historical Basis of Socialism in England* from 1883. Hyndman, who professed his debt to the German school of political economics and its leading figure Karl Marx, and who was fond of issuing vulgar-Marxist manifestos,[103] railed at Rogers for being "a typical bourgeois economist",[104] and criticised his basic description of the forms of agrarian production structures in the fourteenth and fifteenth centuries.

Hyndman has nothing but scorn for Rogers' inflated version of the money and exchange economy in the period. He claimed that the bulk of social reproduction all the way into the fourteenth century was of the primary type. The exchange economy was according to Hyndman a marginal phenomenon tied to a few very large estates.[105] All the same, he thought, like Rogers, that "it may be doubted indeed whether any European community ever enjoyed such rough plenty as the English yeoman, craftsman and labourers of the fifteenth century", and that this wave of prosperity lasted from the end of the fourteenth to the beginning of the sixteenth century.[106]

One is not surprised to see that Hyndman traced the reasons for this happy state to the fact that the Peasants' Revolt of 1381, although it was put down, protected the working classes against personal slavery and arbitrary relations with the means of production and land. He explicitly agreed with Rogers in this view, but had nothing to say about the assumptions on which Rogers based this result of the rebellion. We may recall how Rogers did not think that the Revolt itself was a success, either in the short or long term. If it did prove successful in the longer term, this was to be seen, according to Rogers, in the context of the reduction in the population, the development of prices and wages and the decay of the manorial system.[107]

When Hyndman calls the period "The Golden Age of the People", it is because the Peasants' Revolt ensured freedom - that is, to a great extent, a free contractual relationship between the landlord and the tenant farmer/leaseholder, and between the purchaser of labour and the labourer; and because the corollary of this freedom was not alienation from the means of production and land. The relationship of the primary producers to these necessary means of reproduction was maintained and secured during the period. "The means of production and exchange were alike at the disposal of the individual."[108]

Theoretically, this is an important point, which perhaps might lead to a deeper understanding of the alleged prosperity of the fifteenth century. The point is perhaps not historically correct,[109] but the theoretical distinction between wage labour, feudal personal dependence relationships, and what could be called a semi-petit-bourgeois organization of production is fruitful. At least it opens the way for Hyndman to play down the development of prices and wages as an explanation of the improved conditions of reproduction in the century.

Although he supports his view of the advantageous situation of the wage labourers by using Rogers' prices and wages foundation, it is clear to him that wage labour in the capitalist sense was still not a general phenomenon. The distinction between tenant farmers, leaseholders and farm labourers is still fluid, among other reasons inasmuch as the last of these are not considered to be absolutely wage-dependent. As yet they were not absolutely separated from the means of production and land. "For the labourer himself owned land, and worked upon it for the support of himself and his family....He was a wage-earner for the most part when it suited him to be so: by no means a wage-earner at the disposal of the employing class in return for the bare means of subsistence his life through."[110]

Thus Hyndman finds the socioeconomic background of the prosperity of the working classes in the fifteenth century neither in the development of rents nor of wages and prices, although both factors are considered to be contributory. The background is the quite unique agrarian relations of production in the fifteenth century, which on the one hand secured the relationship of the primary producers with the means of production and land, and on the other ensured them the right to the free disposal and exchange of their labour - a state of innocence before the final fall from grace, before the inception of the duality freedom/alienation from the means of production and land.

I.13 Socioeconomic and technological agricultural conditions

It has been noted that neither Pearson nor Hyndman to any significant extent documented their views empirically, and how it is only possible to a negligible extent to identify the basic source material directly. The case is rather different with Ochenkowksi's book *Englands wirthschaftliche Entwicklung im Ausgange des Mittelalters*. Here the reader finds a detailed apparatus of notes, providing insight into the author's empirical foundation. Ochenkowski's work - an account in about fifty pages of the feudal agricultural structure and its decline in the late middle ages - is built up around two major works, supplemented with other accounts and primary sources less important in this context. In the description of what Ochenkowski calls "die sozial-landwirthschaftliche Verhältnisse", Rogers provides the model, while a work by Erwin Nasse from 1869 forms the backbone of Ochenkowski's account of what he calls "die technisch-landwirthschäft-liche Verhältnisse".[111]

There is therefore good reason to outline some of Nasse's points, even though he does not deal directly with our problem area. Apart from a reference to Rogers' opinion that the wages in agriculture rose as a result of the plague, there is nothing in Nasse's account that indicates a state of crisis in the fourteenth century. The upheaval in agricultural technology in the late middle ages, the object of Nasse's investigation, did not take place under the influence of a state of crisis, but first and foremost as a result of the rise of the money economy. The reason agricultural technology demonstrably changed in England before it did so anywhere else in western Europe is to be found according to Nasse in the geography of the country - in the fact that England was blessed with so many navigable rivers and a long coastal area which created good natural conditions for any kind of commodity exchange, and thus for the development of the money economy.[112]

The development of fixed money rents had a disintegrating effect on the medieval common field system, and led to a polarization, also described by Rogers, in the formerly more socially and economically uniform status of the feudal peasants. Some became leaseholders of large stretches of land, while others were forced to become labourers. This process is described by Nasse as having lasted a very long time. But he does not venture to pinpoint its beginnings - perhaps in the thirteenth century, perhaps in the fourteenth, but at all events in the fifteenth. Two points in his account suggest, however, that the process had started as early as the thirteenth century: first, he found indications that big landholders then preferred to withdraw from the common field system; secondly, certain sources suggest that even then it was being considered whether the division of the common areas would be an advantage. At any rate Nasse concludes that any resistance of the landlords to the division and consolidation of the land disappeared with the cessation of the labour rent.[113]

The parcelling-out of the common and the consolidation of the peasants' plots was technically necessary because the medieval common production system in Nasse's opinion did not provide space for a natural ratio between tillage and stock-breeding - and more specifically because the fixed classification into tillage and pasture areas and the inviolability of the common on the one hand meant that the farms gradually became too small, and on the other that the animal manure could only fall where it would be of benefit if this was arranged by artificial means.[114] So even though Nasse saw the innovations mentioned as derived from an expanding exchange economy, he thus did not think, as Seebohm did, that the expansion of stock farming was an end in itself.[115] It was, or at least it became, a stage in a revolution in agricultural techniques, which, given the climate of the country, was a natural replacement for the artificial three-course system.[116]

Before we began discussing Ochenkowksi's book, and thereby Nasse, our investigation was characterized by the fairly one-sided treatment in the texts discussed of the socioeconomic development of the agricultural sector. With Nasse and Ochenkowski this treatment gains a new dimension: the changes in the technological structures. This does not mean that these works can be judged to be technologically deterministic. On the contrary, we have seen how Nasse posited socioeconomic preconditions for the technological changes. For Ochenkowski this approach is only made explicit by his use of precisely Rogers and Nasse, and by the structural significance of these works for his own account - which must suit him well, as he writes that the agrarian technological structures go hand in hand with the "sozialen Agrarverhältnisse".[117]

Apart from Ochenkowski's theoretical virtues in this respect, he contributes nothing significantly new. He describes the development of socio-economic conditions in the fourteenth century in fairly close conformity with Rogers, and given this background, he keeps just as close to Nasse's view of the innovations in agricultural technology, although in both cases with a certain critical distance from his models.

It is the emergence of the money economy and individual freedom that we witness from the thirteenth to the fourteenth century. The money rent won ground and agricultural labour was released; but this emancipation was "weder allgemein, noch auch rechtlich festgestellt..."[118] It was "dennoch nur eine Tendenz".[119] The barter economy was still the economic form of the agrarian society, and corn production took place as a necessary precondition of the direct reproduction of the individuals in society - it was as yet no "lohnende Unternehmung".

In this socioeconomic form nature was a hard taskmistress, setting limits both to social reproduction in general and more specifically to the development of the money economy, which according to Ochenkowksi characterizes the first half of the fourteenth century. "Als eine Ereigniss, welches den rühigen Bildungsgang der Verhältnisse zu stören vollkommen geeignet war, muss die Pest im Jahre 1349 bezeichnet werden."[120] Nature, that is, interfered with the successive self-development of society, as an

27

exogenous factor which provoked the reaction of the estate-holders against progress. After the plague outbreak "fand man nichts Besseres, als der Landwirthschaft billige Arbeitskräfte durch Zwang und durch Zurückgehen auf die früheren weniger entwickelten Verhältnisse zuzuführen"[121] - a step back into the darkness of the barter economy. But the forces of reaction were overcome. Despite the formal defeat of the peasants, the Peasants' Revolt of 1381 can be regarded as a victory for "die bäuerlichen Klassen".[122] Although Ochenkowksi remains doubtful about Rogers' claims for the extent of the "stock and land lease" system in the latter half of the century; and although the same period exhibited no crucial technological innovations, the way was open for the fifteenth century's upheaval in agricultural technology. The feudal bonds had been loosened and the estate-holders were faced with "die Wahl eines Weges". Which road they chose, we have already seen in Nasse.

I.14 The development of population and agricultural technology

R. Ernle's article on agricultural technology is unique compared with the texts dealt with above; for it is built around eight treatises on agricultural technology from the period from the sixteenth to the eighteenth century. The middle and late middle ages are nevertheless represented in Ernle's account. What his sources are for this part of the article is not immediately clear, beyond the fact that he appears to have read Seebohm's *The English Village Community* and that a number of points are derived from Tusser's, and in particular Fitzherbert's, books from the sixteenth century.[123]

The main content of the article must be described as an account of developments in agricultural technology through eight centuries. Here too the technical innovations are presented in their socioeconomic context. The socioeconomic changes are presented, as we have seen with Ochenkowksi, as the preconditions for the achievement of the technological changes. And Ernle, too, reckoned with one factor or catalyst for the progress of events - the development of the population.

Even though he does not, with analytical rigour, keep his account inside the framework of the paradigm "population development - socioeconomic change - technological change", this is the only possible explanatory cohesion one can extract from his investigation. Thus he writes that "the progress of English Agriculture was in its infancy determined by the growth of population",[124] yet without in any way basing his opinion that the three-course system rapidly gained ground in the thirteenth century on demographic data.[125] It is obviously understood that the reader is aware that there was an expansion of the population up through the high middle ages. But even for the reader who has this knowledge, it would have been interesting to see some arguments for the relationship between this and the spread of the three-course system.

It is similarly understood that the Hundred Years' War and the Black Death brought about demographic changes, for after 1350 the relationship between "owner" and "occupier" assumed a modern form. "Out of the Black Death and the French wars arise tenant farmers, copyholders, free wage-earning labourers."[126] The relations of production were changed, and very radically at that, because of the decline in population. Technically, this change meant that the peasants withdrew from their village community and settled on consolidated farms, with the result of increased yields. Wages rose, and the estate-holders parcelled out the demesnes. "The first half of the 15th century most nearly realized the peasant's dream of Arcadia."[127]

After 1460 the reorganized agrarian relations of production, along with the growth of wool manufacturing and the rocketing price of wool, resulted in the conversion of tillage land to pastures. The landlords withdrew their land from communal production or evicted their tenants and converted their land to grazing grounds for sheep. They forcibly enclosed the former commons, and neither pedantic legislation nor narrow-minded critics could prevent them from doing so. It gave many people pause that the working masses of the agrarian society were once more brought down to earth; that farm labourers were now left unemployed; that tenants were driven from their farms; and that smallholders and cottars had much of their reproductive basis torn from beneath them with the enclosure of the commons. But "fortunately" there were men with a clearer eye for the perspectives of the situation. "Practical agriculturalists such as Fitzherbert saw the advantages of enclosures both to landlord and tenant: advanced freetraders might agree with Raleigh that England, like Holland, could be wholly supplied with grain from abroad without troubling the people with tillage."[128] Apart from thus observing that the peasants were released from the toilsome cultivation of corn, Ernle goes into no more detail about the nature of the benefit Fitzherbert saw for them in this situation. One suspects that it was the

Elizabethan Poor Law. How the landlords and large-scale capitalist tenants throve on the situation is described, however, in detail.[129]

I.15 Summing-up

In the authors discussed above we have seen how the thirteenth and four-teenth centuries are judged to be the period of the decline of the man-orial system, culminating in the second half of the fourteenth century. To the extent one can speak of crisis on the basis of their accounts, it is first and foremost a matter of the crisis of the manorial economy and the lords of the manors, apart from Pearson's opinion that the great landlords came successfully through the crisis. Here we have a profound disagreement with both Seebohm's and Rogers' views. The same is true of their assessments of the primary cause of this crisis. Except for Ochen-kowksi and Nasse they all rely on a population theory that gives the reduction in the population after 1350 the main responsibility for the final dissolution of the manorial system. Yet we have not seen any serious explanation of the possible backgrounds for the drastic population losses. The Black Death and the subsequent reduction in the population are not considered to be rooted in social and economic circumstances endogenous in society. The plague seems rather, through the medium of population losses, to intervene in the social reproduction process as an exogenous natural force.   Regardless of whether the authors, like Seebohm, say that the change in the land-manpower ratio gave rise to a new relationship between the supply and demand of labour, or, like Rogers, compound this view with economic value terms and say that the relation between prices and wages changed, it is the demographic factor that is in the last analysis the determining factor for the fall of the manorial system. Most of the authors who concern themselves with the period before the fourteenth century - Stubbs, Green and Pearson - agree, however, with Rogers that the erosion of the manorial system set in as early as the thirteenth cent-ury. Only Ernle seems to reflect Seebohm's notion of a genuinely revolutionary process after the Black Death, but he is on the other hand in agreement with Rogers about the consequences of the reduction in population for the structure of production. So are the other authors, apart from Green, who inclines to Seebohm's view that the decline of the manorial system immediately led to a general transition from arable farming to sheep-farming. Green and Pearson moreover disagree with

Rogers in his view that the living conditions of the peasants and rural labourers improved in the century after the Black Death - the period that
Hyndman called "the Golden Age of the People".

Notes

1. "1. How is it that England, unlike almost every other country in Europe, is divided by hedgerows into separate fields? 2. How was it that the English peasantry, unlike almost every other European peasantry, in becoming freed from feudal serfdom, became *detached from the land*, instead of remaining rooted to it as peasant proprietors? 3. What was the real cause of the desolation and decayed condition of all the towns and cities in England complained of in the statutes of Henry VIII? 4. If the English poor-laws existed, as they did, before the dissolution of the monasteries, what was their origin?" F. Seebohm, "The Black Death, and its place in English History", in *The Fortnightly Review*, 1865, ii, p. 149.

2. Seebohm had his information for the diocese of Norwich from Blomfield's *History of Norfolk*, (ibid., p. 151).

3. Ibid., pp. 152, 160.

4. Were the parish priests "conspicuous for attending those duties? Was it not rather the mendicant monks who won their way to the hearts of the people by passing to and from amongst the dying, receiving often in reward the death-bed bequest of the men whose heirs were dying by their sides?", ibid.

5. In the case of Bristol, the local historian Seyers is quoted as having observed a drop in the city's tax payments to the Crown from £245 at the time the tax was established in 1225 to £158 in 1377. The chronicler Knighton is quoted as noting that almost the whole populations of Bristol and Southampton died of the plague. In Oxford, the Chancellor of the University Fitzralph noted that the student body in 1357 did not exceed one fifth of the 30,000 students he had before. For London, an inscription on a tombstone at the Charterhouse is cited as saying that more than 50,000 were buried there - according to Seebohm, as a result of the Black Death. Seebohm believes that the city had at least 100,000 inhabitants before the plague. According to the Poll Tax figures of 1377 it had 35,000 inhabitants (ibid., pp. 158-160).

6. Ibid., p. 152.

7. F. Seebohm, "The Land Question - Part II - Feudal Tenures in England", in *The Fortnightly Review* VII, 1870.

8. Ibid., p. 94.

9. Ibid., p. 104.

10. Ibid., p. 97.

11. Ibid., p. 104.

12. In a later work from 1883 Seebohm nevertheless supposes that the three-course system was introduced very early in England. "...Roman improvements in agriculture may well have included the introduction into the province of Britain of the three-course rotation of crops." F. Seebohm, *The English Village Community*, Cambridge 1926, pp. 411, 417.

13. As regards the yield per acre, Seebohm takes this data from Rogers. It is calculated on the basis of the yield in the 13th-14th century. "The Land Question..." op. cit., p. 97.

14. Ibid., p. 104.

15. F. Seebohm, "The Black Death, and its Place in English History - Part II" in *The Fortnightly Review*, ii, 1865, p. 269.

16. Ibid., p. 270.

17. "For, in 1377, we have direct proof of what hitherto we only have hints - that the villeins, like the free labourers, were really engaged in what in modern English we should call a "strike", to obtain just wages instead of being bound to yield to their feudal masters that labour and those services, which by feudal law, their lord could compel them to perform for his benefit. And further, there is clear proof that these strikes were maintained as strikes are now, by combination, and by subscription of common funds, for the support of those who for the common end were refusing to perform their feudal services." Ibid., p. 273.

18, "We have in this ordinance, issued in the very year of the plague, evidence of the commencement of what in modern English we should call a strike, on the part of the labourers, both "bond and free", for an advance in wages corresponding with the actual rise caused by the plague in the value of their labour." Ibid., p. 217.

19. "That the corresponding rise in the value of labour was also to a great extent permanent, is proved by the history of the struggle which took place between the labourers and the landowners, and which ended in favour of the former. In calling attention to this history, it is to be again observed that its main facts rest upon evidence which is not likely to be biassed in favour of the labourer, seeing that it chiefly consists in the recitals of the Acts of Parliament, passed during the course of the struggle, in favour of the landowner, and against the peasant." Ibid., p. 270.

20. Ibid., p. 269.

21. Ibid.

22. Ibid., p. 274.

23. Ibid.

24. Ibid., p. 276.

25. True, Seebohm also argues that the labour and wages legislation continued after 1381. It is a little difficult to see how this can be taken as documentation that the dissolution of villeinage was general around 1400. Ibid., p. 275.

26. Ibid.

27. Ibid., pp. 276-277.

28. Karl Marx, *Das Kapital*, Vol. I, MEW Vol. 23, Berlin 1977, p. 183.

29. "The Black Death...Part II", op. cit., p. 278. Against this, Marx thought, like James Stuart, that the proletarianization of the primary producers only started in the last third of the fifteenth century and the first decades of the sixteenth. Ibid., pp. 745-746.

30. Seebohm makes no bones about his view of other historical social formations than capitalism when he confronts this "new order" with the past and with certain others' visions of a different future. In the above-mentioned book from 1883, *The English Village Community*, where his subject is the genesis and development of, and the identity between the manorial system in eastern England and what he calls the tribal system in western England, he concludes by writing: "Its [the new order's] fundamental principle seems to be opposed to the community and equality of the old order in both its forms. The freedom of the individual and growth of individual enterprise and property which mark the new order imply a rebellion against the bonds of the communism and forced equality, alike of the manorial and of the tribal system. It has triumphed by breaking up both the communism of serfdom and the communism of the free tribe" (Ibid., p. 439) and further: "Communistic systems such as we have examined, which have lasted for 2,000 years, and for the last 1,000 years at least have been gradually wearing themselves

out, are hardly likely - either of them - to be the economic goal of the future." Ibid., p. 441. It should be noted that, as is evident from the last-quoted passage, Seebohm has now, about twenty years after his article on the consequences of the Black Death, become rather more evolutionary in his assessment of the decline of the feudal relations of production.

31. "The Black Death...Part II", op. cit., p. 277.

32. J. E. T. Rogers, "England Before and After the Black Death", in *The Fortnightly Review* iii, 1866.

33. These first two volumes, covering the 1259-1400 period, appeared the same year as the article mentioned was published in *The Fortnightly Review*. Later they were followed by two volumes on the 1401-1582 period and another two volumes covering the 1583-1702 period. Rogers concluded his principal work by publishing the last volume of the series in 1902, dealing with the period from 1703 to 1793.

34. J. E. T. Rogers, *A History of Agriculture and Prices in England*, Vol. I, London 1866, p. vi.

35. J. E. T. Rogers, *Six Centuries of Work and Wages*, Vol. I, London 1884, p. 3.

36. Rogers, *England Before...*, op. cit., p. 191.

37. Ibid., p. 193.

38. Ibid.

39. Ibid., pp. 191-192. Seebohm responded to the estimates Rogers presented as documentation of his view of population figures before the Black Death. Seebohm accepts without further ado the historical basis of the estimates: an average grain yield for all crops of 1/4; that the volume sown and the total cultivated area in England were equal in the fourteenth and nineteenth centuries; and that one quarter of corn was necessary for the reproduction of one individual. On this basis, and given that annual grain production in the nineteenth century amounted to twelve million quarters, Rogers reckons that annual production in the fourteenth century must have been about 3.5 million quarters. Against the background of this result, and with due consideration for the fact that at this time there must have been a number of sparsely populated areas, he concludes that the population cannot have exceeded 2.5 million. Referring to McCulloch's article "Corn Laws and Corn Trade" in the *Commercial Dictionary* from 1859, Seebohm objects that Rogers' figures for *total* corn production in the nineteenth century correspond closely to McCulloch's total for wheat production *alone*. If the other grain types are included, as they purportedly are by Rogers in his estimates, the total yield amounts to about 25 million quarters according to McCulloch. Seebohm further thinks that McCulloch's figures show how wheat only accounted for a small portion of the total cultivated area, and that the crops known in the fourteenth century - wheat, barley, oats, rye, beans and peas - only took up a good half of the total cultivated area - 7,200,000 acres. The remaining 5,500,000 acres were in the nineteenth century used for crops that were unknown in the fourteenth century - potatoes, turnips, rapeseed, clover etc. On this basis, but against Rogers' methodical background, Seebohm thinks his estimate of the population before 1350 justified or at least probable. F. Seebohm, "The Population of England before the Black Death" in *The Fortnightly Review* iv, 1866, pp. 87-89.

40. Ibid.

41. "Despite the statute of labourers, wages nearly doubled, and remained, for causes sufficiently known to students of prices, permanently at these high rates." Ibid., pp. 193-194.

42. "...the Black Death...*I believe*...destroyed not much less than half the population..." *A History of Agriculture...*, op. cit., Vol. I, p. 60. "The mortality was no doubt enormous and appalling. It is *probable* that one-third of the population perished. *Six Centuries...*, op. cit., Vol. I, p. 223 (my emphases).

43. *A History of Agriculture...*, op. cit., Vol. I, p. 60.

44. *Six Centuries...*, op. cit., Vol. I, p. 240.

45. *A History of Agriculture...*, op. cit., Vol. I, p. 61.

46. Ibid., p. 3; *Six Centuries...*, op. cit., Vol. I, p. 24.

47. *A History of Agriculture...*, op. cit., Vol. I, p. 70.

48. *Six Centuries...*, op. cit., Vol. I, pp. 218-219.

49. Thus in 1300 Merton College leased out its estates Ibstone and Gamlingway for 35 and 14 years respectively. The College's northern estates had been leased out since 1280. *A History of Agriculture...*, op. cit., Vol. I, p. 24.

50. "It appears...that in the panic, the confusion and the loss which ensued on the Great Plague, that process which, as I said before, was going on already, the commutation of labour rents for money payments, was precipitated; that the lords readily gave in to compositions; and that even less than had hitherto been demanded in exchange for the service was arranged for the future. The plague, in short, had almost emancipated the surviving serfs." *Six Centuries...*, op. cit., Vol. I, p. 227.

51. *A History of Agriculture...*, op. cit., Vol. IV, p. 2.

52. Ibid., p. 4.

53. *Six Centuries...*, op. cit., Vol. I, pp. 226-227.

54. *A History of Agriculture...*, op. cit., Vol. I, p. 82.

55. "The free labourer, and for the matter of that, the serf, was, in his way, still better off. Everything he needed was as cheap as ever, and his labour was daily rising in value. He had bargained for his labour rent, and was free to seek his market. If the bailiff would give him his price, well; if not, there were plenty of hands wanted in the next village, or a short distance off. If an attempt was made to restrain him, the Chiltern Hills and the woods were near, and he could soon get into another county ....He had slaved and laboured at the farm, and now his chance was come, and he intended to use it." *Six Centuries...*, op. cit., Vol. I, p. 242.

56. *A History of Agriculture...*, op. cit., Vol. I, p. 78.

57. Especially the smaller landlords. "The owners of one or two manors, the small gentry of the time, must have been more severely tried than any other class. They must have been constrained to descend in the social scale, and to live like the tenant-farmers who sprung up about them. The great lords possessed resources which, though narrowed by recent events, were still sufficient for their state." Ibid., p. 676.

58. "England Before and After...", op. cit., p. 196.

59. *Six Centuries...*, op. cit., Vol. I, p. 282. It is interesting to see how Rogers, as far as this question is concerned, is quite in line with Marx in his presentation of "the genesis of the capitalist land rent": "Als eine Übergangsform von der ursprünglichen Form der Rente zur kapitalistischen Rente kann betrachtet werden das Metäriesystem oder Teil-wirtschaft-System, wo der Bewirtsschafter (Pächter) ausser seiner Arbeit (eigner

oder fremder) einen Teil des Betriebskapitals under Grundeigentümer ausser Boden einen Teil des Betriebskapitals (z.b. das Vieh) stellt und das Produkt in bestimmten in verschiedenen Ländern wechselnden proportion zwischen dem Mair und dem Grundeigentümer geteilt wird." *Das Kapital*, op. cit., Vol. III, MEW Vol. 25, p. 811. Erwin Nasse, on the other hand, takes Rogers to task, and thus also Marx, as regards the comparison of "stock and land leasing" with the southern European métairie system, as "...das Wesen der Metairie liegt nicht darin, dass der Halbpächter kein eigenes Inventar hat, sondern in der Anteilswirtschaft, d.h. darin, dass der Grundherr keinen festen Pachtzins in Geld oder Naturalie, sondern einen Antheil am Rohertrag des Pächters bezieht." Erwin Nasse, *Über die Mittelalterliche Feldgemeinschaft und die Eingehung des Sechzehnten Jahrhunderts*, Bonn 1869, p. 55, Note 1.

60. "England Before and After...", op. cit. p. 196,

61. *Six Centuries...*, op. cit., Vol. I, p. 281.

62. Roger's interpretative approach is typically diachronic in the sense that he consistently interprets the economic historical processes in the light of a very positive view of their result. From his point of elevation and distance from the beginning of the fourteenth century he is thus also able to dress up the commutation process in the garb of social partnership. "It was to the interest of both parties that these commutations should be effected." Ibid., p. 218.

63. The reader is at times treated to veritable eulogies of the unique significance of economic liberalism for the progress of the nation - for example in connection with Rogers' explanation of why the dynamic centre of the economy in England since the fourteenth/fifteenth centuries has shifted from the east to the west coast. *A History of Agriculture...*, op. cit., Vol I, pp. 107-108.

64. Ibid., pp. 26, 81. *Six Centuries...*, op. cit., Vol. I, p. 256.

65. And this despite the fact that the revolt began in Kent, where villeinage was said never to have existed.

66. The alliance that arose between craftsmen and peasants, and between the free peasants of Kent and the East Anglian villeins in connection with the 1381 revolt is not incomprehensible in the light of the Statutes of Labourers. But, as Rogers remarks, the revolt should not just be seen as the expression of the peasants' resistance to the attempt to reintroduce villeinage and the repression inherent in the labour legislation. And it should certainly not be seen, as was the French revolt in 1358 (the Jaquerie) as an expression of "the desperate effort of excessive suffering". According to Rogers the English Peasants' Revolt should very much be seen as an expression of the socioeconomic improvements that followed in the footsteps of the Plague for the peasants. *A History of Agriculture...*, op. cit., Vol. I, pp. 79-80, 95. "Such political movements as are organized and developed with any hope of effecting their object ultimately and permanently are always the outcome of times in which prosperity, or at least relative comfort, is general..." *Six Centuries...*, op. cit., Vol I, p. 270. It should further be mentioned that Rogers assigns the religious revolt and the Lollards no little influence, not least on the spread and organization of the Peasants' revolt. *A History of Agriculture...*, op. cit., Vol. I, p. 95. Precisely the organizing of the peasants beyond the local level and their alliance with the craftsmen, religious groups, smaller landholders and labourers in the towns was considered very important - historically epoch-making - by Rogers: "...the serfs entered into what are now called trade unions, and supported

each other in resistance to the law and in demands for higher wages. *Six Centuries...,* op. cit., Vol. I, p. 252. Rogers thinks that the 1381 rebellion had at least a contributory importance for the subsequent further break-up of the feudal relations of production - an importance so great that he is perhaps trying to warn the latter-day labour movement against following this historic example when he writes: "Once in the history of England only, once, perhaps only in the history of the world, peasants and artizans attempted to effect a revolution by force." Ibid., p. 271.

67. *A History of Agriculture...,* op. cit., Vol. I, p. 22.

68. Ibid., p. 63.

69. Ibid., p. 62.

70. The expansion of the manorial system was, as is well known, very extensive. It was first and foremost an expansion of the cultivated area, without any really crucial transformations of the cultivation methods, apart from three-course rotation, which Rogers does not seem to have known of. However, Rogers cannot have meant that the manorial system in the thirteenth and fourteenth centuries had reached its spatial limits. For he writes: "It is probable that in such parts of England as were, for the resources of the thirteenth and fourteenth centuries, fully peopled, not much less land was regularly under plough than at present..." (ibid., p. 34) - after having claimed (ibid., p. 29): "There can be, I think, no doubt, that while ornamental wood was scarce in the thirteenth century, and long after, natural forest was abundant, and occupied considerable tracts or belts."

71. "It was to the interest of both parties that these commutations should be effected. It was a vexation to the tenant that he should be called away from the work of his own holding to do the lord's labour. It is plain, from Walter de Henley's statement, quoted above, that the bailiff had no little trouble in getting the due quota of work from the tenant....Hence, if the lord could get a fair money compensation for the labour, he could spare the cost of the bailiff's supervision over unwilling labourers. And as money was more useful than the work he got, as perhaps more profitable in the end, he would be induced to make liberal terms with the tenants in villeinage, even if he were not morally constrained to take alternative in money, which was prescribed as an alternative in case the labourer, for any cause, made default in the field. At the same time he could, unless he made a special bargain, save the allowances which he made of bread and beer, and the license of every day taking as large a sheaf as the serf could lift on his sickle from the corn crops." *Six Centuries...,* op. cit., pp. 218-219.

72. *A History of Agriculture...,* op. cit., Vol. I, p. 10.

73. *Six Centuries...,* op. cit., Vol. I, p. 217.

74. *A History of Agriculture...,* op. cit., Vol. I, pp. 28-29.

75. *Six Centuries...,* op. cit., Vol. I, p. 221. As for the last question in the quote, Rogers is slightly lost for an answer. In "England Before and After...", op. cit., p. 192, he thus writes, in contrast to this, that "it is known that the Black Death, in England at least, spared the rich and took the poor."

76. "Yet if there be, as some writers have perhaps over-hastily asserted, a science of history, that is a method of analysing facts by which the future of a nation may be predicted, as well as the past interpreted, this will surely be found most fully in that portion of its annals which is economical. The English nation has not been moulded into its present shape by its constitution and its laws, since its history is by no means an

uninterrupted advancement; for both laws and constitutions have been the product of a variety of transient energies, most of them, in so far as they are expressions of the national temper, being derived from economic considerations, or in great part modified by them." *A History of Agriculture...*, op. cit., Vol. I, p. vii.

77. Johannes Steenstrup, *Historieskrivningen i det Nittende Aarhundrede*, Copenhagen 1921, p. 70.

78. *A History of Agriculture*, op. cit., Vol. IV, p. xiii.

79. Ibid., Vol. II, p. x.

80. Ibid., p. xi.

81. "...just as rent, when agriculture improves and a country progresses, is a constantly increasing quantity, so when a serious reverse takes place, when labour is dearer and deteriorates, or capital is lost and scanty, or agricultural profits are otherwise depressed, it is natural, nay, inevitable, that rent should have its reverses, and decline in value and quantity, even for a time to a vanishing point." *Six Centuries...*, op. cit., Vol. I, p. 239.

82. Ibid.

83. *A History of Agriculture...*, op. cit., Vol. I, p. 26.

84. According to Rogers himself, at least the so-called regular farm servants, tables of whose wages he has drawn up in Volume II of *A History of Agriculture...* (pp. 329-334) had their "own" land and their "own" animals (ibid., Vol. I, p. 18). This means that these regular servants, as far as their reproduction was concerned, would not have been absolutely dependent on price fluctuations. And it may mean that the wages Rogers sees in the manorial accounts are perhaps to a certain extent no more than arbitrary or customary monetary expressions of regular labour services, entered in the accounts as an expenditure because they represented a deduction from the rent yield of the land the servants in question held, wholly or partially in lieu of their labour services. Similarly, Erwin Nasse, for example, writes: "Ein freier ländlicher Tagelöhnerstand ist...schon im Mittelalter in England entstanden.... Höchst wahrscheinlich bildeten der Stamm dieser Volkklasse die Cotarii....Ihr kleiner Besitz und die Nutzung der Gemeinweide für ein oder einige Stück Vieh gab ihnen eine feste Grundlage für ihren Erwerb, deren die ländlichen Tagelöhner unserer Tage in England ermangeln." *Über die Mittelalterliche...*, op. cit. p. 52.

85. *Six Centuries...*, op. cit., Vol. I, p. 56.

86. Ibid., p. 251.

87. William Stubbs, *The Constitutional History of England*, Vols. I-II, Oxford 1875.

88. J. R. Green, *A Short History of the English People*, London 1876.

89. Charles H. Pearson, *English History in the Fourteenth Century*, London 1876.

90. W. v. Ochenkowski, *Englands wirthschaftliche Entwicklung im Ausgang des Mittelalters*, 1879.

91. H. M. Hyndman, *The Historical Basis of Socialism in England*, London 1883.

92. R. Ernle, untitled article in the *Quarterly Review*, No. 318, Vol. 159, London, April 1885. This article was later expanded into a book, *The Pioneers and Progress of English Farming*, published in 1888. Published in revised and expanded form under the title *English Farming Past and Present* in 1912. We will later be returning to this work.

93. Stubbs, op. cit., Vol. II, p. 399.

94. Ibid.

95. Ibid., pp. 452-454.

96. Ibid., p. 400.

97. "In England more than elsewhere, constitutional progress has been the result of social development." Green, op. cit., p. vi.

98. "Thus we find the manor of Sandon leased by the Chapter of St. Paul's at a very early period on a rent which comprised the payment of grain both for bread and ale, of alms to be distributed at the cathedral door, of wood to be used in its bakehouse and brewery, and of money to be spent in wages." Ibid., p. 238. It is a general problem with Green that he equates the agrarian feudal relations of production in their general nature with the labour rent. So he connects their general dissolution with the successive abolition of the labour rent. Green does not seem to have realized that the two other forms of feudal rent - in money and kind - in principle involved the same legal and social content as the labour rent.

99. Ibid., p. 240.

100. Ibid., pp. 242, 248-251. An important source for Green's assessment of the conditions of the rural labouring population is William Langland's poem of *Piers Plowman*.

101. Pearson, op. cit., pp. 229-30.

102. According to Pearson, the 1381 revolt happened because the peasants' original rights were trampled underfoot; the commons had been enclosed, and mills had been built where they were forced to have their corn ground. They further demanded that free areas should be set aside where they could hunt and fish. "They were in fact suffering only from the extension of agriculture and from trade monopolies." Ibid., p. 24.

103. "...the manner in which wealth is produced, the power, that is, which man has over the forces of nature, is the basis of the whole social, political and religious forms of the period at which the examination is made. Forms of social intercourse, custom, law, political institutions, and religion no doubt influence even economical methods long after their origin has been forgotten, and constitute the conservative side of human society, keeping back the changes made necessary by the more or less rapid modification of the system of production below..." Hyndman, op. cit., p. vii.

104. Ibid., p. 18. Rogers was accused of quite a few things in his time - purportedly also of being a communist and a champion of violent social upheaval. J. E. T. Rogers, *The Economic Interpretation of History*, London 1887, p. xi.

105. "Competition for farms in our modern sense was unknown. The relations between the various parties interested were in the main personal, and these continued even when the main fabric of feudalism was falling to decay. Such a body of tillers of the soil produced their crops as a whole for the use of their own people. Farming with a view to profit alone was only just beginning. Though England at this time exported her superfluity of grain, wool and hides after the people had been well fed, well clothed, and well shod, only a few large landed proprietors carried on this business with a direct view to commercial gain." Hyndman, op. cit., p. 5.

106. Ibid., p. 1.

107. Hyndman does not, like Rogers and others, present any arguments showing why the revolt had this long-term outcome. At certain points in his work it appears as if the revolt itself was the cause. "The great risings of Wat Tylor and Flannoc (1381) though put down at the moment by treachery and false promises, *really secured* freedom for the mass of the people." Ibid., p. 3. At other points he seems more uncertain, and suspects perhaps that the matter is not quite so simple. "Great risings of the peasantry had

obtained *or confirmed* for the people the freedom from personal slavery and the security for their property..." Ibid., p. 1 (my emphases).

108. Ibid., p. 21.

109. Hyndman makes no great effort to verify his account empirically. Rogers' *A History of Agriculture...* is on the whole the only source he refers to directly in the first section of the book, which is relevant for us.

110. Ibid., p. 6-7.

111. Erwin Nasse, *Über die Mittelalterliche Feldgemeinschaft und de Einhegungen des Sechszehnten Jahrhunderts in England*, Bonn 1869.

112. Ibid., p. 50-51.

113. Ibid., p. 55-56.

114. Nasse speaks of three-course rotation, which, as we recall, was apparently unknown to Rogers, who indirectly suggested that two-course rotation was predominant. Rogers, *A History of Agriculture...*, op. cit., Vol. I, p. 15. Thus Nasse thinks like Seebohm that three-course rotation existed in the high middle ages. On the other hand, they disagree on the origin of this system. Nasse thinks it was probably brought to the country with the Anglo-Saxon invasion. Ibid., p. 65.

115. Ibid., pp. 65-67.

116. The first condition for a field grazing system to exhibit its advantages is "eine starke atmosphärische Feuchtigkeit. Gerade diese Bedingung ist...bei dem Seeklima Englands in hohem Grade erfüllt, so dass es nicht leicht zu erklären ist, wie man überhaupt statt eines so sehr durch örtlichen Verhältnisse angezeigten Feldsystems dort so lange Dreifelderwirtschaft hat treiben können.... Die Bewegung des 16. Jahrhunderts wäre dann nur eine Rückkehr zu den natürlichen Wirtschaftsverhältnissen des Landes gewesen." Ibid., p. 65.

117. Ochenkowski, op. cit., p. 8.

118. Ibid., p. 14.

119. Ibid., p. 11.

120. Ibid., pp. 12-13.

121. Ibid., p. 21. Ochenkowski's compatriot Rudolf Gneist has a similar confirmation of this view of Rogers' in his book *Englische Verfassungsgeschichte*, Berlin 1882, p. 444.

122. Ochenkowski, op. cit., p. 22.

123. A. Fitzherbert, *A new Tracte or Treaty most profytable for all Husbandmen and very frutefull for all other persons to rede*, London 1523 and *The Boke of Surveying and Improvements*, London 1523; Thomas Tusser, *Five hundred Points of Good Husbandry*, London 1812.

124. R. Ernle, op. cit., p. 323.

125. Ibid., pp. 324-25.

126. Ibid., p. 329.

127. Ibid.

128. Ibid., p. 330.

129. Ibid., pp. 331-32.

Chapter II

Change, Crisis or the Status Quo
(1888-1890)

II.1 The Assault

In the previous chapter it was demonstrated that Rogers considered the crisis of the landlords to have been provoked by the rising mortality that followed in the footsteps of the Black Death. He was aware that the manorial system was in incipient decline long before this event occurred, but he attributed importance to it as an accelerating factor for the dissolution of this feudal organization of production, and crucial importance to it for the development of the crisis. Seebohm, who did not himself speak of crisis, on the other hand exclusively saw the manorial system as having been broken down as a result of the population losses after the first plague outbreak, and thus assigned the plague revolutionary historical importance.

We also found disagreement between Rogers and Seebohm in their different assessments of the depth and extent of the changes the plague entailed. Seebohm believed that the reduction in the population was more extensive than Rogers would agree to, and he described the subsequent socioeconomic changes as an abolition of the agrarian feudal relations of production. By contrast, Rogers spoke only of a relative formal change in the relations of production. It can therefore be noted that Seebohm's view was the more radical of the two.

Around 1890, as mentioned earlier, we come to the first serious assault on Seebohm's and even more so on Rogers' authority. The attack was launched from three different sides; from three authors working in different fields, but united in their theoretical and methodological showdown with Rogers. All three directed their assault against most of Seebohm's and Rogers' positions.

The second edition of W. Cunningham's *The Growth of English Industry and Commerce* appeared as early as 1882. But we are to concern ourselves here with the later edition from 1890; not because the author patently changed the plan and aim of the work with this edition, but because the 1890 edition is a complete reworking of the two earlier editions and contains more than double the amount of material. With this highly-esteemed and frequently-used work Cunningham was aiming at a wide public. So was W. J. Ashley with his overview work *An Introduction to English Economic History and Theory*, from which we will look at Volume I, the first book of which appeared in 1888, followed by the second book in 1893. We are faced here with two compact accounts covering a long period: both start with the Dark Ages before William the Conqueror's Domesday Book shed light on the history of England, and both take us up to Tudor times.

The third work is considerably less wide-ranging in terms of period. Its in-depth area - the fifteenth century - is supplemented only by two relatively short introductory sections covering respectively the three centuries of the high middle ages and the fourteenth century. W. Denton's *England in the Fifteenth Century* was published posthumously in 1888, and further differs from the above works in concentrating on the dominant production sector in society - agriculture. As the titles suggest, Cunningham's and Ashley's works cover all the production sectors of feudal society as well as commerce.

There is wide agreement among Denton, Ashley and Cunningham on the one side and Rogers on the other that the thirteenth century was an age of growth. The disagreements among them are to be found in their evaluations of the fourteenth and fifteenth centuries. Seebohm and Rogers and their immediate followers took the view that the boom period of the high middle ages continued up through the first half of the fourteenth century, and that the plague triggered off the crisis of the manorial system as an exogenous factor. Against this body of opinion we find a general opposition in the present three authors, which cannot however be unified into a common view.

II.2   Change

Cunningham thought that the fourteenth and fifteenth centuries could not be characterized simply as a period of decline. This was rather a period of social and economic transformation which logically had to follow the boom of the thirteenth century:[1] a boom which he acknowledges as having been stopped in the fourteenth century by the Hundred Years' War and the plague.[2] Cunningham sees the late medieval centuries as primarily typified by political, constitutional and socioeconomic structural changes that would have made their impact even in the absence of the Black Death. The constitutional efforts of the Edwards were for him synonymous with the strengthening of central power, and were reflected in social and economic changes. These efforts were necessary before crucial social and economic changes could make headway.

Cunningham was not blind to the fact that these structural changes involved certain demographic preconditions.[3] However, he did not concern himself much with the causes of these demographic factors, apart from observing that the Hundred Years' War was in itself so destructive that it would have been disastrous for population growth, even if Europe had not had to undergo the Black Death. As for the causes of the Black Death, he relied heavily on Hecker's monograph.[4] So we can conclude that Cunningham's very sparse deliberations in this respect are by and large identical to Denton's.

II.3   Crisis

The latter regarded the fourteenth and fifteenth centuries as one long period of crisis. The death of Edward I in 1307 was in his view a turning-point in the history of England. These dark centuries are described by Denton bearing in mind that the country was torn by wars - the Scottish War of Independence, the Hundred Years' War, the Wars of the Roses - by famine, cattle murrain and recurrent epidemics. On the face of it, Denton's account makes it appear that the death of Edward I was a primary cause of all this misery.[5] That his view was more complex beneath the surface will appear from the following.

Denton was concerned with two general types of crisis. On the one hand he describes a political/constitutional crisis, and on the other what he calls material and cultural crisis. The former of these seems to have been of primary importance for him insofar as it was seen as the triggering factor for the material and cultural crisis. And correspondingly, the political and constitutional changes brought about by the Tudor reigns seem to him to have been the groundwork for the road out of the material and cultural crisis he believes existed under this dynasty. So according to Denton the material and cultural déroute extended from the death of Edward I to the accession of the first Tudor king in 1485, while the political and constitutional crisis persisted until the beginning of the seventeenth century, until the accession of the first Stuart.[6]

Just after the death of Edward I, the Scottish War of Independence was unleashed by his unworthy and weak son's foolishness.[7] This war, which led to between three and four centuries of strife and lawlessness in the border areas between Scotland and England, caused great destruction in the northern shires of England, followed by the harrowing of the east, south and west coasts by Scottish and French pirates. As if this were not enough, these ravages were followed by famine and epidemics - "the fruits of war".[8] During most of Edward II's reign, then, the country was marked by a situation of scarcity where the suffering of the peasants was immense.

The situation was not improved under Edward III. Denton also held this king responsible for plunging the country into war - a war which because of Edward's perhaps reasonable claim to the French throne was to prove almost interminable.[9] For more than a century the country was subjected to sufferings like those mentioned above, compounded by crippling taxation of the people.

Somewhat simplified, Denton's theory of crisis can be stated as follows: incompetent monarch - war - famine - epidemic. Thus we find the determining factor at the political level. The warfare of the incompetent kings after Edward I resulted for the country in general social dissolution and material scarcity. This situation he considered the sufficient condition for the spread of the epidemics, and the epidemics triggered off further social turbulence and material shortages, which again could form a basis for the resurgence of the epidemics.[10]

All other things being equal, Denton thought that the late medieval crisis was brought on by factors inherent in society;[11] factors primarily to be found in the political field. Thus the crisis of the late middle ages was in Denton's interpretation set off by factors endogenous in the feudal social system, but factors that, compared with the agrarian relations of production, must be termed exogenous. It was a socially-created crisis, although it did not originate in conflicts in these socioeconomic structures, but rather separately from them, in the political structures of society.

II.4   The status quo

The most radically different from all the previous readings of the late
medieval centuries we have discussed was Ashley's view of the period. He
only hints at crisis, and even denies in polemics with Rogers that there
was any change in economic and social relations in agriculture in the
fourteenth century. For Ashley this century was apparently fairly uninte-
resting. At all events the development of the agrarian relations of
production in the century are simply incorporated into his general account
of "the Manor and Village Community" as they developed throughout the
high middle ages.[12] Moreover, Ashley only mentions the Black Death and
the reduction in the population in the context of his critique of Rogers'
views. For him the Black Death and the population reduction were of no
crucial importance for the "Agrarian Revolution" to which he devotes a
chapter of his book.[13]

Despite the fact that we find no sign of crisis in Ashley's work, he thinks
that the process of change in the agrarian sector leading to "modern
conditions", which lasted well nigh four centuries, included two periods of
a revolutionary nature. The first he dates from 1470 until 1530.[14] This
view implied that until 1450 "no vital change had taken place in the
organization of the manorial group".[15] But this does not mean that there
had been no changes at all. In particular the thirteenth century stands
for Ashley as a century in which the agrarian relations of production
changed significantly.

II.5   Commerce and rent types

In discussing Ochenkowski we noted the distinction between the money
and barter economy applied to the study of English feudalism. Ashley
takes up this thread, since he thinks that precisely the transition from a
natural to a money economy can be traced throughout the high and late
middle ages. But how does this transition take place, and to what extent
does it transform the agrarian relations of production?

"Such a change implies two conditions: first, the existence of an adequate currency; and secondly, the existence of markets..."[16] But even if the
two conditions exist, and money according to Ashley penetrates the
agrarian sector as early as the eleventh century either as a means of
preserving and measuring value, or as a medium of exchange, he does

not think that the money economy effects substantial changes in the relations of production before the fifteenth century. In the first place, he sees commutation as a very long-lasting process extending all the way into his own century.[17] Although he thinks that commutation became more general in the thirteenth century than before, he is very cautious in his assessment of its extent. Thus he warns against Rogers' interpretation of the manorial accounts in this respect. Just because in the sources we find money expressions for various services, this does not necessarily mean according to Ashley that commutation of the services in question had taken place.[18]

The money items in the manorial accounts are not necessarily an expression of transactions mediated by money, at least not between lord and peasant. He finds an example of how money as a measure of product rent yield gained some ground as early as the twelfth century in the account *Dialogus de Scaccario*, probably written in 1178. In this it is "especially important to observe...that it shows that a currency can be used as a common measure of value, long before it is actually employed in everyday transactions as a medium of exchange".[19] Thus we cannot, because we come across services priced in money amounts in the sources, conclude that in the cases in question commutation had necessarily taken place. Nor can we conclude against this background that the transactions between the manorial tenant and the lord of the manor had become purely money relations.

In the second place, commutation for Ashley meant no crucial change in the relations of production. Regardless of the form rents took, they were fixed by custom.[20] So in Ashley's view commutation had no influence worth mentioning either on the productive or social structure of the village community.[21] This is a view that appears to conflict with his view that commutation primarily took place among the better-off peasants, and that it also required the existence of wage-dependent agricultural labourers.[22]

Thus Ashley did not believe commutation in itself was an expression of a qualitative change in the agrarian relations of production. This and his cautious assessment of the extent of commutation before the fifteenth century made up the primary premiss for his view that the late medieval agrarian revolution only came at the end of that century.

The second premiss was his opposition to Rogers' interpretation of the relationship between population decline, rising wages and the reintroduction of labour rent/villeinage, and his claim that this relationship led to a Peasants' Revolt that in the end emancipated the peasants.[23] "In spite of what Mr. Rogers says as to the real success of the movement, the burdens against which they revolted only very gradually disappeared; and their final modification into the innocuous curiosities of copyhold was the result of the transition from arable to pasture during the next two centuries."[24]

We recall that Rogers based his view that the Peasants' Revolt was successful on the idea that the Revolt could not be seen as an expression of the misery of the peasants; that on the contrary it was to a great extent an outcome of the improved conditions that arose for the peasantry after the great population decline. Ashley completely disagreed. "The Black Death, one cannot help thinking, did more harm to the mortality of the people than good to their material prospects. It shook them out of the habits of their lives and the customs of their village".[25] Thus the causes and consequences of the Revolt are seen in a quite different light from Rogers'. The social demoralization after the Black Death and the teaching of Wycliff was enough to trigger off the Revolt.[26] Neither the plague nor the Peasants' Revolt were of crucial importance to the economic history of England. The feudal agrarian relations of production persisted undisturbed until socioeconomic factors outside the agrarian sector had assumed enough importance to affect its organization.

II.6   Enclosures and wool manufactures

For Ashley the crucial factor in the transformation of the agrarian relations of production was cloth production[27] and the rising demand for wool it gave rise to as it gradually, throughout the fourteenth and fifteenth centuries, dissolved the ties of craftsmanship. In the wake of the textiles production based on domestic industry and small and middle manufacturing came the enclosure movement - the fencing-in of manorial land, free peasants' land and parts of the common with a view to using these areas as pastures for sheep. Where manorial land had not already been withdrawn from the common field system, as seems to have been the case on many estates since the thirteenth century,[28] the enclosure of manorial land disturbed the three-course system.[29] The same was true when the free peasants withdrew from the common field system and enclosed their fields for sheep farming. First and foremost, this meant smaller pasture areas for the village community with fields lying fallow. The same thing happened when the lord enclosed parts or the whole of the common.

"From 1450 to 1550, enclosure meant to a large extent the actual dispossession of the customary tenants by their manorial lords. This took place either in the form of the violent ousting of the sitting tenant, or of a refusal on the death of one tenant to admit the son who in earlier centuries would have been treated as his natural successor".[30] Only from this time on did the enclosure movement mean an upheaval in the agrarian relations of production: the loosening of the primary producers' ties with the land - the process we have called above, with Adam Smith, "previous accumulation". But how could this dispossession of the tenant farmers take place. Were they not secure in their tenancies?[31]

In principle Ashley did not think that the tenant farmers had any rights beyond those that had been determined by the needs of the lord. Custom ruled, but was broken in cases where the situation of the lord so dictated. Throughout the middle ages the primary producers' land holdings were at the arbitrary sufferance of the lord. But it was no more arbitrary than that the peasants kept their holdings as long as they observed their customary duties to the lord; and, historically, as long as the reproduction of the lord depended on granting the peasants these holdings. As long as this situation lasted, the lord only exceptionally exercised his right to evict the peasants from the land. This became custom, a custom that through time fossilized into law. But when the economic situation of the lord changed, as the demand for wool made sheep farming more and more lucrative, he broke this customary law and fell back on his old, but in reality always-existing, right. He dispossessed the primary producers from the land.[32]

With the enclosure movement, the small customary tenant farms disappeared, and instead a tenancy system was introduced which Ashley calls "leases for copyhold". But this change in the structure of the relations of production did not in the first instance significantly alter agricultural techniques in the arable cultivation that was left. The connection between the spread of sheep farming and developments in arable farming that we have previously seen stressed by Nasse and Ochenkowski in discussing the early enclosure movement is pushed forward in time by Ashley. He believes that the first wave of enclosures, from 1450 to 1550, involved a simple transition from arable to sheep farming, and that this transition was necessary for the next wave, from 1750 to 1830, to achieve its aim: "to introduce a better system of arable cultivation".[33]

In this view that the enclosure movement originally primarily only led to a transition from arable to sheep farming, Ashley agrees with Seebohm. But as regards both the dating of this socioeconomic phenomenon and its causes he takes a quite different view. Ashley's dating of the first enclosure movement has its origins in three things, two of which we have already touched on: his very strong playing-down of the demographic and economic consequences of the Black Death and his unequivocal emphasis on the importance of wool manufactures. However, neither of these affords a real basis for his very rigid dating. In reality, his dating of when the movement came into effect in earnest is documented by just two sources, both of which mention the enclosure movement as a generalized phenomenon.[34] The very precise dating we are presented with is thus not too convincingly supported.

II.7 Villeinage - its definition and cessation

The end of villeinage is an event of central importance in the accounts we are working with. All of them offer datings for the disappearance of villeinage. One of the reasons these efforts have such different results is that there is rarely any attempt to define the concept of villeinage; indirectly, the authors assign it different contents.[35] We see a good example of this problem when we compare Denton's and Cunningham's approaches to villeinage and its departure from English history.

For the former, the villein is a person who holds land "at the will of his lord".[36] This means a person whom the lord, all else being equal, can drive off the land at any time, and from whom the lord, also all else being equal, can demand any amount of rent in the form of labour. Although the term "tenantry in villeinage" still existed in the model reign of Edward I, the villein had now become a "copyholder"; and although he still legally held land "at the will of his lord", the lord now no longer had the power to remove him from the land. As long as he carried out the labour services imposed on him or paid the commuted money rent, if such existed, he could not be chased off the land he "held by copy of court roll".[37]

For Denton the end of villeinage meant a fixed rent, commutation and security of tenure. In addition there was the legal and political acknowledgement he thought the former villeins were given under Henry III.[38] All that was left of the former ties of villeinage were certain personal ties related to things like education and marriage, but "most of these restrict-ions the humble manorial tenants shared with the most powerful noblemen in the land".[39]

By contrast, Cunningham thought that when Richard II succeeded to the throne a large number of the English peasants were "serfs"; when Henry of Richmond defeated Richard III villeinage was fast dying out.[40] We note that Cunningham uses the term "serf" for the villein, but apart from this terminological difference, how does he manage to date the end of villein-age almost two hundred years after Denton, and about a century later than other datings we have come across than Ashley's? It is connected with the fact that Cunningham, like Ashley, simply ties in the dissolution of villeinage with the transformation of the agrarian relations of production into "something like the simple cash nexus of modern times".[41] For Cunningham villeinage is synonymous with the feudal agrarian relations of production in general. For this reason, among others, he can-not admit either that the Black Death and the subsequent social revolt led directly to the goals set up by the leaders of the revolt.[42]

Neither Ashley nor Cunningham occupies himself with the process that led to the dissolution of villeinage; they deal rather with the ultimate dissolution of the relations of production. Unlike Ashley, who described this change in the light of the expansion of the wool trade, Cunningham was particularly interested in certain political and constitutional factors leading to the modification of the relations of production towards capitalistic conditions.

II.8   Constitutional change

In contrast with Rogers, who thought that he could find free wage form-ation in the period before 1350, Cunningham believed that wages, labour services and rents had always been regulated; and that the wage regula-tion introduced by the Statutes of Labourers was simply the expression of a formal change. It was now Parliament that tried at the national level to regulate wage developments and working conditions in general, as oppo-sed to the practice of earlier times, when the local unit - the manor - was responsible for a specifically local wage regulation. For Cunningham the labour legislation was part of an incipient mercantilist policy that gained strength under Edward III. Hitherto, the custom of each manor and city guild had been enough in itself to regulate not only wages but also other conditions like working hours. "But in the presence of the terrible plague which swept over England in 1349, the frame of society and the ordinary instruments of social authority were entirely shattered and it was necessary for the central government to interfere. This is the principal case, during the reign of Edward III, in which parliament took over a department of regulation that had been hitherto left to local bodies."[43]

The Statutes of Labourers thus seemed first and foremost to have been part of an incipient state takeover of the social and economic regulation which in the local context was typical of the feudal social system. This was a constitutional change which occurred with and as a condition of economic changes in general, but specifically in connection with the Black Death, because the plague led to a dissolution of local social and economic regulation. For this reason Cunningham had difficulty in understanding Rogers' and Seebohm's moral censure of the labour and wages legislation as an attempt to circumvent free competition - all the more so as it had turned out that the free competition that had developed in the nineteenth century was not to the advantage of the economically worst-off.[44] We see here how Cunningham was picking at Rogers' most important theoretical imperative - economic liberalism.

In Cunningham's view, then, the wages legislation did not have a repressive effect on the conditions of the wage labourers, because it suspended the free formation of wages in a situation where labour scarcity would otherwise have forced wages up; this although he acknowledges that the Black Death led to a labour shortage.[15] The extent to which it had a repressive effect, and the real reason why it was not observed, he finds in the fact that the inflationary effect undermined the money wages frozen by the Statutes.[46] This reasoning would have been valid if there had been inflation due to a strong development in prices. But according to Rogers' account of price development this does not seem to have been the case. As is evident from the chronicles with which Cunningham documents the development of inflation,[47] it was due to the Crown's constant debasements of the coinage.[48] This must have had an equally serious effect on wages and prices. In addition, these state regulations of wages were in fact followed up by corresponding efforts to regulate price developments.

One main reason for Cunningham's sabotaging of Rogers' liberal prices and incomes theory is that he thought Rogers had dated commutation too early and neglected the fact that the labour rent was predominant right up until the fifteenth century.[49] This does not mean that Cunningham thought there had been no increasing transition to money rents towards the close of the thirteenth century;[50] only that commutation had not become so widespread that the money rent had become the dominant rent type. This apparently never happened in the feudal mode of production in England, since the labour rent seems to have disappeared at about the same time as the dissolution of this mode of production began around 1500.

Inasmuch as commutation is not considered to have been a general phenomenon, it follows that it cannot be assigned the general explanatory force we have seen earlier. This also means that the wage-dependent class of agricultural labourers was not so crucially important in the manorial system, and on the whole that the money economy and market-mediated reproduction had not assumed dimensions that could justify the historian in using the development of prices and incomes as anything but an indication of specific socioeconomic phenomena. This, along with the fact that Cunningham thought, as did Ashley above, that the personal ties between lord and peasant persisted throughout the fourteenth and a large part of the fifteenth century, is a major reason why we see the developments in the structure of production in their works in particular in another light than in those of Rogers and his successors. It also sheds a different light on the issue of who suffered from the crisis or the changes in the structure of production, and how they were affected.

Where Cunningham speaks at all of signs of crisis - direct constitutional and demographic consequences of the Black Death - it is characteristic that, unlike Rogers, he talks of general "rural difficulties" - not just the crisis of the manorial system. Nor does he mention that one direct consequence of the Black Death was an improvement in the conditions of the primary producers. For him, of course, the high mortality resulted in a thinning-out of these productive individuals in society,[51] but this was not to say, although he speaks directly of labour scarcity, that their conditions were improved. The primary producers were not better off in their relations with the class of unproductive landlords just because there were fewer of them, precisely because Cunningham did not like Rogers see the relations between these major classes of society as something regulated by anonymous market forces.

The personal relationship between lord and peasant remained, as a customary, and now also state-regulated, relationship between landholder and producer. In principle and in general Cunningham thought that the estate-holders' lands were still cultivated by peasants with labour rent obligations, supplemented by wage-dependent agricultural labourers, whose wages he did not see as directly derived from the lords' money rent incomes.

II.9   The preservation of the relations of production

The high mortality in the second half of the fourteenth century of course entailed, from the point of view of the estate-holders, a labour shortage - a problem that Cunningham thought was faced in two ways. First, there was a tightening of constraints on the peasants with labour rent obligations. The land-holders attempted to force the rent up to the extreme limits that custom could bear, or demanded labour rent from those peasants from whom, by ancient usage, rents could be claimed either as money or labour.[52] Secondly, it was attempted to cultivate the deserted land by the introduction of the stock and land lease. This meant that new tenants were allowed to hold land and livestock against payment of an annual rent, and so some of the deserted tenant farms began producing again. But for Cunningham this system did not mean, as it did for Rogers, a transformation of the tenancy conditions of former times, since "the yeoman farmers, or tenant farmers...probably sprang from the class of free labourers, as the surviving villeins, who already had their own holdings, would not be easily able to offer for a portion of the domain land which the lord desired to let."[53]

Even if "the manorial system was doomed from the time of the Black Death", there was no decisive change in the relations of production before the accession of the first Tudor. The stock and land lease and the demands of the peasants during the 1381 Revolt for the end of villeinage were foretastes of the agrarian revolution that spread slowly but surely over the realm, until it came into full flower when Henry VII ascended the throne.[54] It is important to be clear about what Cunningham means by decisive changes, since it might otherwise seem self-contradictory, when we have noted that he saw the late medieval events first and foremost as a structural change in the relations of production, while at the same time we are asserting that the epoch-making change only came at the end of the fifteenth century. As suggested before, the crucial change in the relations of production for Cunningham was nothing less than the dispossession of the primary producers and the emergence of the capitalist leasing system.

To sum up, Cunningham's view of the development of the agrarian relations of production in the late middle ages can be described as follows. Towards the end of the thirteenth century the money rent gains ground, but the labour rent is generally the dominant type of rent all the way up to the end of the fifteenth century. Villeinage continues to exist throughout the period, undisturbed by the Peasants' Revolt of 1381 and the advance of the money rent at the end of the 14th century. Cunningham found the most telling change, also for the subsequent transformation in the direction of capitalist leasing and the dispossession of the primary producers, in the stock and land lease.[55]

Cunningham did not occupy himself much with the technological structures of agriculture. He had his sights first and foremost on the development of the relations of production, and considers the technological composition of the agrarian production process only to the limited extent that he speaks of the transition from arable to sheep farming. He dates this historical shift, as we have already seen, more or less to the same period as Ashley. And it was a shift that meant, in the interpretation of both scholars, that sheep farming expanded at the expense of arable farming.[56]

II.10   Political crisis versus agrarian crisis

Denton, on the other hand, was concerned both with the development of the relations of production and with changes in the technological structures in the agrarian production process. As we have already established, he connected the development of the relations of production and agricultural technology with fluctuations and events in the political field. Let me now just recall how, in view of the model reign of Edward I, he said that the villein had become a copyholder already at this point, with all that implied of secure tenancy and a fixed money rent, before we turn to the fourteenth and fifteenth centuries.

It will also be recalled that both these centuries were typified by a long succession of miseries that began with the warfare of incompetent kings and continued with the strife between rival families for the Crown. These events again led to famine and repeated epidemics, all of which resulted in a decline in the population. The mortality due to the Black Death was overwhelming enough in itself.[57]

Denton, like Rogers, thought that wages rose with the decline in the population. But the actual rise was not only due to this. Commutation, too, and the accompanying growing need for wage labour in the manorial system, forced wages up. Given, too, that the commuted money rents from the copyholders tended to lose their nominal value, Denton understood perfectly why the land-holders were placed in a situation where they were forced on the one hand to demand the reintroduction of the labour rent, and on the other to attempt to force the money rent up - which they did, after the Black Death and the recurrent epidemics had intensified the problem so much that their reaction was justified.[58]

Regardless of the justice of the estate-holders' action and regardless of Denton's view that the causes of the Peasants' Revolt in 1381 were many, he concluded, like Rogers and for that matter like Cunningham, that the main cause of the Revolt was to be found precisely in the landlords' attempts at counter-action.[59] Of all the many reasons for the Revolt, Denton specified only one besides the one mentioned: that is, the Poll Tax, which, unlike Cunningham, he saw rather as having been the grievance that made other social classes than the peasants take part in the Revolt than as an event that united the scattered dissatisfaction of the peasants in a general uprising.[60]

Also unlike Cunningham, Denton took the view that "the movement in itself was in the end successful"; not only because he saw the various Statutes of Labourers the same way as Cunningham, as "a confession that the days when lords of manors could require the personal services of their tenants in return for the lands they held had gone by..."; for after the Peasants' revolt had been quelled, "the large landowner hastened to adjust the rights of the tenants and their own claims by means of acts of parliament"; but also because the last remains of villeinage now disappeared. The primary producers had won the right to move freely, even into urban trades, and they could now marry without the permission of the lord.[61]

The liberalization of the relations of production in agriculture thus comes in two waves. Commutation, the fixed rent (copyhold) and the politico-legal recognition of the former villeins are said to take place at the end of the thirteenth century. A century later, partly as a result of this, but just as much as a result of the Black Death, the landlords' reaction and the Peasants' Revolt, the last remnants of the bonds of villeinage fell away. By the close of the fourteenth century, the agrarian relations of production thus appeared on the one hand as a relationship between

55

feudal landlords and wage-dependent agricultural labourers, and on the other as the relationship between the landlords and the tenant farmers, who in return for the use of an allotment of land were obliged to pay a fixed money rent. Despite the fact that Denton does not mention the stock and land lease or other types of leasing, his view leads one to think that the leasing system and free wage labour came into force around 1400. He speaks of fixed money rent, political and legal security of tenure or lease and free mobility, and these characteristics indicate the presence of impersonal "cash nexus" relations.

Even though the fourteenth and fifteenth centuries, as stated above, are described by Denton as centuries of general crisis, he does not think that the crisis affected all classes of society equally. In contrast with Rogers, who thought that the crisis first and foremost affected the landlords and the bigger free peasants, while the period after 1330 heralded general prosperity for the wage-dependent class in- particular, but also for the peasants in general, Denton thinks that "the tenant farmers of England, who on the whole were less affected by foreign war than either the landowners or the agricultural labourers, were rising as a distinct and important class in the community."[62]

The documentation for this view he found in the first instance in the fact that villeinage ended, and that the primary producers became tenant farmers; and in the second instance in the problems this development brought for manorial production; and thirdly and finally, apparently conflicting with this, in the fact that the wage-dependent agricultural workers suffered under the various Statutes of Labourers. Denton thus, unlike Rogers, thought that state regulation of the conditions of the wage labourers had an effect, which was why manorial production did not run into difficulties due to rising wages. The problem was simpler than that: that is, a sheer scarcity of labour.[63]

II.11   The soil exhaustion theory

But what significance did the changes in the agrarian relations of production now have for the development of agricultural technology? Denton thinks like Ashley that the technological changes first appeared at a time when the capitalist conditions of production became dominant, that is at the end of the fifteenth century.

This does not mean that in the preceding period there had been no breakthroughs in agricultural technology at all. Thus he thought that soil fertilization in the thirteenth century had developed into a systematic practice, and that "the neglect of manure was at once a symptom and a cause of the decline of agriculture at the end of the Middle Ages".[64] The attention which up to the beginning of the fourteenth century had been paid to marling the ground had declined during the time of war, pestilence and famine".[65]

The breakdown of manorial production, the changes in the agrarian relations of production, and the general labour shortage caused by war, famine and epidemics meant that the earlier manuring of the soil was neglected; and this to such a degree that "it was clear at the end of the fifteenth century that the fertility of the arable land of England was well-nigh exhausted."

The landlords were therefore forced, if for that reason alone, "to lay down their lands in grass, and to turn their attention to sheep-farming", which was why "the chief part of the sixteenth century was one long fallow for the old, exhausted arable land".[66]

It may give us pause that Denton introduced a view that later came to play a role in the interpretation of the causes of the crisis in the fourteenth century - the theory of the exhaustion of the arable land. But in contrast with the later version of this theory, which we shall make the acquaintance of later, Denton dated the exhaustion of the English arable land to the end of the fifteenth century, not as in later interpretations of the theory as a consequence of the population growth of the high middle ages. The exhaustion of the soil is for Denton a consequence of the late medieval crisis, not its cause - among other reasons because he thought, as did his contemporaries in general, that the population growth of the high middle ages "had been too slow to encourage to any considerable extent the conversion of the woodlands into pasture or tillage grounds, or even to check the growth of these forests....When land was broken up, or assarted and tilled for corn, it was because the forest had been cleared for fuel, and not because it was required for wheat." Similarly, he thought that the growth of sheep farming in the high middle ages was due to poor chances of commercial profit from the sale of corn, "because the land under cultivation was more than equal to the wants of the country in good seasons".[67]

Thus the crisis of the late middle ages led to an erosion of agricultural technology, to the breakdown of one of the agricultural innovations that

57

had arisen in the middle ages:[68] a fact that for Denton was a direct reason why large areas were converted to pasture, with all the social consequences this entailed - consequences that in the longer term created the basis for a more fundamental development of agriculture.

According to Denton there were two obstacles to the development of agricultural techniques in the later middle ages. One was the exhaustion of the soil. The other was the medieval village community. The enclosure movement around 1500, that is the general abandonment of arable farming and the dispossession of significant numbers of the primary producers, was therefore a precondition for the innovations in arable farming that he considers first arrived in the eighteenth century.

The long fallow period, the growth of stock farming, and the enclosures themselves were technical and socioeconomic preconditions for the development of a more dualistic type of agriculture - one which could function in a biological balance between arable farming and stock breeding. The enclosure movement had the result that the village community and the common field system broke down, and this provided the direct opportunity for the restoration of the arable land.

II.12 Summing-up

Rogers proclaimed a revolt against the classical political economists' "disregard for facts", their "strangling...with definitions" and their "economic laws".[69] He argued for and practised a pragmatic, empirical approach to the study of economic history, while his empirical method involved a priori positions that flawed it with a general historiographical problem - a problem both Ashley and Cunningham directly warned against.

Both pointed out the connection between the economic form of an epoch and the ideas that arise about economics. Modern economic theories are therefore not universally true; they are not true in relation to a past where the relationships they postulate did not exist, and they are therefore not true either in relation to a future in which the same relationships have changed.[70] Political economics is therefore not a body of absolute truths developed in recent centuries.

Rogers would in fact have agreed with this manifesto. The problem was that he was unable to practise it. Rogers' critique of a personally, legalistically and militarily orientated historiography prior to the middle of the nineteenth century drove him out on a limb of what was well nigh economic determinism, which despite his unwillingness had not surprisingly dropped him into the arms of the political economists he was so opposed to. Rogers' empirical method was certainly not emancipated from universal economic dogmas. In particular, it built on the economic law of supply and demand, and he applied this politico-economic law to the study of the late medieval centuries. At the same time he poured empirical material into a theory whose general value he denied. He made it universally applicable and allowed it to become not only the structuring element in his empirical method, but also the crucial cognitive element in his investigation of the economic fluctuations of the period.

Denton's, Ashley's and Cunningham's break with Rogers is compounded of two interdependent elements: in the first place, the pointing out of the problems of anachronism in historical research; in the second the rejection of economic determinism.

In the introductory section of his book Cunningham tries to drive a wedge through the previously very rigid view of the domain of economic history. Without attempting a general definition of the concept of economy he ventures to assert "that economic and political circumstances constantly re-act on one another"; and that "wars and revolutions, court intrigues as well as religious movements, have all had an industrial side".[71] This dialectical view of the relationship between economy, politics, social conditions and the formation of consciousness, to which Ashley also adheres, is however successfully demonstrated neither by Cunningham, Ashley nor Denton in their works. Ashley despite everything assigns the economy - the money economy and wool manufactures - crucial importance. And Denton and Cunningham in the last analysis consider political and constitutional development to be the determining factor.[72] Looking at the composition of the three works, it strikes one that only Cunningham makes any attempt to present these dimensions as interrelated.

Cunningham took the view that economic history is only one particular approach to the history of the nation,[73] and that each historical epoch was perfect in itself. For him history was not just a course of development, but a gradual course of development alternating with decline.[74] In this he differed fundamentally from Rogers, who in his optimistic belief in progress saw the past as a more or less imperfect prelude to the present.

Cunningham's effort to solve what for him was the greatest problem of historical research - anachronistic interpretations of the residue of the

59

past[75] - was therefore made in an attempt to investigate history in the light of its own premisses. This was why he thought that an economic investigation in the modern sense of his area of study was impossible;[76] and not only because of the shortcomings of the sources in this respect. We must therefore, wrote Cunningham, whether we want to interpret specific phenomena or reconstruct a satisfactory picture of society as a whole, "begin from the general influences and actual forces in each epoch".[77]

We must thus decide in advance the determinants of the historical process we want to investigate. "And since the growth of industry and commerce is so directly dependent on the framework of society *at any one time*, it may be the most convenient to take periods which are marked out by political and social rather than by economic changes. This will give...a clear picture of the economic conditions of each period, and a clear understanding of the reasons for the changes that ensued."[78]

It is hard to see how this manifesto in itself solves the anachronism problem, but one must acknowledge that it is a break with every kind of economic determinism in keeping with Denton's effort. It is certainly a break with the unconscious use of the laws and categories of classical political economy in the study of economic history. But it is also to some extent an expression of a return to the historiography Rogers opposed.

In contrast, Ashley's assigning of a high priority to the determining importance of the money economy and wool manufactures for the medieval and late medieval process of development is more in keeping with Cunningham's manifesto for the solution of the problem of anachronism. On the other hand, of course, it does not break with economically deterministic interpretations of this historical process.

These theoretical and methodological aspects are the background against which Ashley, Cunningham and Denton shed a rather different light over the late medieval period and the crisis problem in particular than we have come across before. Ashley neglects the crisis problem, Cunningham reformulates it, and Denton backdates the impact of the crisis.

Ashley can neutralize the crisis problem because his theoretical premiss is that agrarian and other socioeconomic developments have their "motor" in the money economy and wool manufactures. This a priori assumption means in the first place that the changes in the agrarian reproduction complex are dated in accordance with the successive development of the money economy and wool manufacturing; and in the second place that any agrarian crisis had no crucial influence on the change in the reproductive relations.

Cunningham reformulates the crisis question, because he also thinks, a priori, that political and constitutional development are determining factors for the structure of the economic relations of reproduction, The crisis is transformed into a political and constitutional crisis. The consequences of the Hundred Years' War and the plague break down the authority of the local institutions, and the state must take over the functions of these local institutions. But this constitutional change does not lead directly to changes in the agrarian relations of production.

Similarly, against the background of a preconception about the determining role of the political dimension, Denton dates the impact of the crisis earlier. The monarchs of the fourteenth and fifteenth centuries from Edward II on, with their warfare, prepare the way for famine and pestilence to have their disturbing effect on the agrarian reproductive cycles.

Notes

1. W. Cunningham, *The Growth of English Industry and Commerce*, Cambridge 1890, pp. 334-35.

2. Ibid., pp. 275-76.

3. In considering Cunningham's view, one should make allowances for the fact that he was writing a history of industry and commerce. These sectors of the economy appear in a rather different light in the period than the agrarian sector. In the fourteenth and fifteenth centuries English textile manufacturing took off seriously, and according to Cunningham, mercantilist policy began in England with Edward III. Ibid., p. 276.

4. Ibid., p. 303; J. F. C. Hecker, *The Epidemics of the Middle Ages*, 1844,

5. The latter half of the thirteenth century was a flourishing period. But "all this, however, depended, humanly speaking, too much on the life of one man, and the untimely death of Edward, and the dark days which followed, served but to show the instability of material wealth and prosperity when resting upon so insecure a basis. They show, however, at the same time, how much England owed, and still owes, to the genius, the honesty of purpose and inflexible will of its great king." W. Denton, *England in the Fifteenth Century*, London 1888, pp. 61-62.

6. "...in material wealth...England at the accession of Henry VII was far behind the England of the thirteenth century." Ibid., p. 120. "Though greatly inferior in intellectual and moral worth to the great Plantagenet, Henry [VII] was the successor of Edward I, rather than of Richard III." Ibid., p. 124. "Not until the passing away of the Tudor dynasty (1603) had the constitutional institutions which the genius of Edward I fostered into existence their full sway." Ibid., p. 126.

7. Ibid., p. 66.

8. Ibid., p. 91.

9. Ibid., p. 79.

10. As suggested before, Denton bases his epidemiological insight on Hecker's seasoned work. The epidemics in the fourteenth and fifteenth centuries he considers to have been of the typhoid type, which he thinks need particular environmental conditions to thrive. "Disease was mostly of a typhoid character [Hecker]. The undrained neglected soil, the shallow stagnant water which lay upon the surface of the ground; the narrow unhealthy homes of all classes of the people...predisposed the agricultural and town population alike to typhoid diseases and left them with little chance of recovery when stricken down with pestilence." Ibid., p. 103.

11. At certain points Denton also assigns exogenous factors some significance for the spread of the epidemics, and thus of the crisis, for example on p. 96: "These years of disease were often preceded by frequent rains, and sometimes by large inundations..." Cf. Hecker again.

12. W. J. Ashley, *An Introduction to English Economic History and Theory*, Vol. I, Book I, Chapter I, London 1888.

13. Ibid., Book II, London 1893, Chapter IV.

14. Ibid., pp. 285-86.

15. Ibid., p. 264.

16. Ashley posits two quite empty premisses. In the first place it is incorrect to give money and the market equal status as conditions of the development of the money economy, since money is only separated out from the mass of commodities as a general equivalent, as a medium of exchange, with the expansion of the exchange of commodities and thus of the market. In reality, too, it is wool manufacturing in particular, and the wool trade, that are the preconditions of the money economy for Ashley. The history of money is given a very summary account in relation to the development of commerce (ibid., Book I, pp. 48-49 and pp. 163-78). In the second place it is completely meaningless to claim that the presence of money is a condition of the money economy. It explains nothing about where the money comes from, how it is separated out as the medium of exchange, or what the functions of money in general are. In the third place Ashley's positing of the market as a condition of the money economy is just as empty of content since one can claim that the market is also a condition for barter. This is however also gobbledygook, because it is on the contrary barter, the simple exchange of product for product, that gives rise to a market. When a product can be exchanged for a product different in kind, the market exists. But the necessary condition for such an act, and thus for the existence of a market, is that products have lost their use value for their respective possessors, and that they appear as useful values for others - in other words that surplus products have an exchange value.

17. Ibid., p. 46.

18. "We find in many of the customals of the thirteenth century that, even where the labour is not generally commuted, each item of it - a day's work of each sort - is precisely valued. In some manors no labour-dues had been commuted; in others, some had some had not." Ibid., pp. 29-30.

19. The King's need of hard cash, his poor possibilities of consuming his rent on the spot, and the grumbling of the peasants over having to bring the product rent to the King entailed that cash settlements were introduced between the King and his stewards. And this type of settlement affected the relationship between the steward and the peasant, as a "common measure of value" with no change in the real rent type. Ibid., pp. 47-48.

20. Ibid., p. 42.

21. Ibid., pp. 41-43.

22. Ibid., pp. 31-32.

23. To refute this interpretation and demonstrate how dubious it is empirically, Ashley documents how Rogers, without presenting new empirical material, manipulates the above-mentioned hypothesis from the first volume of *A History of Agriculture*..., pp. 81-83, into an absolute fact eighteen years later in *Work and Wages*, p. 253, and then four years later notes with pride in *Economic Interpretations*..., p. 30, how historians "have accepted in silence the proof which I published more than twenty years ago as to the causes and consequences of the insurrection". Ashley, op. cit., Book II, p. 265, n. 3.

24. Ibid., pp. 266-67.

25. Ibid., p. 338.

26. Ibid., p. 266.

27. "It is when we turn to the industrial conditions which took the place of the guild system, and the to agricultural changes with which the transition was accompanied, that we begin to understand how great a part the woollen manufacture has played in English social history". Ibid., p. 222.

28. Ibid., pp. 269-70.

29. Ibid.

30. Ibid., p. 273.

31. For example, Seebohm claimed that terms of tenure were gradually secured by statute from about 1300 on. F. Seebohm in "The Land Question - Part II - Feudal Tenures in England" in *The Fortnightly Review*, 1870.

32. Ashley, op. cit., Book II, p. 281. However, the lord had to take certain measures. Of these, increases in the entry fines were not as frequently used in Ashley's opinion as the lord's right to mete out and collect fines. Another approach was to replace a copyhold with a lease. "In many cases a lease was but a stepping-stone to tenure at will" - among other reasons, because leasing terms were purportedly not subject to the protection of state legislation until 1529. Ibid., pp. 283-285.

33. Ibid., p. 286.

34. Statutes 4 Henry VII, cc 16, 1488, mentions that many villages on the Isle of Wight had been deserted; cc 19, 1499, contains an order to restore the houses that had fallen into decay all over the country in the preceding years. An anonymous pamphlet, probably from 1519, says that between 400 and 500 villages had been deserted in central parts of the realm. Ibid., n. 64.

35. Paul Vinogradoff shows in his book *Villainage in England* from 1892 how the concept in the first instance can be interpreted in different and sometimes directly contradictory ways on the basis of legal and customal sources respectively, and in the second how the meaning of the concept develops in the course of the feudal epoch. Ibid., pp. 211-12 and pp. 217-18. From a later work by Vinogradoff, *The Growth of the Manor* from 1905, we can extrapolate an attempt to define the concept as it appears in the legal sources of the thirteenth century - for example from the many cases of trials of villeinage. The villein has no legal rights vis-à-vis his lord. He possesses only what the lord puts in his possession, and the lord may employ and dispose of the villein as he wishes. On the death of the villein, a duty must be paid in cattle - the heriot - and he must ask for the lord's permission to marry off his daughters, and on acceptance must pay a duty - the merchet. The villein is tied to the lands of the manor. The lord can impose arbitrary taxes and labour rents (ibid., pp. 343-51). The arbitrary labour rent disappeared, according to Vinogradoff, as a sign of villeinage around 1300 (*Villainage*, p. 315). The distinction between arbitrary and fixed rent is solely a formal legal phrase which has no significance in practice. It only occurs exceptionally in the manorial accounts (ibid., pp. 82 and 167). On the other hand, labour rent, contrary to what Vinogradoff writes in *The Growth of the Manor*, cannot be found as a proof of villeinage in the legal sources, "but in actual life...all stress is laid on the distinction between land held by rent and land held by labour." Up until 1300, that is (*Villainage*, p. 167).

36. Denton claims that the term "villain" is actually connected with this aspect - the will of the lord. "Its real meaning is best gathered from the word will." Denton, op. cit., p. 36, n. 4.

37. Ibid., p. 37.

38. Ibid., p. 38.

39. Ibid., p. 40.

40. Cunningham, op. cit., pp. 336-37.

41. Ibid., p. 339.

42. "It has been generally, but too hastily, assumed that the villains were practically successful....If the revolution had really been successful on its social side, it is hard to see why it was so inoperative politically...there is incidental evidence of the very general continuance of serfdom long after the time of the revolt." Ibid., p. 360.

43. Ibid., p. 303.

44. Ibid., pp. 308-09.

45. Ibid., p. 305.

46. Ibid., pp. 308-09.

47. Ibid., p. 308.

48. The silver content of the shilling fell regularly between 1300 and 1500 according to Seebohm's article in *The Archaeological Review* 1889, "The rise in the value of silver between 1300 and 1500". This means the cash price of silver was rising. The article was written during a polemic with Rogers, who thought that the price of silver was stable up until Elizabeth's reform of the coinage (Rogers, *The Economic Interpretation...*, op. cit., 1888). The inflationary development in the late middle ages is hesitantly confirmed by Seebohm, but firmly opposed by Rogers. So Cunningham and Rogers were in opposition to each other on this issue too. See Cunningham, op. cit., pp. 300, 486.

49. Ibid., p. 356, n. 4. This rather crucial view is supported in the first place by a general remark that labour services were not always itemized in the manorial accounts, secondly by a demonstration that on the one hand the labour services can sometimes only be deduced indirectly from the accounts, and were on the other hand temporarily commuted, only to return later to their original form. The term *opera vendita* is said to refer to such a temporary sale of work, a practice which according to Maitland was current at Wilburton and surrounding manors throughout the whole fifteenth century (ibid., p. 357, n. 1). The same is thought to have been the case in 1326 at Symondshide in Herts. Examples of indirect information about labour services are presented from three manors in the fifteenth century (ibid., p. 221). Finally, the persistence of the labour rent is documented by an example of its existence at the end of the sixteenth century (ibid., p. 476).

50. "At the end of the thirteenth century there were three different classes of tenants; those who had commuted all their services for a definite money rent; those who paid either actual service or gave the value of the services in money according as the lord preferred, and those who still performed their obligations either in whole or in part in the form of actual service." Ibid., pp. 218-21.

51. As for the issue of the demographic consequences of the Black Death, Cunningham is by and large in agreement with Rogers. His assumption "that nearly half of the population was swept away" (ibid., p. 304) corresponds fairly accurately with Rogers' calculations in the article "England Before and After the Black Death" from 1866, and is only a little more than Rogers' modification of this calculation from *The Economic Interpretation of History* from 1888, where Rogers thought that the Black Death "probably killed a third of the population" (ibid., p. 22). By contrast, Cunningham does not agree with Rogers on the numerical assessment of mortality. He does not believe that Rogers in the controversy with Seebohm refutes convincingly enough Seebohm's assumption that the population of England before the plague was five million. Rogers' premiss for his refutation of Seebohm is wrong, Cunningham thinks. "His conviction that the populace lived practically on wheat...oatmeal and other cereals than wheat were commonly used

for food....The area of food-producing land may therefore be taken as much larger than that which Professor Rogers assumes." Cunningham, op. cit., p. 304.

52. Ibid., p. 356.

53. Ibid.

54. Ibid., pp. 336-37.

55. Thus summing-up of Cunningham's views on the development of the relations of production must be read with the reservation that one can easily find utterances in the text that directly contradict this concentrated version. Sometimes it is difficult to tell what Cunningham actually means - a problem not peculiar to him, but fairly common among nineteenth-century historians. For example, as far as villeinage is concerned, and in direct contradiction of what has been stated above, we can read in connection with the Peasants' Revolt of 1381 that "it was possible to force back the villains into nominal serfdom" (ibid., p. 361). Similarly, in connection with the Revolt, we can read that the pressure on the peasants was increased with the recurrent plague outbreaks "until it at last resulted in the general outbreak of the peasants' revolt in 1381" (ibid., p. 357). On the very next page this absolute position is softened so that it was only a source of scattered, local dissatisfaction, while the Poll Tax, and in particular its method of proportional taxation "was enough in itself to render it unpopular, and this was the occasion which brought the separate and local discontents into a single focus" - despite the fact that "we cannot assign one cause only..." since Wycliff's teachings gave religious and ideological support to the smouldering rebellion (ibid., p. 358).

56. Ibid., p. 393.

57. "It had been computed that far more than half the population of England died during this terrible year. Sufficient proofs remain that this computation is far below the truth, though we may dismiss as an exaggeration the assertion that only a tenth part of the people of England remained alive". Denton, op. cit., p. 98. Denton is thus in line with Seebohm. It is interesting to see how different authors, more or less against the background of the same source types, can reach widely varying results. Cunningham criticized Rogers' source material, but reached the same mortality percentage as he did on the basis of source material said to be of the same type as Seebohm's: that is, ecclesiastical chronicles supplemented with court rolls. It should be noted however that Cunningham had not worked at first hand with the ecclesiastical sources, but got his information from Jessopp's *The Coming of the Friars*. Against this, Denton was able to reach the same result as Seebohm, apparently without using either the same primary sources or Seebohm's results, but using almost exclusively secondary sources. Apart from Knighton's dubious contemporary account and *Chron. Anglia*, Denton used a number of older, but non-contemporary, historical accounts. Although Denton thought that the decline in the population set in after the death of Edward I (Denton, op. cit., pp. 65, 79) he confirms that the Black Death had a hitherto unrealized drastic effect on the development of the population of England.

58. Ibid., pp. 106-108.

59. The difference between Rogers' and Denton's positions is on the whole simply that the former only speaks of the landlords' attempt to reintroduce the labour rent, while the latter also cites the landlords' attempts to force the labour rent up.

60. Ibid., pp. 108-109.

61. Ibid., p. 113.

62. Ibid., p. 234.

63. "Between the time of the Norman Conquest and the Battle of Bosworth Field the progress of agriculture in England was almost as imperceptible as the growth of its population". Ibid., p. 146.

64. Ibid., pp. 149-53.

65. Ibid., pp. 155-56.

66. Ibid., pp. 153-54.

67. Ibid., pp. 138-39.

68. There is no empirical documentation here for this, nor for the exhaustion of English arable around 1500. The closest he comes to verification is when he quotes Fitzherbert's *Surveying*, Chapter XXVII for the view that inadequate manuring of the soil in the fifteenth century was connected with miserable conditions of tenancy. Denton's thesis is grounded solely in a logical summing-up of the consequences that the development of the relations of production and the erosion of the manorial system in the fourteenth and fifteenth centuries must have had for the treatment of the soil.

69. Rogers, *The Economic Interpretation...*, op. cit., pp. viii-ix.

70. Ashley, op. cit., Book I, pp. x-xi.

71. Cunningham, op. cit., pp. 6-7.

72. "...politics are more important than economics in English history". Ibid., p. 9.

73. "Economic history is not so much the study of a special class of facts, as the study of all the facts of a nation's history from a special point of view....Nor should we be justified in contending that the special point of view from which we look at these changes is the one which gives us the most important and adequate survey of the national history". Ibid., p. 8.

74. Ibid., pp. 14-15.

75. "The chief problems which have to be faced are far less due to want of information than to the difficulty of interpreting the facts which lie to hand; there is a danger of reading modern doctrines into ancient records..." Ibid., p. 20.

76. Ibid., p. 18.

77. Ibid., pp. 18-20.

78. Ibid.

Chapter III

Epidemic, Commutation and Enclosure
(1884-1915)

III.1 Gaps in research

Research on the socioeconomic history of the late medieval centuries in the latter half of the nineteenth century left certain central fields of study to its immediate posterity - fields in which more detailed research was extraordinarily important for the further development of the more accurate and unified understanding of the period for which Seebohm and Rogers had prepared the way.

Although their pioneering work had its point of departure in a discussion of the socioeconomic consequences of the Black Death - a discussion that continued in the subsequent decades - the plague was not immediately made the object of a more exhaustive research effort. Interest was focused on placing it in its general sociohistorical context. To this end new source material was constantly taken up.[1]

At that time there existed no other monograph works on the Black Death than the above-mentioned book by J. F. C. Hecker - *Schwarze Tod* from 1832.[2] Not only did this study suffer from being far too general in the English historiographical context; it was now also considered obsolete.[3] In English research, there had as yet been no attempt to determine the epidemiological nature of the Black Death and relate it to the factual spread of the disease or diseases in society, or to its consequences for that society. But the standing controversy about the significance of the plague throughout almost half a century had created fertile soil for such an effort.

Not only that: the same controversy also led to more exact determinations and datings of the erosion of villeinage and the enclosure movement - two salient phenomena in the late medieval transformation process that had been discussed by historians of the latter half of the 19th century. From just before and after 1900 we have a group of monographs which purportedly concern the disappearance of villeinage, but which in reality are rather contributions to the debate on the course of the commutation process. It is characteristic of this literature that the upheaval in the agrarian relations of production is seen as having been determined by one crucial factor: the end of villeinage and the manorial system is, in all its complexity, concentrated in one theoretical fixed point - commutation. This approach is defended by Paul Vinogradoff in an article from 1900, where, besides generally defending the use of a few dominant concepts in investigating the structures of feudal society, he argues that labour services were the most crucial, general indications or proof of villeinage;[4] and, consequently, that commutation was the same thing as the end of villeinage.

There was one major reason why commutation, and thus the development of the feudal rent types at this juncture, took on enough importance to become the object of specialized empirical studies: not only had commutation assumed a very important position in the interpretations of the general course of economic history that we have reviewed; also, and perhaps more importantly, the transition from labour to money rents, and the latter type's genuine relationship with the change in the social nexus, had never been discussed in these earlier contributions.[5]

The linking of the enclosure movement with the late medieval transformation process, and in particular the rather obscure treatment this problem complex had been subjected to in the more general works from before the turn of the century, were in themselves sufficient grounds for the numerous studies of this historical problem that we encounter in the first decades of the twentieth century. But they were not the only grounds. A general interest in changing the time perspective in research on agrarian history also played a role.[6] And the growing interest in this subject must also be seen as a culmination of an intense discussion in the last decades of the nineteenth century of current problems in English agriculture and the conditions of the English rural labourer in particular - i.e. as a historical clarification of these problems.

This connection makes its appearance in W. Hasbach's book on the English rural labourer from 1908, in Lord Ernle's agrarian history from 1912, to which we shall return in another context,[7] and is directly described in the series of monographs on the enclosure problem from the period between 1900 and 1915. All of these relate the enclosure movement to the development of the specific structure of English agricultural landholding and to the position of the English rural proletariat.

Despite the fact that there were also other motives behind this research effort, it had direct results for the development of the view of the socioeconomic and technological changes in the late medieval centuries:

in the first place, because it led to a discussion of the chronology of the enclosure movement; in the second, because it provided insight into formal regional differentiations and temporal shifts in the process; and thirdly, because it supplemented existing knowledge of the motive forces behind this salient, much-discussed problem in English historiography.

III.2 The "epidemiological" theory

In 1891 there appeared the first volume of Charles Creighton's two-volume work *A History of Epidemics in Britain*,[8] a standard work which, with the necessary amendments, was reprinted as recently as 1965. Creighton understood perfectly that certain economic historians, as we have also seen, either neglected or doubted the socioeconomic importance of the Black Death.[9] And he also understood why the same historians were more or less forced to consider the Black Death as a crucial factor for the understanding of historical development, without necessarily being interested in the causes of the epidemics. For Creighton the Black Death was "part of the great human drama" that made its influence felt for about three centuries after the first plague outbreak in 1349.[10] The development of society in the fourteenth century could only be understood in relation to the Black Death, and vice versa. "The coming of the great plague was part of that movement, organically bound up with other forces of it, and no more arbitrary than they".[11] But in what sense was the Black Death a factor in the development of society?

Against the background of contemporary accounts he has difficulty in deciding whether mortality was particularly high in the lower social classes. For this reason among others, Creighton rejects any connection between the general reproduction level in society and the plague's ability to spread. He distinguishes between epidemics caused by famine, and epidemics without this connection. Thus he believes that the reasons for most medieval epidemics apart from leprosy *before* the Black Death must be sought in ordinary famine.[12]

Like various other diseases, such as Asiatic cholera, yellow fever and typhus, "bubo-plague also is a soil poison".[13] This theory, widespread as Creighton himself states in the sixteenth century, he seeks to verify partly with historical data, partly with studies from the nineteenth century of plague-stricken regions in the Arab world. He supports these with the assumption that there is a connection between the development of the plague and "cadaveric decomposition in circumstances of peculiar aggravation and on a vast or national scale".[14] The plague thus developed according to Creighton in China in the period up to 1342, where millions of people had perished in floods and famine, and their bodies had become the breeding-ground of the plague.[15] From here it spread by trade routes to a Europe whose soil was prepared for the explosive growth of the plague.[16]

However, Creighton is not quite sure that the spread can be attributed exclusively to the soil's potential for receiving and disseminating infection and unhygienic conditions in general. There was apparently also infection from individual to individual; but this aspect of the matter is rather more obscure. The source of infection, all other things being equal, was identical to the medium through which it spread. The plague arose and was spread as a poison in the soil, which in certain circumstances had a special affinity for the results of decomposed cadavers. He therefore understands excellently why the plague, when it hit England, found many victims among men of the church - those who literally lived on the graves of the dead.[17]

Creighton's view of the Black Death differs in general from the view predominant among the economic historians before him, even though a few of these sought to connect the rise and development of the epidemic with socially-created circumstances - for example Seebohm (see Chapter 1, p. 3, n. 6). But it is not a new and original approach to the problem we find here. As mentioned before, the main element in his theory was several centuries old. Creighton tried to develop this theory and in particular to verify it with historical and contemporary data from Arab plague regions.

III.3 The plague - a "neglected" historical turning point

A few years later the hitherto most comprehensive monograph on the Black Death in England appeared: Francis Aidan Gasquet's *The Great Pestilence*.[18] Gasquet attributed the Black Death extraordinary importance for the understanding of English history in the late middle ages. "The Black Death inflicted what can only be called a wound deep in the social body, and produced nothing less than a revolution of feeling and practice....In truth, this great pestilence was a turning point in the national life. It formed the real close of the Medieval period and the beginning of our modern age".[19] He therefore saw it as his task to contribute an account of the Black Death in its entirety. He did not think Creighton had done so in his work, which was announced while Gasquet was gathering material for his own.

One must concede Gasquet that his work is far more exhaustive than Creighton's, where the Black Death was only one epidemic among many. In Gasquet we find a far more detailed description of the spread of the plague and in particular of its socioeconomic and spiritual/theological consequences. Then again, he enriches us with no unified theory of the causes of the plague and its possible connection with the general conditions of life in late medieval society. In general, the plague appears in his account as an exogenous natural force, making an impact on society that is of extraordinary importance for the reproduction of its members and its socioeconomic and religious structures.

All the same, it is possible to piece together a picture of the nature of the plague and its relationship with social conditions, to the skimpy extent that he deals with these issues. In the first place he provides a description of the symptoms of the disease which does not differ substantially from Creighton's.[20] Secondly, he deals with the impact of the disease on various social strata, and the extent to which it was possible to protect oneself from the contagion. Thirdly, he notes the relationship between hygienic conditions in various sectors of society and the risk of contagion and tries to identify the vectors of infection.

For Gasquet the plague is a natural force whose spread is connected with particular climatic circumstances.[21] The infection came from direct and indirect contact with the victims,[22] and consequently the plague spread quickly in densely populated areas.[23] This was probably why the mortality was higher among clerics than laymen,[24] and why the cities must have been hit relatively harder - also because the sanitary standards here were in general very poor.[25] Despite the fact that personal hygiene in the middle ages was considerably better than normally assumed, and that it is thought that personal hygiene might have functioned as a precaution against the spread of the plague,[26] no social group escaped the plague; but the poorest were hardest hit.[27]

Gasquet's epidemiological insight and his information on the relationship between the Black Death and contemporary living conditions do not extend much further than what we found in certain of the economic historians. Our knowledge of these subjects is no further enriched by the two articles on the Black Death in East Anglia that were published by Augustus Jessopp in 1884 and 1885. Jessopp's articles concentrate on presenting and dealing with source types relevant to the clarification of the consequences of the plague in this area,[28] and allow only for a supposition that the Black Death "probably was a variety of the Oriental plague".[29] Jessopp's description of the disease goes no further than Hecker's account of its symptoms as described in the sources. And he completely neglects to relate the plague to social conditions.

Nor is our knowledge in this respect increased by Arthur Dimock's article from 1897, "The Great Pestilence".[30] The article, which bears the misleading subtitle "a neglected turning point in English history",[31] tells us only of the symptoms of the plague - cf. Gasquet - and that neither climate nor physical constitution had any influence on the impact of the plague. On the other hand, he thinks that the rich were less vulnerable than the poor, judging from the continental sources, and that poor hygiene and bad food stimulated the spread of the plague.[32]

In all the four works investigated, we find a description of the disease, and an outline of its spread. But we are told a limited amount about the causes of the epidemics and their relationships with social conditions. In fact only Creighton presents an actual epidemiological theory - and for good reasons, as the epidemiological study of the plague in the last decade of the nineteenth century stood just before the threshold of the most important discoveries for the scientific understanding of the nature of the disease.[33]

III.4 The population losses

The issue of the effect of the plague epidemics on the development of society plays a large and crucial role in the four works by Creighton, Gasquet, Jessopp and Dimock. They deal with many of the same problems raised by the economic historians. But do we find new material and new approaches to these in the monographs?

It is striking that these works use a far more systematic, developed body of source material to illustrate the demographic consequences of the epidemics than we have seen presented before. Thus we find Seebohm's study of mortality among priests in East Anglia confirmed by Jessopp's more detailed, extensive use of the Institution Books of the dioceses.

Jessopp also includes a type of source not previously used for this purpose: the court rolls, with whose help he thinks he can register the succession of certain tenancies in a number of manors. But the source is difficult to use, as many of the registers are missing precisely for the years 1348-1350.[34]

Against this background Jessopp estimates that between 1349 and 1350 the Black Death exterminated more than half the population of East Anglia.[35] He believes there was the same mortality in the towns, but that no documentation can be presented for this conjecture.[36]

Creighton on the other hand thinks that for a number of cities, including Leicester and London, there are mortality figures "which have an air of precision".[37] The chronicler Knighton's estimate of the mortality in Leicester is such a source.[38] The inscription, mentioned above and criticized by Rogers, on a stone monument in Charterhouse Churchyard is another which is used, along with the remarks of the chronicler Avesbury, to back up Creighton's assessment of the mortalities in London,[39] estimated at 50%, more or less the same as the cities of Norwich and Yarmouth.[40]

Creighton assesses the mortality among churchmen partly by means of a selection of monastic chronicles reproduced by various historians,[41] partly by comparing Seebohm's and Jessopp's studies of vacant benefices in the year of the first wave of the plague. So it is not surprising that he arrives at a confirmation of Seebohm's estimate of the mortality in this population group.[42]

Creighton ventures no general assessment of mortality among tenant farmers and rural labourers, although he had studied a series of court rolls with the express purpose of clarifying this.[43] Nor does he embark on a general assessment of mortality, either numerically or in percentages. Gasquet, however, attempts both: he thinks that at least half the population of England was wiped out as a direct result of the Black Death.[44] And although he admits that the source material available to him then was not "sufficient to enable us to form the basis of any calculation worthy of the name", he does attempt a numerical estimate of the development of the population between the Black Death and the Poll Tax in 1377 - with figures that correspond closely with those of Seebohm. Gasquet repeats Seebohm's view that the population figures before 1348 were somewhere between four and five million, and that in 1377 they had been reduced to less than 2.5 million, without mentioning Seebohm, but with due acknowledgement of Cunningham, who, as we recall, backed up Seebohm in the controversy on this point with Rogers.[45] It is noteworthy to see how Gasquet, like Cunningham, agrees with Seebohm on the numerical estimates of population development after the Black Death; while, also like Cunningham, and for that matter like Jessopp, he inclines more towards Rogers on the issue of mortality percentages.

Gasquet bases his assessment of population development on source material that represents a larger geographical area than we have seen before in any study. And although he thinks that only the abundance of the religious sources enables us to judge indirectly the mortality among the lay population groups,[46] he does not refrain from using non-religious sources - court rolls and *inquisitiones post mortem*.[47] He finds the diocesan registries in particular - the "institution books" - valuable; but supplements them with some monastic chronicles and a source type we have not seen used before: the so-called "patent rolls".[48]

It has been demonstrated how difficult it was to assess population development in the fourteenth century on the basis of the available material. The sources used were relatively few, and most of them refer to particular social groups, especially the men of the church. The most serious combatants in the controversy over the demographic consequences of the Black Death were therefore fairly cautious in their estimates. Creighton did not venture to make a general estimate of mortality, and Gasquet only pronounced on the issue with some uncertainty.

Less cautious estimates are made by Dimock. He ventures to draw up mortality rates distributed over social groups, inclining heavily towards Jessopp's and Gasquet's material - this despite his doubts about the representativeness of the material: around half of the servants of the church and the London middle class fell victim to the plague, while only a third of the secular lords succumbed. True, he is unwilling to determine mortality among tenant farmers, labourers and craftsmen,[49] but still considers himself able to claim that as far as general mortality is concerned "the conveniently round number of one-half is about correct".[50] But let us not spend too much time on Dimock's superficial and popularizing article, although it may have had some communicative value in its time. Let us instead look at the consequences of the high mortality.

III.5 Social, economic and moral consequences

Apart from Jessopp's, Gasquet's and Dimock's great emphasis on the description of the consequences of the decline in population for the church, we do not find much new material here.[51] In the socioeconomic and technological fields most of the views could, with modifications, be traced back to Thorold Rogers and some of his opponents - Cunningham, Ashley and Denton. There is thus a broad consensus that wages rose,[52] villeinage ceased,[53] and the manorial system was given its death blow in the latter half of the fourteenth century. There is agreement that the effort of the state to regulate wages with the Statutes of Labourers failed, that the manorial system was replaced by new forms of production and new tenancy conditions,[54] and that the enclosures were a sign of these developments, as was the increasing tendency in the fifteenth century to change from arable to sheep farming,[55] because of falling rents, rising wages and rising prices on non-agricultural products.[56]

It should therefore be obvious that the time after the Black Death, as Gasquet put it, "was a hard time for the landowners".[57] Jessopp was not convinced that the situation of the landlords could be described so simply. The smaller estate-holders were perhaps in difficulties because of the shortage of labour and rising wages,[58] but the class as a whole in his opinion gained considerable immediate advantages from the high mortality. The estate-holders not only collected large extraordinary duties - heriots - but also came into possession of areas that had hitherto been uninterruptedly tenanted. Jessopp also thinks that the mortality led to a concentration of "the landed property of the country". The land "came into fewer hands; the gentry became richer and their estates larger".[59] The surviving landlords profited not only from the high mortality among their tenant farmers, but also from the mortality within their own ranks.

Although Jessopp does not document these relationships convincingly, it is interesting to see these arguments used precisely in the consideration of the situation of the landlords after the Black Death. Hitherto we have only seen them used to argue for the advantageous position of the peasants and rural labourers and the crisis of the estate-holders in the second half of the fourteenth century.[60] But Jessopp goes further and uses the arguments inductively: "the country was producing less, it may be; but the people, man for man, were much richer than before".[61] This meant that the population crisis in general was to the advantage of the survivors.

We recall that Cunningham saw the acceleration of the dissolution of local justice that had already begun as the most serious social consequence of the Black Death. This theme was touched on by both Gasquet and Creighton, but in particular by Jessopp, who looked at what one could call the moral habitus of the population after the plague.

Although he found terrible examples in the Norwich court rolls he studied - a priest who took up highway robbery, more violent behaviour from the estate stewards, peasants' refusals to pay rent, and increasing strife and bloodshed among them - he still did not think that he had found proof that the time after 1350 was typified by widespread and growing lawlessness. Apparently, the drastic mortality did not cause panic.[62] The decentrally organized societal structure ensured, according to Jessopp, that calm and the general rule of law prevailed.[63]

III.6 Summing-up

It was only natural that the monograph works surveyed above enriched debate on the late medieval centuries with a more unified impression of the history of the Black Death. They helped to clarify the routes by which the epidemics spread - especially Creighton and Gasquet - and showed some possible aspects of their consequences that the economic historians had not dealt with, and which we have not touched on either, as they fall outside the scope of our study - for example spiritual and theological aspects.

On the other hand the monographs - apart from a few exceptions mentioned here - contributed nothing substantially new on the socioeconomic and technological changes in society. Much was drawn from the results and approaches of the economic historians, and from their assessments of the population development in the second half of the fourteenth century. Yet it should still be stressed that in the monographs we meet a more differentiated and representative body of source types for the demographic problem than before - source types that to a reasonable extent allow us to assess the mortality among both laymen and clerics, lords and tenants, rural labourers and townsmen.

This is not to say that the source material presented was sufficiently representative, either socially or regionally. Even after the publication of the monographs on the plague epidemics at the end of the nineteenth century we must note that the demographic consequences of the epidemics could still be most reliably documented for the *men* of the church. And, no less important, the southern and eastern parts of England were still over-represented, despite the inclusion of sources from Cornwall and Lancashire, for example.

None of the four authors speaks directly of crisis in the fourteenth century. The plague epidemics were for Gasquet and Dimock the starting-point for a general socioeconomic transformation process. Against this, Jessopp and Creighton saw the plague epidemics rather as factor accelerating social changes already in progress. For Jessopp, Gasquet and Dimock the plague was part of this economic process - but at the same time separate from it. The plague was exogenous to the development of society in general. Only Creighton attempted to make a direct connection between the plague outbreak and conditions inherent in society which made it an organic part of the social process and the stage of social reproduction which prevailed before 1348.

III. 7 Commutation, 1325-1450

We have already noted that the discussion of the end of the institution of villeinage in England around and after 1900 centred on one crucial point - the commutation process. The composition of Thomas Walker Page's article "The End of Villainage in England" from 1900 is a clear example of this. The article falls into two parts, of which the first must be considered a general framework for the actual object of the study presented in the second part. The first part describes the institution of villeinage at second hand, as Polloch and Maitland had done in the *History of English Law,* and as Vinogradoff had tried to do in *Villainage in England* in particular.[64] Then, in the second part of the article, Page turns to an empirical account, relying on primary sources, of when the commutation process took place.

With this aim in mind, he had reviewed a large number of "ministers' accounts" in the Record Office, at the British Museum and in the collections of St. Paul's Cathedral, and to a lesser extent had consulted some currently published and edited manorial archives.[65] The material covered Crown estates as well as vassal and ecclesiastical holdings. Of these, the last category is clearly over-represented. Geographically, the material represents an area of south east and central England up to a line drawn from Southampton through Gloucester and Warwick up to Lincoln, but with a representative focus in the eastern part of the country. For good reasons, Kent is excluded, since villeinage never gained a foothold there. Page ordered his material in three tables. The first shows the proportion of manors on which the labour rent had respectively been abolished, half abolished, almost abolished or maintained in the period between 1325 and 1350. Table Two specifies the same ratios for the period 1350-1380, while Table Three covers the period between 1380 and 1450.[66] From these constructions it emerges that commutation first gathered speed in earnest after the Black Death, and that the labour rent persisted on many manors right up until the fifteenth century. This applied first and foremost to Ramsay Abbey's holdings in Huntingdonshire, Cambridgeshire and Norfolk, which made up a large proportion of the collected material.[67]

III.8 Commutation and population losses

On this basis Page thinks that the plague must have been the factor that triggered off the commutation process - a rather far-reaching conclusion, given his professed familiarity with Rogers' work and Rogers' observation that precisely the religious estates, particularly the monastic ones, seem to have had a deliberate anti-commutation policy and kept up commutation longest. But how does Page argue for his view otherwise?

He does not let it stand or fall with the empirical documentation. He presents what he himself feels is a "logical" explanation of why, as his empirical studies showed, commutation first began in earnest after 1350. Unlike Karl Marx, who (as we saw in note 5) both thought that the money rent had helped to dissolve the feudal social synthesis by stimulating trade and manufactures, and assumed that this dissolution was already in progress even then, Page believed that commutation quite simply only began when trade, manufacturing and the money economy had gained a foothold in society.

He considered that this situation had been latent just before the Black Death left its mark on England. The money economy flourished around London, and "the spoils of Calais and the new cloth industry increased the amount of money in circulation after that time; but it was the Black Death, which, by destroying nearly half the population while leaving the available capital and the medium of exchange as great as ever, hastened the transition from a system of barter to a system of money payments.

Before that event it does not seem that the change had gone far enough to render possible a complete abolition of the predial services of villains".[68]

It is not easy to grasp what Page means by his statement that the amount of capital in society was constant, and impossible to understand what significance this is supposed to have had for the implementation of commutation in the context of the declining population figures. We can also note that he does not pursue this line of reasoning. For the development of commutation is not explicitly claimed to have its roots in the development of trade and manufacturing - only in the increased money supply in society. It emerges that the money rent simply gains ground because the Black Death "by destroying half of the population...doubled the amount per capita of the medium of exchange; and at the same time by causing...a greater fluidity of the surviving population it made the use of money more familiar in the country districts".[69] This means that the relative rise in the money supply along with the increasing mobility of the population constituted, according to Page, the immediate causes of commutation.

But the fact that the Black Death resulted in a crisis for the landlords also played a role. The reduction in the population led to a long-term shortage of labour in agriculture which lasted until the middle of the fifteenth century, by which time the countermeasures of the landlords against this unfavourable situation for them, the introduction of new terms of tenancy, and the transition to sheep farming, had made such an impact that they had regained the upper hand over the rural population. In the meanwhile the villeins took their chance. "Instead of permitting the lords to lay heavier impositions upon them, the villains seized the opportunity to lighten those they already bore".[70] Against Rogers' theory of the causes of the 1381 Revolt, Page claims that the lords did not even attempt to reintroduce or enforce villeinage, and thus the labour rent, in the period after 1350.[71]

Francis G. Davenport agrees in principle with Page about this sequence of causes and effects, and on the general dating of the beginning of the commutation process. Davenport's study of manorial accounts and court rolls, mainly from the manor of Forncett in Norfolk - incidentally reasonably free of inductive inferences - by and large confirms Page's dating of the commutation process.

The source material from before 1300 is very scanty and discontinuous, but it is Davenport's view that there was no clear trend towards the replacement of the labour rent in this period. But between 1306 and 1376 "the economic position of the serfs, the administration of the demesne, and the whole organization of Forncett Manor were revolutionised".[72] The revolution began after 1349, when "the decrease in the amount of labour rents available which must have been felt immediately after the Black Death, the high price of hired labour, and the restless and refractory spirit of the tenants, of which there is abundant evidence in the rolls of Forncett and of Moulton, as well as of other Norfolk manors, were doubtless the causes that finally induced the lord to give up the cultivation of the demesne."[73]

Thus Davenport thinks that the commutation documented in the fourteenth century at Forncett and other manors was mainly set off by the decline in the population after 1349. But his study does not preclude that the process had already begun by this time.[74] It should however be pointed out that both Davenport and Page directly connect the commutation process with the population decline after the Black Death, even though Page's explanatory model is permeated by a vulgar-monetarist theory, and though they consequently date the process to the latter half of the fourteenth and the first half of the fifteenth century.

III.9 Commutation and population growth

Edward P. Cheyney utterly rejects this, taking the view that commutation and the end of villeinage was a process that extended over a period of more than 250 years - from the end of the thirteenth far into the fifteenth century.[75] Cheyney reached this position by studying a highly differentiated body of sources consisting of historical accounts, contemporary chronicles and reprinted primary sources, first and foremost consisting of edited court rolls - material which, at least geographically, seems more representative than both Page's and Davenport's.

The gradual erosion of the manorial system and the dying-out of villeinage are said by Cheyney to have been caused by two processes, both of which began around 1300: on the one hand, the commutation of labour rents to money rents; and on the other the leasing out of the demesne.[76] And these two processes seem to have influenced each other. For it emerges implicitly from his account how the abandonment of the labour rent seems to have required that the demesne lands formerly run on the basis of this type of rent now had to be leased out, and vice versa. He

81

does not consider a factor that had been stressed by historians in this context since Rogers' day - the emergence of a class of wage-dependent rural labourers - or that it may have been because of the efforts of this class that the manorial land could be maintained after commutation.

These views, and the fact that Cheyney can actually present examples of demesne land being leased out even before 1349, explain why he attributes no great importance to Rogers' idea that the stock and land lease was introduced after the Black Death. "Whatever may have been the case, and whatever the exact dates, the silent revolution was in progress during the fourteenth and fifteenth centuries".[77]

Cheyney is not particularly interested, as is evident from the above quote, in the reasons for the process he describes and tries to verify. He neither attempts to explain the course of commutation or the increasing leasing-out of demesne land. He observes and contrasts these processes without otherwise clarifying the connections between them. Given this, it is striking to see how close Cheyney is to presenting a new and radically different explanation of the fall of the manorial system and the progress of commutation.

Unlike almost all previous historians, he thought that the shortage of labour in the latter half of the fourteenth century after the plague can hardly have failed to delay the commutation process, "although actual testimony to that point is scarcely available".[78] Not too far from Rogers' theory of the origins of the 1381 Revolt, Cheyney thought there were manifest grounds for the landlords to turn against a general abolition of villeinage in connection with the Revolt.

Cheyney's view that the population decline after 1350 acted as a check on the dissolution of villeinage that was already in progress bears within it the understanding that the population growth may have stimulated commutation - a relationship touched on in another context by Paul Vinogradoff in his review of Page's article. In the same volume of the journal in which Cheyney's article was published, Vinogradoff wrote: "The treatment of surplus population was one of the weakest sides of the manorial arrangement, and it has to be reckoned with in any attempt to explain the gradual change in economic conditions". There were limits on how much population growth the manors could absorb without bursting out of their original form, because "to put it shortly, villain service was in main fitted to the holding and not to the persons".

The extent of labour services was, all other things being equal, determined by the fact that "a certain quantity of labour should be performed from certain tenements". If the number of persons associated with a given tenancy rose, this in the first place precluded "any thorough exploitation of the personal element" - that is, of a growing volume of labour. In the second place it meant that "under ordinary circumstances" it was relatively easy to leave the manor. The ties of the villein with the land "held good as long as conditions were not only positively but negatively against migration" - that is, as long as labour services were reasonably moderate, and as long as there were not "the agricultural, industrial and commercial conjunctures close by for the tillers of the virgate to forsake it and seek better wages and better profits in other employment".[79]

If this interpretation is accepted it must be remarked that the population growth not only loosened the bonds of villeinage that Vinogradoff was referring to. It also created an incentive and conditions for the replacement of the manorial system, either through the leasing of demesne land or its continued cultivation using hired labour; in both cases with commutation as the result. The incentive - a different and more adequate "exploitation of the personal element", and the preconditions for this - an emancipated surplus population - were according to Vinogradoff derived from the population growth. Incidentally, Vinogradoff also thought that as early as the thirteenth century a certain amount of the labour rent had already become "bad business" for the lord. The so-called "boon-days", work imposed on the peasants on the busiest days of the year, involved according to the documentation presented counter-payments in goods to the peasants by the lord. The relative rise in the value of this payment caused by the generally rising level of prices gradually undermined the value of these special labour services to the lord.[80]

It is important to keep in mind these hints of an explanation of the view, often expressed before, that commutation, the leasing-out of demesne land, and the dissolution of the manorial system in general, were already in progress before the Black Death; not only because there are very few, rudimentary attempts at explanation to be found in the literature before 1900, but also because here for the first time we are seeing the demographic factor used in the exact opposite way from what was usual at that time. It must however be stressed that there were only hints, and that we have not yet been presented with more elaborated attempts to explain the socioeconomic changes in the period before the Black Death. Interest was still primarily concentrated on the documentation of the chronology of the process.

III.10   Commutation, 1320-1350

Up to this point, the most comprehensive empirical survey of the extent of commutation just before the Black Death could be found in an article by H. L. Gray from 1914.[81] Gray had no great interest in the causes of commutation - only in providing documentary proof of its spread in the last decades before 1350. The article partly takes the form of a polemic against Page, of whose study he is highly critical - among other reasons because Page's sources mainly referred to church estates, with eastern England greatly over-represented.[82]

Gray, however, wished to study both church and secular estates spread all over England. He therefore also included the so-called *inquisitiones post mortem*, which could shed some light on secular estates all over the country. Gray thought that this type of source was far more suitable for illustrating the extent of commutation before 1350 than the manorial accounts, which he had to use however for the diocesan estates.[83]

Gray classified the information he found in the *inquisitiones post mortem* into 1) appraisals of the manor's buildings, fields, meadows, pastures and woods; 2) statements of incoming money dues (consisting of a. *redditus assisae* or fixed rents, b. *auxilium* or money rent, c. fines, and d. mill charges); and finally 3) statements of *opera* or labour rents. His use of the source, in all simplicity, is that in every single case he compares points 2 and 3 with one another.[84] The method can be used because point 3 in the source - the labour rent - is almost always converted to money and therefore directly comparable with point 2 - the money rent and other money dues.

Gray categorizes the material in four groups: A) where labour dues were still rendered in full, point 3 should make up at least half of point 2; B) where labour dues were rendered at certain times of the year, they should account for between about a quarter and a third of the money dues; C) where the labour rent is only of the *precariae* or "boon" type, or where only a few peasants rendered full labour services, it must not exceed one eighth of the money dues; and, finally, D) where the labour rent has been fully replaced by money dues.

The survey of the secular estates covers the period from 1333 to 1342 and a total of 521 estates.[85] Of these, 212 are distributed over the northern and western counties, while the remaining 309 are distributed over the central and southeastern counties. In Norfolk, Suffolk, Essex, Hertfordshire and Sussex the estates are predominantly of type A or B. In Cambridgeshire, Buckinghamshire, Middlesex and Hampshire types C and D are in the majority, and the farther we move to the west, the more type D dominates. In direct contrast to Page, who thought he could prove that commutation was most advanced around London, Gray's scheme thus shows that commutation in the secular estates had progressed furthest in the northern and western regions.[86] But at the same time Gray supports Page's study within the limited area he had looked at. For Gray concludes that "in that part of England lying south and east of a line drawn from Boston to the mouth of the Severn, full or considerable services were still rendered on about one-half of all the manors, the burden heaviest in the south-eastern counties, Kent excepted".[87]

Five dioceses in northern and western England and four in southern and eastern England had been studied. Gray approached the task as methodically as possible, analogously with his approach to the secular estates: that is, he attempted to compare money dues and labour services. He could make these comparisons when there were notes in the manorial accounts on labour services rendered in the so-called *grangiae*. When this was the case, the items listed could be compared with the more frequently-itemized *opera vendita* in the accounts in question. In cases where the method was not feasible he compared, as an alternative, the *opera vendita* with the itemized expenditure on the purchase of labour.

The trends noted above are confirmed by these methodologically more difficult and uncertain constructions. The study of the diocesan estates shows that commutation was almost accomplished before 1350 in the northwestern shires and in Norwich, where the commutation process seems to have been further advanced than on the secular estates. In the south and east labour rent was still widespread on the Winchester and Chichester estates. On the estates of the Ely diocese the commutation process had hardly begun.[88]

The review of a relatively large number of "extents" from 227 monastic estates scattered over the country, but with the highest concentration in the southern and eastern areas, further supports this impression. A comparison of "opera" and "assize" rents from these extents in the last three decades before the Black Death leads Gray to the conclusion that commutation on these monastic estates appears to have been at least as advanced as on the secular estates in this period. "What stands out clear is the general agreement between lay and monastic lands, not only in the contrast of the north-west with the south-east, but in the identity of the counties where services were most burdensome".[89]

In sum, it can be concluded that Gray's investigation of documents from more than nine hundred estates spread all over England showed that north west of a line drawn from Boston to Gloucester, insofar as labour rent had existed in this area previously, commutation had almost been accomplished before 1350. South east of this line the labour rent still existed after this time, but to very varying extents from estate to estate and region to region.

III.11 Summing-up

One must grant Gray that this result created a basis for understanding how Rogers and Page could arrive at such different views of the extent of commutation before 1350. Rogers' position, that commutation was in progress before the Black Death and that the process was accelerated by it, was split in two in the discussions between Page and Davenport on the one side and Cheyney and Gray on the other; but it was reconstructed in the latter's account, in his demonstration that the apparently rather incompatible views on the chronology of the commutation process had an empirical background in regional differences.

Methodologically, it is true not only of Gray but also of Page, Davenport and Cheyney that, like Rogers, they developed their positions from interpretations of primary source material - a method we have not seen used very much on our field of research in the period between the publication of the first volume of Rogers' major work and the 1890s. Just as we found that the monographs on the demographic aspect of the Black Death used different supplementary source types from those used by Seebohm and Rogers, we can note here that Rogers' one-sided use of the manorial accounts was complemented by other types of sources for the commutation process - extents and court rolls.

The methodological model is thus not taken directly from Rogers. What is probable, however, is that F. W. Maitland's study of the history of the manor of Wilburton was a source of methodological inspiration, at least for Gray. Maitland's article was not the first monograph on the economic history of a single manor or manorial complex,[90] but it was the first work to take up the commutation issue in the light of empirical data from different source types applied to a delimited local unit.[91]

There are grounds for seeing this work as the precursor of a whole school of thought, as a series of articles and books appeared in the following period whose content seems traceable to it; either to Maitland's subject - the commutation process - or to his method. The four articles on the breakdown of villeinage we have looked at in this section were only part of a tendency in historical theory and methodology around, and particularly after, 1900. A number of other works might have been considered, but - like Maitland's - they are not all equally relevant to the clarification of our problem area.[92]

In the introduction to this chapter it was said that interest in one of the most manifest structural features of late medieval England - the dissolution of villeinage - was concentrated in this period on a particular theoretical approach to the problem: the commutation of feudal labour services to money dues was seen as constituting the end of villeinage, although there was full awareness that this feudal institution involved a number of other important legal, social and economic characteristics than just the labour services, and that the money dues were not in themselves fundamentally incompatible with the maintenance of the institution.

The most obvious argument for developing the links in this theoretical chain is that there was much to indicate a real, specific, historical succession leading from commutation to the decline of the other characteristics of the institution of villeinage in the late medieval centuries. But more general theoretical reflection also seems to have played a role. Thus we saw how Page returned to a previous paradigm where the overall development of trade and the money economy was linked with the spread of the commutation process. Page saw the reason for commutation in increasing trade and a growing money economy in general, and in a relative increase in the money supply after 1350 in particular.

Although, apart from this example, we have not come across direct, and certainly not developed, theoretical attempts to interpret the connection between commutation and the general form of social nexus in the four authors dealt with here, it seems reasonable to see their approach to the understanding of their subject in the light of a more general theoretical concept which encapsulated the feudal economy in a network of primary reproductive structures, and where the breakdown of these structures was heralded by the development of socioeconomic relations of production borne up by money - a development, in Ochenkowksi, Ashley and others' terms, from the natural to the money economy.

Within this theoretical framework commutation can be interpreted as an indication of the dissolution of the feudal social and economic order in general, or of aspects of it, as in the present case, where commutation was identified with the end of villeinage.

III.12 The chronology of the enclosure movement

Enclosure in the historical context means the fencing-in of a piece of land with hedges, fences, dikes or other barriers preventing the free passage of animals or people. In terms of agricultural technology, the enclosure of a piece of arable land in an open-field system was the first step towards converting it to pasture. Land was enclosed to keep cattle or sheep in, while the fencing-round of a common pasture area was the first step towards the cultivation of this piece of land. The fencing was set up to hold the cattle out. In socioeconomic terms both types of enclosure meant the end of the former collective usufruct and common cultivation, if such there was. With enclosure, usufruct was individualized. The enclosed land was removed from the public domain.

We recall how the prevalent view among nineteenth-century historians was that the enclosure movement in England took place historically in two tempi. The transition from tillage to cattle and sheep farming was thought to be a major agricultural result of the social and economic unrest of the late middle ages, while individual arable farming was thought to have appeared later.[93] Ashley, for example, spoke of two intensive enclosure periods - from 1470 to 1530 and from 1760 to 1830.[94]

Both this classification into phases and, on the whole, the close connection between the late medieval transformation process and the enclosure problem that we found in the literature from before 1900 must be directly doubted in the light of the extensive literature dealing with the problem in the first decades of the twentieth century.[95] The bulk of this concentrated on the period from the seventeenth to the nineteenth century,[96] while the researchers R. H. Tawney and E. F. Gay addressed themselves to the historical significance of the enclosures of the sixteenth century.[97]

Against this background, the enclosures in English agriculture seem more associated with the period from the sixteenth to the nineteenth century than our period - the thirteenth to the fifteenth centuries. This is an impression that is further reinforced by the documented fact that enclosures first gathered speed in the eighteenth and at the beginning of the nineteenth century. Thus Tawney concludes that enclosure in his period - the sixteenth century - was only sporadic.[98] He applauds J. L. and Barbara Hammond's observation that "at the time of the great Whig Revolution, England was in the main a country of commons and of common fields", since "the whole of the cultivated land in England in 1685 did not amount to more than half the total area, and of this cultivated portion three-fifths was still farmed on the old common-field system".[99]

The determination of the chronology of the enclosure movement was of course to a great extent due to purely documentary efforts. But the relative shift in the temporal perspective of the movement that most of the literature after 1900 gave rise to, compared with previous research's close linking of parts of the enclosure movement with our late medieval issues, must also be assessed in the light of a number of other factors, some of which we have already mentioned in the introduction to this chapter. A significant factor in this respect is the view of the reasons for the enclosure phenomenon found in the literature. The continuing discussion of this issue sheds light on the chronological aspect and - especially interesting to us - provides further clarification of the relevance of the enclosure movement to our subject.

III.13 Enclosures and wool manufacturing

We recall how the early enclosures were linked - by Ashley, among others - with the growing demand for wool and rising wool prices. The expansion of wool manufacturing along with the rising wages in agriculture were thought to have led to extensive conversion of arable land to sheep pastures. This interpretation was questioned from several sides, in the first place because doubt was cast on the price statistics constructed by Rogers;[100] and secondly because the now-established opinion of the chronology of the enclosure movement, faced with this interpretation, raised the question of why the first enclosures only took place about a hundred years after wool manufactures began to boom. This question led supporters of the prices and wages theory either to argue that the agrarian changes were rather due to low corn prices than high wool prices,[101] or to seek the reasons in rising wages.[102]

But these objections did not involve a blanket rejection of the theory of the relation between the commercialization of society, in particular involving the expansion of wool manufacturing, and the agrarian enclosure movement. Eminent historians like Tawney and Gonner still saw an obvious connection between the growing social division of labour connected with the growth of wool manufacturing and the agrarian innovations. With this division of labour, they thought, the demand not only for wool, but also for food and labour, grew. There arose a division of labour between countryside and town which had a dissolving effect on the agrarian subsistence economy.[103]

Tawney viewed the problem dialectically. The enclosure movement was a precondition of the development of wool manufacturing.[104] But then wool manufacturing and the rising demand for wool, labour and food in the urban trades were preconditions or perhaps rather motives for enclosure, for more extensive sheep farming, and for the consolidation and individualization of arable farming.[105] Thus, apart from considering mercantilist state policy's importance for the process from Tudor times up to the seventeenth century, Tawney also linked the socioeconomic changes in late medieval agriculture with the enclosure phenomenon.

So did Gonner, although he did not exhibit the same dialectical thinking. He thought that the enclosure movement either took place as a necessary development of the original common cultivation system, as an incorporation of new land outside the framework of this system, or as a measure which toppled the common cultivation system[106] - a necessity which became pressing as a result of commercial and urban development. But he also stressed the importance of certain natural preconditions for arable farming.[107] Gonner, who thought that the enclosures of the fifteenth and some of those of the sixteenth century were primarily a matter of creating conditions for expanded sheep farming with the resulting attrition of the village commons,[108] therefore arrived at a partial confirmation of Denton's soil exhaustion thesis. The rising demand for wool, combined with "a disturbance in the labour market" and soil exhaustion "in many places" led to "the partial abandonment of cultivated land to sheep farming".[109]

Drawing the inferences from these assumptions that there was a close connection between the development of trade and wool manufactures and the progress of enclosure over the English shires, one might reasonably expect that the areas lying near the centres of commerce and wool manufacturing in eastern, southeastern and western England would be enclosed first. According to Leadam's and Gray's studies this inference is correct, inasmuch as the *early* enclosures were said to have affected precisely these areas, and inasmuch as the sixteenth century's admittedly modest enclosures first and foremost took place in central England.[110] The inference is also confirmed by Slater's map of the eighteenth and nineteenth century enclosures, which shows that the open field system survived longest precisely in the central area.[111] On this basis, there is thus some indication that in the case of the *earliest* enclosures one can trace a connection between the areas affected and their locations near the wool manufacturing and commercial centres. But this connection is not absolute. There are also examples of the open field system disappearing just as early in areas that were not near these centres.

III.14 Enclosures and field systems

The conflicting evidence is examined in H. L. Gray's standard work *English Field Systems* from 1915, where Gray views the enclosure problem in a quite different context than we have seen before, and which also sheds light on the chronology and geography of the process.

Gray thought that "agricultural progress [in the middle ages] was bound to take one of two directions. It was necessary either that the unenclosed arable of a township should be brought under better tillage continuing to lie open, or that it should be enclosed and given over to convertible husbandry".[112] In the central area the first direction was taken - first with a general transition from a two-course to a three-course system in the thirteenth and fourteenth centuries;[113] then with a division of two fields into four or a change in the open fields proper through partial consolidation of the peasant's lands or partial enclosures in the period from the mid-sixteenth to the mid-eighteenth century.[114]

That this direction was taken in central England is connected according to Gray with the fact that the open-field system only existed within this area - from Durham in the north to the Channel in the south - and because for Gray it is an important point that the central English field systems were fundamentally incompatible with enclosures. The development of the open-field system here arrested the enclosure movement so long that when it did begin, it was very drastic and revolutionary.

An equally important point in Gray's book is that the medieval field systems which existed outside this area exhibited structural elements that can be seen as related to aspects of the enclosure movement. It is thus attempted to explain the early enclosures in Devon, Cornwall, Cheshire, Lancaster, Suffolk, Kent and Essex[116] by saying that the medieval field systems in these shires among others had an intrinsic affinity with the progress of the enclosure movement. Thus Gray presented a theory that explained the chronology and geography of the enclosure movements in the light of the original medieval field systems' varied structures in England.

Besides the two-course and three-course systems in the central area, Gray thinks in principle that one can speak of three other field systems: the Celtic-inspired in the north west, west and south west; the Kentish; and the prevalent system in East Anglia.[117] What the three systems had in common was that the peasant's plots had originally been more or less consolidated, and that the later splitting-up was an effect of split inheritances.

While the central English "virgates" consisting of subdivided strips of land represented an original state of affairs according to Gray, as they were first described in the thirteenth century, he attempts to demonstrate how the splitting-up of the peasants' lands took place in parts of the "Celtic" area as a reflection of early Celtic laws on the equal division of land among the heirs to a holding.[118] In areas where tillage was predominant this led, because of cooperative ploughing,[119] to the development of the so-called "runrig" system, where the infield was redistributed every year in land plots that were non-contiguous for the individual peasant.[120] This, along with the fact that the consolidated land represented the original historical situation, might suggest that the semi-Celtic field system made it relatively easy for the landlords to reconsolidate the peasants' land as the leasing system gained ground.[121] In the sense of consolidation of the individual plots, enclosure in the semi-Celtic area was equivalent to a return to the original form of the field system.

In areas dominated by cattle and sheep farming the split inheritances led on the other hand to fenced-in, consolidated pastures, since the splitting-up of the land through inheritance did not stand or fall as much, according to Gray, with the quality of the soil with cattle and sheep breeding as with tillage. "Probably this is what happened at times in Wales. There in the sixteenth century township after township consisted of closes, those of a holding being frequently contiguous".[122]

Thus Gray saw the early enclosures in Devon, Cornwall and Cheshire, that is the early reconsolidation of the peasants' plots, and the enclosure of areas either for keeping cattle or sheep out or in, as "a normal manifestation of the Celtic system".[123] He was unable to say, however, that the same applied to the northwestern areas, which were otherwise also considered to be under Celtic influence, since there was no certain evidence here of the splitting of inheritances. However, he does claim that it applied to Kent and East Anglia.

In Kent all cultivated land was split into more or less rectangular areas called *iuga*, *dolae* or *tenementa* - a classification which in the middle ages was mainly for taxation purposes. The peasants' plots were now often scattered over several of these units. According to Gray, they were not originally split up, but became so through split inheritances.

A more or less equivalent system was used in East Anglia. The fiscal unit known as the *tenementum* in the fourteenth and fifteenth century suggests to Gray that the peasants' original plots must either have been completely consolidated or at least must have consisted of a group of pieces of land that were not widely scattered.[124] It is his thesis that there was an original unit of possession which, like the Kentish *iuga*, was gradually subdivided through inheritance, but that the manorial system and the collection of rents preserved the unit for administrative reasons.[125]

The rotation of crops in Kent, Gray thought, was variable. Sometimes there was alternation between cultivation and fallowing, but often the tendency was towards successive cultivation without fallow periods. And "the absence of a three-course rotation, and especially of a large compact fallow field, made easily possible the reconsolidation of scattered parcels as soon as the tide turned in that direction....Toward...enclosure the flexible field system contributed in no negligible degree".[126]

An important reason why the "large compact fallow field" was not used in Kent was that here, as in the "Celtic" areas, one often had great forests, heaths or moors that could be used for pasture.[127] Such resources were not available to the same extent in East Anglia, and for Gray this, along with the traditionally widespread sheep farming in this area, and the competition for sheep manure, formed the background against which a field system that had originally been identical to the Kentish one developed special "pastoral arrangements".

Nor did East Anglia develop a proper two-course or three-course system with regular fallowing of a compact area. What did happen was that the individual peasants, and even more so the lords, fenced in their part of the land in fallow periods, ensuring that they could duly exploit the manure of their own or others' animals. These specific pastoral arrangements - the fold-course system - cannot thus be explained exclusively in the light of the prevalent field system. In that case they could perhaps also have developed in Kent. They are thought to have been particularly associated with large sheep flocks and the competition over rights to the sheep manure.[128]

Gray claimed, then, that a number of features of the field systems that were predominant outside the central area made enclosure more or less inherent in these systems, or at least eased the progress of the system. In the first place he thought that he could demonstrate how there had originally been consolidated land areas for every farm in all the areas mentioned. Secondly, it was claimed that flexible field systems were used in Kent and East Anglia that, most importantly, did not entail the use of the two-course or three-course rotation systems' compact fallow areas. Thirdly, both the semi-Celtic and the East Anglian field systems involved temporary or permanent enclosures as early as the middle ages. Fourthly, the subdivided parcels were redistributed every year in the "Celtic" system.

III.15 Summing-up

On this basis it might seem that there are grounds to play down the significance of the enclosure phenomenon as a problem in our field of study. The primary reason for this is that research around and immediately after 1900 placed the focus of the movement centuries after the late medieval era. But this does not mean that we can conclude that our subject has no significance for the development of the enclosure movement. On the contrary, the social and economic upheavals of the late middle ages seem to have been a crucial initiating factor.

We have seen a confirmation of the general assumption of nineteenth-century historians that in the rather longer term the population decline from the mid-fourteenth century on forced a general transition from tillage to sheep farming. We have seen how it was thought that the expansion of English wool manufacturing created a basis for this process, and how the spread of sheep farming underpinned the development of wool manufacturing. Beyond this, we have been presented with the view that the development of wool manufacturing in particular, and the commercialization of society in general, pressed agrarian innovation in the direction not only of an expansion of sheep farming, but also towards a rationalization of the necessary tillage technology by means of enclosure - consolidations of the individual farms and the introduction of new crop rotations to meet the growing need for exchangeable agricultural products and labour caused by the increasing social division of labour. In the period following the thirteenth to fifteenth centuries we saw how it was thought that enclosure was only implemented in certain areas of England. We first found the answer to this regional heterogeneity among scholars whose views corresponded with the views current in the nineteenth century - that is, that the regional nature of the early enclosure movement, its early impact in the eastern, southeastern, western and southwestern parts of the country, should be seen as connected with the location of the wool manufacturing and commercial centres. This answer neglected the fact that there had been agricultural development in the central area in and after the late medieval centuries.

This neglect was remedied by Gray, who thought that agricultural development in England since the late middle ages took place either because of the enclosures or, as in central England, because of the development of two-course and three-course rotation. Gray sought the explanation in the original structures of the various field systems and offered an argument that, if it is linked with the demographic fluctuations of the high and late middle ages - which Gray did not do - sheds an interesting light on the enclosure problem.

Gray spoke of a "necessary" development of agricultural technology, but he went into no detail about the circumstances that made it necessary. However, his view that the originally consolidated land areas in the Kentish-influenced eastern and southeastern areas and in the "Celtic" western and southwestern parts of the country were split up by inheritance during the high middle ages corresponds strikingly with the general view also current then that there was a rise in the population in these centuries. On the other hand, his thesis concerning the reconsolidation - enclosure - of the original farming units between the fifteenth and seventeenth centuries seems, at least as far as the first of these centuries is concerned, to be explicable in terms of the population reduction from the mid-fourteenth century.

It is further worth stressing that certain aspects of enclosure, according to Gray, were not incompatible with the prevalent medieval and late medieval agricultural practice in those areas where the enclosure movement proper first made its impact - which does not however alter the fact, we must note, that he too thought that the process took place after the late medieval centuries.

We have already mentioned certain facts that may have influenced the general account of the chronology of the enclosure movement in the first decades of the twentieth century. One important feature in this respect was the available sources and the limitations they imposed on research. Gray's method illustrates this problem excellently.

He uses what one could call a retrospective method - a method necessitated by the very nature of the sources for the history of the English field systems. His main sources are thus manorial surveys and village land registers from Tudor and early Stuart times. He uses these source types with a kind of historical hindsight, as a tool for the interpretation of the earlier, far more fragmentary source material, which for the most part says very little directly about the prevalent field systems.[129]

A striking example of Gray's retrospective method can be found in his account of the "Celtic" field system, which is mostly written on the basis of a number of reports on Scottish farming in the eighteenth century.

This method of course leads one to exercise the greatest caution when assessing the validity of statements originating in it. But in the first place Gray claimed to be driven by necessity because of the shortage of adequate sources for the early period; and in the second, he was very careful in his shifts between using the more complex later sources and the more sporadic, less adequate early sources - for example, when he applies his general account of the "Celtic" system to developments in the "Celtic-influenced" areas, and when he describes the special East Anglian pastoral arrangements. In the third place, Gray's theory seems so reasonably consistent that it can be supposed capable of countering much of the detailed criticism of his method that would take more time than the occasion warrants.

Notes

1. See for example A. G. Little, "The Black Death in Lancashire" in *The English Historical Review*, Vols. V, 1890 and VI, 1891.

2. This was published under the title *The Epidemics of the Middle Ages* along with two other works by Hecker, *The Dancing Mania* and *The Sweating Sickness*, by the Sydenham Society in 1844.

3. Hecker's account, about seventy pages long, describes the disease, its symptoms and its course, with its causes and spread, mortalities and moral consequences for society, and with the attitude of contemporary medical science to the disease; but "to indicate scientifically the influences which called forth so terrific a poison in the bodies of men and animals exceeds the limits of human understanding". Ibid., p. 17.

4. "The medieval period, taken broadly, is also capable of being brought under a few dominant conceptions. One of those would undoubtedly be the notion of service" (p. 308). "There can be no doubt that villain service meant agricultural service" (pp. 317-18). "...the material distinction [between villein and freeman] from which the lawyers started, was that of agricultural service, not more and not less. They assumed that agricultural service, as such, was villain service and a presumption of villain tenure, unless proof of the contrary was forthcoming" (p. 320). P. Vinogradoff, "Agricultural Services" in *The Economic Journal*, Vol. X, London 1900.

5. In the third volume of Karl Marx' principal work, *Das Kapital*, published posthumously in 1894, there is such a theoretical study in Chapter 47, Section IV. From this it emerges that money rent, even though it only constitutes a change in the form of product or labour rent - as it has the same basis as these forms of rent, that is the customary holding of land by the primary producer against an equally customary consideration to the lord - on the one hand makes for the dissolution of this basis and the whole nature of the mode of production; and on the other hand requires a relative development of trade and manufacturing in society. The money rent is thus in Marx' interpretation both a consequence of, and crucial to, the process whereby impersonal money relations and the exchange economy displace the former personal, directly reproductive social nexus. MEW, Bd. 25, Berlin 1977.

6. Since "the agriculture of the Middle Ages has often been ably described, I have devoted the greater parts of this work to the agricultural history of the subsequent period, especially the seventeenth, eighteenth and nineteenth centuries", writes Curtler, for example. W. H. R. Curtler, *A Short History of English Agriculture*, Oxford 1909, p. iii.

7. W. Hasbach, *A History of the English Agricultural Labourer*, London 1909; Lord Ernle, *English Farming Past and Present*, London 1912.

8. Charles Creighton, *History of Epidemics in Britain - from A.D. 664 to the Extinction of the Plague*, Vol I, Cambridge 1891; *History of Epidemics in Britain - from the Extinction of the Plague to the Present Time*, Vol II, Cambridge 1894.

9. "It is perhaps inevitable that scholars accustomed to deal only with obvious human causation, should look with some distrust upon large claims made, in the way of moral and social consequences, for a phenomenon which has been apt to be classed with comets and earthquakes". C. Creighton, *A History of Epidemics in Britain*, Vol. I, 1964, pp. 141-42.

10. Ibid. "Not the least of the effects of the Black Death upon England was the domestication of the foreign pestilence on the soil. For more than three centuries bubo-plague was never long absent from one part of Britain or another". Ibid., p. 215.

11. Ibid.

12. Ibid., pp. 13, 65.

13. Ibid., p. 162.

14. Ibid., pp. 173-74.

15. Ibid., pp. 155-56.

16. Ibid., p. 175.

17. Ibid.

18. The first edition from 1893 bears as its title this original term for the plague in 1348-49, while the second edition from 1908 was published with the title *The Black Death*. This term is not known from the contemporary sources, but probably first arose later in Sweden or Denmark, perhaps influenced by the black buboes which were one of the symptoms of the disease. At least, the term "the Black Death" has a later origin. In *Rerum Danicarum Historia* from 1631, J. J. Pontanus speaks on p. 476 of "atra mors".

19. Gasquet, *The Great Pestilence*, p. XVI.

20. Ibid., p. 7.

21. Gasquet remarks that the summer and autumn of 1348 had extraordinarily high rainfall. "In such a season, naturally unhealthy, the sickness, of its own nature most deadly, found every condition suitable for its rapid development". Ibid., p. 74.

22. Ibid., p. 21.

23. Ibid., pp. 52-53.

24. Ibid., p. 75.

25. Ibid., p. 95.

26. "Contrary to the opinion entertained by persons of repute there is evidence to show that bathing was common and much used especially among the lower classes, and that even small villages had their public bath places".

27. Ibid., pp. 195-96. Gasquet documents this view with a reference to Rogers' article in *The Fortnightly Review*, 1866, p. 192; but in a note he tones down his and Rogers' view of the reproductive conditions of the peasants as follows: "This is, of course, true, but without qualification might give the reader a false impression as to the condition of the English peasantry in the Middle Ages. Most of what Mr. Thorold Rogers says is applicable to all classes of society. Dr. Cunningham [*Growth of English Industry and Commerce*, p. 275] takes a truer view: "Life is more than meat, and though badly housed the ordinary villager was better fed and amused.""

28. Augustus Jessopp, "The Black Death in East Anglia" in *Nineteenth Century*, April 1885 and December 1884.

29. Ibid., p. 918 (1884).

30. Arthur Dimock, "The Great Pestilence" in the *Gentleman's Magazine*, Vol. CCLXXXIII (283), July-December 1897.

31. It was an unreasonable claim that the Black Death was "a neglected turning-point" at the time the article was published - all the more so as the article must be regarded as a popularized account of Jessopp's and Gasquet's results, which is not explicitly evident from the article.

32. Ibid., pp. 170-71.

33. In 1894 the plague germ was identified, and more than a decade later, in 1908, it was successfully proved that this microorganism was carried by the rat flea. L. Fabian Hirst, *The Conquest of Plague*, Oxford 1953, pp. 106, 172.

34. According to Jessopp, both source types abound with other problems. The court rolls only provide information on the death of the tenant, saying nothing about the rest of the household or cause of death. Jessopp, op. cit. Dec. 1884, pp. 928-29. With the institution books there is a risk of misconception which Seebohm did not take into account. Some of the vacant benefices were probably vacant for several years. Ibid., p. 927.

35. Ibid., p. 932.

36. "...the statements made of the mortality in the towns will not bear examination - they represent mere guesses - nothing more." Ibid., p. 933.

37. Creighton, op. cit., Vol. I, p. 124.

38. Ibid., pp. 124-25.

39. Ibid., p. 128.

40. Ibid., pp. 130-31.

41. Ibid., pp. 131-32.

42. "...some two-thirds of [the clergy] were cut off by the plague in Norfolk, Suffolk, in Yorkshire and Shropshire, and probably all over England". Ibid., pp. 133-34.

43. From the manors of Winslow in Buckinghamshire and Lessingham in Lancashire, and Jessopp's material from Norfolk and Suffolk. Ibid., pp. 136-39.

44. Gasquet, op. cit., p. 194.

45. Ibid., pp. 195-96.

46. Ibid., p. 75.

47. *Inquisitiones post mortem* are, as mentioned before, stocktakings of manors whose lords had recently died. We will return later to this type of source.

48. Patent rolls in this context are registers of donations by the Crown to churchmen working under the Crown

49. Dimock, op. cit., p. 178.

50. Ibid., p. 179.

51. Gasquet thought in fact that the disorganization and subsequent rebuilding of the church was the most important consequence of the Black Death. Gasquet, op. cit., p. xvi.

52. "...wages everywhere rose to double the previous rate and more". Ibid., p. 197. "...the plague accelerated the rise in wages". Jessopp, op. cit., April 1885, p. 619; Creighton, op. cit., pp. 140, 184-85; Dimock, op. cit., p. 180.

53. "...the total extinction of villenage [was] hastened by the Pestilence..." Ibid., p. 183. Gasquet and Creighton agreed that villeinage ceased in practice with the Revolt of 1381, as they agreed with Rogers that the Revolt was due to the attempt of the landlords to reintroduce the labour rent. Gasquet, op. cit., pp. 199-200; Creighton, op. cit. pp. 184-85.

54. Creighton, ibid., pp. 141, 182, 192; Gasquet, ibid., pp. 197, 199-200; Dimock, ibid. pp. 181-82.

55. Creighton agreed with Rogers and disagreed with Seebohm on this issue. The fifteenth century was primarily characterized by new conditions of tenure and "the rise of the yeoman class". The transition to sheep farming only began at the end of the fifteenth century. Ibid., p. 196. Gasquet, on the other hand, was not very clear about his view of the issue. Ibid., p. 201.

56. Creighton, ibid., p. 192; Gasquet ibid., p. 196.

57. Gasquet, ibid.

58. Jessopp, op. cit., April 1885, p. 620.

59. Ibid., p. 618.

60. "The obvious and undoubted effect of the great mortality among the working classes was to put a premium upon the services of those that survived". Gasquet, ibid., p. 185. Creighton: "The labourers although the lowest order on the manors were...masters of the situation". Ibid., p. 185.

61. Jessopp, ibid., p. 620.

62. Ibid., p. 615.

63. "A calamity so sweeping, so overwhelming, its consequences upon the whole social fabric would be incomparably more disastrous [today] than it was in times when centralization was almost unknown and practically impossible". Ibid., p. 611.

64. Nevertheless, Vinogradoff had objections, three in all, to Page's account of the institution of villeinage in his review of Page's article in *The English Historical Review*, Vol. XV, No. LX, 1900.

65. The preliminary work for Page's article made up his doctoral dissertation, defended at the University of Leipzig. Thomas Walker Page, *Die Umwandlung der Frohndienste in Geldrente*, Baltimore 1897.

66. Thomas Walker Page, *The End of Villainage in England*, Publications of the Economic Association, 3rd Series, Vol. I, No. 2, New York 1900, pp. 45-46, 60-64, 78-82.

67. See Vinogradoff's criticism, op. cit., p. 780.

68. Page, ibid., p. 44.

69. Ibid., pp. 57-58.

70. Ibid., p. 54.

71. Page refuted Rogers' theory with a number of arguments we have seen before from Ashley - see notes 23 and 25, Chap. II. Page also believed that the liberalization already in progress formed the background for the ideological social equality doctrines of the Revolt, and had a self-reinforcing effect. "Many villains had already acquired freedom, why should others remain in subjection?" Ibid., pp. 69-71.

72. Francis G. Davenport, "The Decay of Villeinage in East Anglia" in *Transactions of the Royal Historical Society*, New Series, Vol. XIV, 1900, p. 126.

73. Ibid., p. 129.

74. Davenport did not distinguish between the transition to the money rent to leasing. This was also the case with Page for the period up until 1450.

75. Edward P. Cheyney, "The Disappearance of English Serfdom", in the *English Historical Review*, Vol. XV, No. LVII, 1900, p. 20.

76. Ibid., pp. 33-34.

77. Ibid., p. 35.

78. Ibid., p. 34.

79. Vinogradoff's criticism of Page, op. cit., p. 778.

80. Vinogradoff, "Villainage in England", op. cit., pp. 174-75.

81. H. L. Gray, "The Commutation of Villein Services in England before the Black Death", in *The English Historical Review*, Vol. XXIX, No. CXVL, 1914.

82. Ibid., p. 629. "Mr. Page's evidence earlier than 1350 thus proves little more than that villein services upon the manors of certain ecclesiastical foundations of south-eastern England were at that date largely uncommuted. This is far from showing that similar conditions prevailed throughout England or even upon lay manors in the region dealt with. Ibid., p. 630.

83. In the first place, it is claimed that the accounts of the ecclesiastical estates were only kept when the estate in question was unenfeoffed and under the Crown. The accounts were therefore sometimes kept for periods of less than a year, or at intervals of many years. Unless accounts are available for an uninterrupted series of years it is impossible to assess the extent of commutation, since one cannot check whether this was a temporary or permanent situation. Secondly, statements of labour rent were often omitted. And thirdly, the estate accounts according to Gray were not so numerous in the first half of the fourteenth century. There are greater numbers of them from before 1300 and after 1350. Ibid., pp. 630, 636-37.

84. By contrast, Page did not attempt in his investigation to compare the extent of labour rent with that of money rent and the other money dues. He simply observed the extent to which labour services were rendered, and whether the peasants subject to labour rent did all the necessary work on the estate.

85. Page's study only covered 81 estates.

86. Gray, op. cit., p. 634.

87. Ibid., p. 636.

88. Ibid., p. 648.

89. Ibid., p. 650.

90. One of the earliest examples of a monograph on the history of a single manorial complex is G. Poulett Scrope's *History of Castle Combe* from 1852.

91. Maitland's point of departure is two extents from 1221 and 1277. In the subsequent period up to and some way into the fifteenth century, when the commutation process got under way at Wilburton, the study is based on non-synchronic comparisons of information from court rolls and account rolls. There is a large lacuna in the source material between 1320 and 1380. F. W. Maitland, "The History of a Cambridgeshire Manor", in the *English Historical Review*, July 1894, reprinted in *The Collected Papers of F. W. Maitland*, Vol. II, 1911.

92. However, one should mention N. Nielson, *Economic Conditions on the Manors of Ramsay Abbey*, 1898; and K. G. Feilings, "An Essex Manor in the Fourteenth Century" in the *English Historical Review*, Vol. XXVI, April 1911 - both of whom thought they could prove that the Black Death only had a tangential effect on the commutation process.

93. This disregards the fact that it was assumed that there was already a trend for manorial land to be taken out of the common field system as early as the thirteenth century. It should also be added that Erwin Nasse had earlier called the general view into question. Nasse claimed that enclosure in its early phase not only meant a transition to sheep farming and the dispossession of peasants, but also a consolidation of the individual user's land holdings and the opportunity for individual innovation with respect to rotation and crops - measures which were not only directed against the primary producers on the initiative of the lords. Erwin Nasse, *The Land Community of the Middle Ages*, 1871, pp. 81-91.

94. Ashley, op. cit., pp. 285-86.

95. One can find a basis for refuting Ashley's classification into phases very early in T. E. Scrutton, *Commons and Common Fields, or the History and Policy of the Laws relating to Commons and Enclosures in England*, Cambridge 1885; and later in E. M. Leonard, "The Inclosure of Common Fields in the Seventeenth Century" in *Transactions of the Royal Historical Society*, 1905.

96. H. Levy, *Entstehung und Rückgang des Landwirtschaftlichen Grossbetriebes in England*, Berlin 1904; G. Slater, *The English Peasantry and the Enclosures of Common Fields*, London 1907; J. L. and Barbara Hammond, *The Village Labourer, 1760-1832*, London 1912; E. C. K. Gonner, *Common Land and Inclosure*, London 1912.

97. R. H. Tawney, *The Agrarian Problem in the Sixteenth Century*, London 1912; E. F. Gay, "Inquisitions of Depopulation in 1517, and the Domesday of Enclosures", in *Transactions of the Royal Historical Society* XIV, 1900; E. F. Gay, "The Midland Revolt and the Inquisition of Depopulation of 1607", ibid., XVIII, 1904; E. F. Gay, "Inclosures in England", in the *Quarterly Journal of Economics* XVII, 1903.

98. Tawney, op. cit., p. 401.

99. Ibid.; Hammond, op. cit., p. 26.

100. See the discussion between Leadam and Gay on Rogers' constructions for the fluctuations of the price of wool in *Transactions of the Royal Historical Society*, Vol. XIV, op. cit., pp. 259-63; and Gustaf F. Steffen's deliberations on developments in prices in the sixteenth century in *Studien zur Geschichte der Englischen Lohnarbeiter*, Bd. I, Stuttgart 1901.

101. W. Hasbach, op. cit., pp. 31-32.

102. See for example A. H. Johnson, *The Disappearance of the Small Landowner*, 1909, pp. 55-56.

103. Gonner, op. cit., pp. 116-18.

104. "The development of the textile manufactures, which for two centuries were the chief of English wealth, could not have taken place without the production of cheap supplies of raw materials, and the growth of the towns was dependent on the saving of labour from agriculture". Tawney, op. cit., p. 179.

105. "The revolution in the technique of agriculture when sucked into the vortex of expanding commerce is, in fact, simply an early, and owing to the immobility of sixteenth century conditions, a peculiarly striking example of that reaction of widening markets on the methods of production, which is one of the best established of economic generalisations". Ibid., pp. 196-97.

106. Gonner, op. cit., pp. 107-115.

107. Ibid., pp. 115-18.

108. Ibid., pp. 132-36.

109. Ibid., p. 121.

110. Gray, *Transactions of the Royal Historical Society* XIV, op. cit.

111. Slater, op. cit., p. 73.

112. H. L. Gray, *English Field Systems*, Cambridge, Mass., 1915, p. 8.

113. Even though there are traces in all the regions of central England of both two-course and three-course systems, Gray thinks that the two-course system was most prevalent in the south west and the three-course in the north east. Gray can see no fundamental technical difference between the two systems. The difference was simply a

matter of the proportion of the total area that lay fallow annually. The "choice" between one or the other system depended on natural conditions, primarily the quality of the soil and the pressure of population, which meant, with the gradual progress of the three-course system around the thirteenth century, that the development of certain areas was set back by natural causes. Ibid., pp. 72-73.

114. Ibid., pp. 74, 79, 81-82, 88, 125, 406.

115. We have remarked before that Nasse was the first to speak of the existence of the three-course system in the thirteenth century, and how Seebohm confirmed this view, despite Rogers' apparent opinion that the two-course system was dominant in that century. Vinogradoff considered that both systems were widespread at that time (*Villainage in England*, op. cit., pp. 224-25). Despite these nuances, Gray broke with a generalization, since he rejected the assumption that the two and/or three-course system predominated throughout England in the middle ages.

116. Gray, op. cit., p. 404.

117. Gray also operated with the idea that there were field systems in the Lower Thames Basin which could be compared to a greater or lesser extent with the central English, Kentish and East Anglian systems. We will go no further into this area here, since our interest is concentrated on the fundamental features of the field systems and their relationship with the enclosure phenomenon.

118. Ibid., pp. 198-99.

119. Ibid., p. 199.

120. The infield was under continuous cultivation. This was possible partly because of intensive manuring, partly because it was cultivated in a simple rotation system where it was divided in three parts sown with barley, oats and oats in succession. The outfield was divided in two unequal areas. The smallest, about a third, was again divided into ten parts, some of which were cultivated every year, after having been enclosed with a low turf wall the previous year, within which the cattle and sheep spent the night and a couple of hours a day. The bulk of the outfield was not manured. One tenth of this, too, was sown every year. Ibid., pp. 158-60.

121. Gray expresses some doubt as to whether the annual redistribution was really kept up until the first enclosures, as Slater thought must have been the case in some places (Slater, op. cit., p. 174-75). So even though he could find specific documentation for this, he refrained from drawing a generalized conclusion. "What generally gave the first impetus toward consolidation was not the practice of annually re-allotting strips, but the falling-in of the lease and the action of the landlord". Gray, op. cit., pp. 170-71.

122. Ibid., p. 203.

123. Ibid., p. 269.

124. Ibid., p. 340.

125. Ibid., pp. 351-52.

126. Ibid., p. 303.

127. Ibid., p. 405.

128. Gray showed how the peasants in certain parts of East Anglia were obliged to gather and spread manure on the demesne lands; and how they had to set up and move wattles for pasturing and manuring the fallow fields of the manor. But the enclosure of the fallow fields was also a privilege extended to the peasants or at least some of them. Early extents contain lists saying whether each peasant had the right or not to his own enclosures, and whether they might let their own sheep graze on their own fallow fields

(*sua falda*). The competition for the sheep manure was not only between lord and peasant. It was also competition among two or more manors, since the East Anglian villages were often a complex of the landholdings of several manors. Gray thinks this came into conflict with the grazing and manuring practices prevalent under the two-course and three-course systems. The special pastoral arrangements (fold courses) of East Anglia must also be understood in this light. Ibid., p. 350.

129. Ibid., pp. 13-14.

Chapter IV

Evolution and Crisis Theory
(1900-1930)

IV.1 The Historical significance of the Black Death

The significance of the Black Death for socioeconomic development in and after the fourteenth century was one of the most controversial issues in English historical research before and around 1900. Although few historians simply thought that the plague was a turning-point in English medieval history, one constantly encounters authors in the period who mounted an attack on this view.

So far we have not seen the attack carried off successfully. What we have seen is that in monographs on the Black Death - not surprisingly perhaps - and in Page's account of the chronology of the commutation process, there were many arguments for assigning the plague revolutionary significance. Given this, one might expect that this reading was widespread in the period just after 1900. Yet this was not the case, judging from those general works which also dealt with the late medieval centuries and which were published between 1905 and 1915. The general scepticism towards assigning the Black Death too overwhelming historical significance that had developed, and which had been strengthened since the debate between Seebohm and Rogers, also typified these accounts. Let us look at some examples: Volumes III and IV of *The Political History of England*, published in 1905 and 1906; E. Lipson's *An Introduction to the Economic History of England*, from 1915; and A. H. Johnson's *The Disappearance of the Small Landowner*, from 1909.

In the first of these works the fourteenth century is split between Volume III, covering the period from 1216 to 1377, and Volume IV, from 1377 to 1485. Both volumes venture into the economic area as regards two issues of interest to us. In Volume III, T. F. Tout describes the Black Death and its economic consequences in particular. And C. Oman devotes a chapter in Volume IV to a description of the Peasants' Revolt in 1381, with an account in terms of economic history of the causes and consequences of the Revolt.

According to Tout, the Black Death hit a flourishing English economy with its full force. "Medieval England seldom enjoyed greater well-being and tranquillity than during the first eighteen years of the personal rule of Edward III". But "a rude ending to this period of prosperity was brought by the devastation of the pestilence known to modern readers as the Black Death".[1]

The young and healthy were affected just as much as the old and ailing, but "comparatively few magnates died". "The poor, the religious, and the clergy were chief sufferers" among the approximately fifty per cent of the population carried off by the plague.[2] The immediate consequences of the high mortality were falling prices and rising wages despite the attempts of the Crown - natural enough in their time - to regulate them.[3] Wages continued to rise, as did the prices of all manufactured goods, while corn prices stagnated. "Thus the labourers enjoyed the benefit of the scarcity of labour, while the employers suffered the full inconvenience of the change".[4] The peasants profited from a steep drop in rents, from the growth of commutation and from the advantages of the leasing system.[5]

But despite "these undoubted results of the pestilence" the Black Death was not "the one great turning-point in the social and economic history of England". "In truth the Black Death was no isolated phenomenon. There were already in the air the seeds of the decay of the ancient order, and those seeds fructified more rapidly in England by reason of the plague. It is only because of the impetus which it gave to change already in progress that the pestilence had in a fashion more lasting results in England than elsewhere".[6]

The Black Death was thus an accelerating factor. It stimulated the seeds of social and economic change sown in the preceding period. But Tout's history says nothing about the origins of these seeds. They were present or had arisen in a period with stable economic progress which was disturbed by the plague.

This does not mean that Tout saw the period after 1350 as one of want and general crisis. It was an age which contrasted with the uneventful first years of Edward's reign because of the social unrest and increasing mobility that followed on the heels of the plague. "The war of classes, which was beginning, sprang not so much from material discomfort of the poor, as from what unsympathetic annalists called their greediness, their pride, and their wantonness. The wage-earner was master of the situation and did not hesitate to make his power felt".[7]

106

Although attempts had been made to prove that the period just before 1381, for example in East Anglia,[8] was marked by the dissolution of villeinage; and although it was repeatedly claimed that the Peasants' Revolt was the event that gave English villeinage its death-blow, Oman thought that the manorial archives presented the best proof that the manorial system and the labour rent continued after 1381.[9]

Thus the gradual, partial decay of villeinage in the half-century following the Revolt should not be directly attributed to it. "It was not in direct consequence of that rising, but...a result of the rural economic revolution of the fifteenth century. The Lords preferred, more and more, to work their estates in pasturage rather than in arable, and this being so, they had less and less interest year by year to exact the old servile corvees. Villeinage disappeared by slow degrees and from economic causes. It was not killed once and for all by the armed force of rebellion in June 1381".[10] But what these "economic causes" were, Oman nowhere says directly.

He rejects the possibility that it was commutation, falling rents and rising wages that made the transition to sheep farming an economic necessity for the landlords. In the first place, he says, unlike Tout, that it was because the Statutes of Labourers had the desired effect;[11] and in the second, because in his view of the causes of the 1381 Revolt, commutation was neither very widespread nor a salient demand from the villeins before and after the Revolt.[12]

So Tout and Oman both reject the idea that the Black Death and the 1381 Revolt had crucial socioeconomic consequences - yet without attempting to answer the questions this rejection raises. Tout simply passes over the problem, while Oman in general seems to be swayed by the widespread nineteenth-century theory of the influence of commerce and the money economy on the change in agrarian reproductive structures.

IV.2 The progressive decline of the manorial system

Nor did Johnson think the Black Death was a turning-point in English economic history.[13] The change in the agrarian reproductive structures began before and finished long after 1350. In this process the plague played a secondary role, although it had certain serious immediate consequences. "The effect of the great plague was different on different

107

manors",[14] and the high mortality led to no permanent changes. The commutation process was halted,[15] wages rose, and the mortality thus thinned out the villeins so much that the landlords either had to leave much land fallow, lease it out or convert it to pasture.[16] But "many of the changes which have been attributed to [the Black Death] had begun before; some of these it checked, others it accelerated slightly, and that is all. A few years after the visitation and the peasants' revolt the manors assumed their old aspect".[17] The 1381 Revolt had no great significance either, if we are to believe Johnson.

"Nevertheless, the disintegrating influences of the plague and the Statute of Labourers were great".[18] The customary, personal relation between lord and peasant was replaced by an impersonal cash nexus relationship: in the first place, "the only serious result" was that the number of villeins fell either because of the plague or because they fled the estates - "In this way the peasant was divorced from the soil and went to swell the class of landless but free labourers";[19] and in the second place "the Statute of Labourers introduced the agents of the king, and the law entered the sacred precincts of the manor".[20]

But these serious consequences would not have become permanent "had the economic conditions of the country remained the same". Commutation, which began again between 1390 and 1440, the leasing system, agricultural wage labour and the transition to sheep farming achieved a permanence "due primarily to that industrial revolution caused by the transition from an agricultural to a trading and manufacturing country".[21] "It was this revolution which was the real solvent of the manorial system and which prevented its reconstruction after the shock of the Great Plague".[22]

Yet these not uncommon arguments do not explain, according to Johnson, how the late medieval transformation process started. Relying heavily on Pollack and Maitland's *History of English Law*, and Vinogradoff's *Villainage in England*, he described the beginning of the process in the thirteenth century. On the one hand he thinks that the villein's socioeconomic conditions of life, originally laid down at "the will of the lord", gradually fossilized into customary law. This meant for one thing that the former arbitrary rent was changed to a fixed customary labour rent, which began to be commuted in the period before 1350.[23]

The reason for this early commutation Johnson finds in two factors, both of which can be traced to Vinogradoff's writings. In the first place, "this [commutation] may have been caused by the fact that with the increase of the number of villeins, more labour was due than the lord required". In the second place, "in the case of boon, or occasional services because the dues paid in kind to the lord in return became with the rise in price of commodities more valuable than the boon service itself".[24]

IV.3 The evolutionary view of history

The population growth and the rising price trend in the thirteenth century was the initiator, the Black Death was the accelerator, and the flourishing of commerce and manufactures was the final, stabilizing factor in the agrarian transformation process. This ranging and ordering of the factors describes excellently some of the crucial elements in the discussion of the crisis of the manorial system or agrarian transformation process in the latter half of the nineteenth century.

Johnson's ordering paints a picture of an evolutionary process, showing how the description of the economic history of the late medieval centuries became less and less determined by more or less discrete dramatic events. Confirming this evolutionary view of history and the whole approach in Johnson's book, A. F. Pollard wrote in 1910 in his overall account of the birth of the capitalist era:[25] "Whatever factor we take in the making of that change from medieval to modern history...we find that the same thing is true about all. They have their roots stretching back far into the past, and buried far out of sight....The dramatic events which catch the eye and the ear, and by which we date the process or backsliding of mankind, are... but the outward and visible manifestation of causes, working without rest, without haste, without conscious human direction in the making of the history of the world".[26]

The evolutionary historical perspective was maintained in Lipson's introductory book from 1915. Although, influenced by Page's study of commutation in East Anglia, he acknowledged that "the Black Death proved to be an economic catastrophe of supreme importance",[27] he took the many warnings against assigning it too great importance to heart. "On the one hand, it did not set in motion the tendencies towards commutation; on the other hand, the progress of the movement was most rapid at the end of the fourteenth and during the fifteenth century".[28] And this was not due to the Revolt of 1381. "The Revolt itself was but one symptom....The end of villainage in England was not due...to the Peasants' Revolt....The dislocation which foiled all attempts to reconstitute on a stable and permanent basis the old manorial order, and the alienation of the demesne or its conversion into a sheeprun were the real forces which dissolved the economic fabric of medieval serfdom".[29]

Lipson partly rejected Rogers' view of the chronology of the commutation process - in the first place with the support of Page's study and his monetary explanation of the causes of commutation;[30] in the second place with a source-critical objection to Rogers' seeing every money amount itemized in the thirteenth-century manorial accounts in connection with peasants' rents as an indication that commutation had taken place.[31] Thirdly, he considered that there was no basis for assessing the extent to which wage-dependent agricultural labourers were working within the manorial system until after 1350;[32] and according to Lipson the existence of agricultural labourers was one of two necessary conditions for commutation.

The transition from labour to money rent either meant that manorial cultivation continued with the help of wage labour or that the landlords abandoned the cultivation of the demesne and leased it out. Commutation before 1350 was thus related to the leasing system, for which Lipson found indications as early as the thirteenth century,[33] and to the development of a class of free wage-earning labourers which he believed must have already existed before 1350 despite the silence of the sources.[34]

As an argument for his early dating of commutation, he referred, like Rogers, to the irrational content of the manorial system and labour rent. In a very brief passage Lipson listed a number of general reasons for early commutation. As Walter of Henley had remarked, "customary servants...neglect their work and it is necessary to guard against fraud". The labour rent was an obstacle to agricultural progress because of the unwillingness of the peasants and their control of the production process. The transition to the use of wage labour was in this respect a step forward, as it enabled the estate-holder to organize the work more independently. On the other hand, leasing meant absolute savings on management. The peasants, too, drew benefits from commutation. They could now work more continuously on their own land, and "in the long run the peasantry profited by the change from services in kind to services in money", because the change did not entail a variable rent determined by the market, as Rogers had thought, for example. The commuted money rent was fixed once and for all. With the general development of inflation, this meant that rents had a tendency to fall.[35]

Given this background, and even though Lipson partly takes Rogers to task over his view of the chronology of the commutation process and the causes and significance of the Peasants' Revolt,[36] his account must be seen as an underscoring of the evolutionary elements in Rogers' view of the background of the dissolution of the manorial system.

IV.4 The resistance of some "Winchester manors"

In the period between 1916 and 1920 we find three studies of the development and decline of the manorial system in fairly well-delimited parts of England. E. Robo's article from 1929 is a study of a single manor - Farnham[37] - while a 1916 treatise by A. Ballard takes the form of a comparative analysis of the manors of Witney, Brightwell and Downton, extending from the eleventh century to the period after the Black Death.[38] Most interesting for our purposes, however, is A. E. Levett's study, which covers several manors.[39] All three studies are mainly based on data from one of the most complete source collections for late medieval England:[40] the so-called Pipe Rolls of the Bishop of Winchester - a collection of manorial accounts running from 1208/09 to 1455.

Levett looked at a sample of accounts from before 1346, but concentrated her efforts on the period between 1346 and 1356, where the chronology of the sources is complete. Less intensively, she studied the 1376-1381 period. Geographically, her sample represents manors in two main groups around Winchester and Taunton, as well as a third group in Northamptonshire and Berkshire. Against this background Levett issued sensible warnings against attempts at inductive generalization, stressing that her study only aimed to clarify certain local consequences of the Black Death,[41] and that the effects of the plague may have varied greatly with regional conditions and the status of the landlords. Thus, although she directed stringent criticism at Gray's method,[42] she accepted to a great extent his general conclusions.

She accepted that the Black Death had widely varying consequences in different parts of England, and her own work showed directly how the differences were also evident within the Winchester domain - something which was confirmed by Ballard's study. "At Witney the Black Death marks the change between a barter economy and a coin economy", since the labour rent and the manorial system were here replaced by the money rent and leasing, while "the manor of Brightwell was practically unaffected by the Black Death" and "in spite of the large death-roll, the bishop proceeded with his dominical farming at Downton in exactly the same manner as he had done before the Plague".[43]

Levett concluded that on the estates of the Bishop of Winchester there was no revolutionary upheaval either in the relations of production or in agricultural technology in the fourteenth century. All the changes that could be registered, she thought, had to be attributed to a continuous, evolutionary process. The Black Death did not lead to immediate chaos and permanent change. The general increase in commutation she could register after 1360-70 was not "immediately due to the Black Death".[44] She took the same view of the increase in the leasing of demesne that she thought took place from about 1376 on.

There were apparently no great difficulties involved in tenanting, on the usual terms, the land that had been deserted as a direct result of the high mortality in 1348/49: less than 20% of the tenancies remained unfilled.[45]

This observation was confirmed by Robo. Although the mortality among the Farnham peasants in 1348/49 was very high (Robo estimated that between a third and half of the peasants died), "nearly every holding soon found a tenant. Only the poorest and least profitable land probably remained vacant".[46] "On the demesne the land was tilled and weeded; harvesting, winnowing and sowing went on as before; the same acreage was kept under cultivation".[47] The labour rent was rendered uninterruptedly and to the full regardless of the inroads of the Black Death into the manor's villein labour reserves.

Despite Levett's reservations about the representativeness of her material, it was characteristic of her as well as Ballard and Robo that they saw their contributions as major attacks on the accentuation of the historical importance of the plague that we have seen in certain of the earlier accounts.[48] It might seem that the attack achieved exemplary success; earlier views of the demographic results of the first plague outbreak in 1348/49 seem to have been countered by the example of the Winchester estates, where there were no crucial changes in the existing socioeconomic conditions.

There seems to have been a large enough labour base on the Winchester estates to compensate for the immediate results of the high mortality in 1348/49 - enough manpower to check the development of rents and wages and to cultivate successfully more or less the same area as before the Black Death.

On the other hand, all three studies appear to confirm that the subsequent plague outbreaks had certain effects. Levett thought, as mentioned before, that commutation began to gather speed from about 1360-70 on. Ballard concluded that "labour services were entirely commuted at Witney in or before 1376" and that commutation must have started at least at Brightwell in the same period.[49] Finally, Robo thought that wage rises first began in the 1360s, and that "it was only the second appearance of the plague in 1361 which accentuated and settled the new economic condition".[50]

As Levett formulated the problem, she was able to present documentation that convincingly refuted the socioeconomic importance of the plague on the Winchester estates. But it is important to note that she only considered the demographic consequences of the first outbreak in 1348/49 in her arguments. She did not explicitly take the repeated outbreaks and the continuing excess mortality in the subsequent decades into account. For this reason among others, she was able to explain the breakdown of the manorial system in keeping with the evolutionary tradition we have seen develop.

Levett considered that the development of the exchange economy and the internal irrationalities of the manorial system were the factors which led to the decay of the system. The increasing trade in the thirteenth century, the Crusades and the royal conversion of vassal military service to money dues reinforced the lords' need for cash incomes, a need which was intensified after the outbreak of the Hundred Years' War,[51] and which gave rise to the replacement of labour services by money dues.

To this was added the increased money circulation following on inflation, which on the one hand undermined the relative value of the labour services rendered against payment in kind, since these payments from lord to peasant rose in real value with the progressive development of prices.[52] On the other hand the effect of inflation was that already-commuted fixed rents lost their relative "value" to the landlords and left many peasants in so advantageous an economic situation that Levett thought that much of the basis for the development of manufactures was created through the opportunities of these peasants for accumulation.[53]

The Winchester estates, like other church holdings, were subject to these conditions; yet not to the increasing demand for cash income imposed by the Hundred Years' War on the lay lords. So Winchester in the fourteenth and fifteenth centuries was better able to maintain the manorial system; and so, thought Levett, commutation on the Winchester estates in particular was concentrated in those periods when the bishopric was unenfeoffed and in Crown hands,[54] and in the period after 1367 when William of Wykeham had the fief. William of Wykeham, the man "wise of castlebuilding", apart from his building activities, organized subsidies for needy students, among other ways by founding an Oxford college, and these activities increased the bishopric's needs for a cash income. Perhaps "the process of commutation on the Winchester estates was more seriously affected by William of Wykeham's magnificent projects than by that traditional parent of all economic development, the Black Death".[55]

IV.5 Summing-up

Research on the economic history of the late medieval centuries from the latter half of the nineteenth century until 1930 was characterized by intense concern with the socioeconomic changes in the agrarian sector during the period, and rather less by work in the field of agricultural technology. However, in the following sections we will see how the latter area attracted considerable interest in the second and third decade of the twentieth century.

Researchers found the key socioeconomic structural transformation in the breakdown of the manorial system: a transformation which gave rise at an early stage to talk of the crisis of the landlords or of the manorial system. They tried to date the decline of the manorial system empirically by studying the chronology of commutation, agricultural wage labour, the leasing system and the enclosure movement. This documentary effort was extraordinarily important for the evaluation of the most salient problem of the period: the issue of the significance of the Black Death for the decay of the manorial system.

We recall Thorold Rogers' assessment of this issue: he did not quite simply think, as some historians have maintained, that the Black Death was a turning-point in English agrarian history; rather that the plague accelerated certain existing tendencies. This general view of the historical importance of one of the late middle ages' most conspicuous phenomena was still prevalent between fifty and sixty years later, but now reinforced and underpinned by the research results of the intervening period.

In an attempt to collate Page's, Gray's and Levett's rather varied results, E. E. Power concluded that over three centuries the manorial system gradually disintegrated "from within", yet in the belief that "the growth of a money economy was its great solvent", and that "neither the Black Death nor any other cataclysm worked a revolution in its history....The truth seems to be that the Black Death came upon a rural world which was already changing" - a rather bold formulation, given Power's adoption of Page's monetary views, his idea that an increasing money supply combined with the high mortality stimulated commutation, and Power's opinion that "the Black Death gave rise to a general discontent which was favourable to change".[56]

As for the breakup of the manorial system before 1350, Power thought that the exchange economy left early traces in the organization of agricultural production in the form of commutation, "rent-paying tenancies" and a growing class of agricultural wage labourers.[57] But in reality Power did not root the simultaneous development of these factors in the advance of the money economy; they rather seem to be symptoms of it. The real causes were sought more in a growing labour base on the estates.[58]

The link with Rogers is even clearer in Helen Robbins' essay from 1928. Robbins tried to give a synthesizing account of the development of prices and wages in England and France in order to compare the economic and organizational consequences of the Black Death in the two countries. Rogers' works were important sources for this account.

Robbins had no doubt that the price rises just after 1350 derived from the labour scarcity after the Black Death. "The prices of 1348-51 are clearly the results of the devastation following the Black Death; but they are high peaks in the steadily rising price level".[59] For other factors were involved. The whole fourteenth century was marked by many hard winters, droughts and famines, and it was well known that there were other plague outbreaks after 1348/49.[60] The general price rises throughout the thirteenth and the first half of the fourteenth century further had to be related to the drop in silver production and the subsequent inflation.[61] "Hence the monetary rise in prices in the years 1348-51 was an accentuation of the current tendency rather than an isolated phenomenon".[62]

After the Black Death labour became more valuable because more scarce, and because its bearers, now realizing their strength, refused to render labour services and abandoned the estates to which they had been tied by the bonds of villeinage. This allowed the remaining peasants to secure commutation where it did not already exist, and left the landlords in a situation where they were forced either to lease out the demesne or to try to cultivate it with help of expensive wage labour.[63]

Robbins claims that the reduction in the population caused by the Black Death had a permanent effect on the ongoing commutation process and led to a drop in rents.[64] But "one must be cautious in attributing this development entirely to the Black Death".

The fall in rents "seems directly attributed to the Black Death and scarcely to any other operating factors of the period". But "the plague seems to have been the occasion for the commutation...not its cause. The English villein, lured by the prospect of high wages in neighbouring towns, must sooner or later have deserted his manor...the movement towards the town had already received its impetus in the growing industrial development in England".[65] "This is the evolution which was probably a more important cause for the breakdown of the manorial system in England than was the Black Death".[66]

The accelerating effect of the Black Death on the medieval transformation process was defended in a number of other accounts from the 1920s: for example in G. C. Coulton's *The Medieval Village* from 1926;[67] M. E. Seebohm's *The Evolution of the English Farm* from 1927;[68] and William Rees' *The Black Death in England and Wales*.[69] Given this, it is tempting to think that research on the history of late medieval agrarian England had made no particular progress between the 1860s and the 1920s.

Yet our survey has shown that this was not the case. It has shown, among other things, that Rogers' rather ambivalent assessment of the significance of the Black Death, which very often prompted his successors to attribute to him a revolutionary view of the fall of the manorial system,[70] was increasingly replaced by a much more definitively evolutionary view of this process. Especially research from around and after 1900 strengthened this view. This research effort formed the background against which Rogers' most general and elaborated assessment of the historical importance of the Black Death could still be confirmed in the 1920s.

In the first place it had been empirically documented that the disintegration of the manorial system was a long-term historical process that speeded up after the Black Death - if not immediately afterwards, at least in the rather longer term. Secondly, a number of important theoretical developments had shed light on one main issue in Rogers' interpretation where he himself had only hinted at an explanation. Before 1920 a number of elements had already been adduced to explain the incipient breakdown of the manorial system before 1350.

116

The generally disintegrative effect on the manorial system of trade and manufacturing had been cited earlier, but had never been pursued seriously and convincingly for the early period before 1350. Political factors and a political crisis at the beginning of the fourteenth century had been postulated, with no real explanatory value for the incipient breakdown of the manorial system. The most important attempts to explain this later involved three factors, two of which were directly traceable to Rogers' work.

We saw his rather diffuse thesis of the irrationality of the manorial system developed and linked with his general price development constructions for the thirteenth and the first half of the fourteenth century. Price rises derived from, among other things, the development of inflation[71] were thought to have undermined the value of certain labour services on the manors and provoked their commutation. We further saw the population growth of the high middle ages linked with the incipient erosion of villein adscription, the growth of a class of agricultural wage labourers, and a related growth in commutation and the leasing of demesne and land under new cultivation, since "the treatment of surplus population was one of the weakest sides of the manorial arrangement". In conditions of population growth it was thought that the manorial system was not conducive to "any thorough exploitation of the personal element".

Besides this there emerged in the second decade of the twentieth century, as we shall see in the following sections of this chapter, attempts to revise Denton's soil exhaustion theory in order to explain the agrarian structural reorganizations. These efforts too must be seen as elements in the intensified work to identify the reasons for the breakdown of the manorial system, now that it had been documented that the process had been set in motion long before 1348.

The playing-down of the historical importance of the Black Death must be one of the reasons why we very rarely encounter utterances after 1900 which prompt us to see the fourteenth century as an age of crisis. There was no further talk of the crisis of the manorial system; not even in connection with the commercial and manufacturing expansion claimed to have made its impact in the course of the fifteenth century. This only set its seal on the long-term erosion of the manorial order.

Nevertheless, on the basis of the available research, and within the terms of our definition of the concept of social crisis, there are reasons for considering the late middle ages as a period of crisis. First, the period is described as a watershed between feudal and capitalist organizations of production, a period when the feudal agrarian form of organization was

117

transmuted into the capitalist form, primarily by the development of the wage labour and leasing systems, secondarily because agriculture, among other reasons because of the expansion of sheep farming, increasingly moved towards a social division of labour between countryside and town. It must be objected to this, however, that around 1930 it was still not clear whether leasing and wage labour really were more widespread than customary tenancies against money rents in the fifteenth century; nor was it clear how far the labour rent withstood its rival. Secondly, the progressive playing-down of the importance of the Black Death in this process meant that the dissolution of the feudal agrarian relations of production was not seen as derived from extra-societal influences. The crisis of the manorial system which manifested itself in what was perhaps a general dissolution of the feudal relations of production made its impact because of internal contradictions in this socioeconomic mode of reproduction, and as a result of specific economic factors in other sectors too. In this sense one can with more and more justification, given the general development of research between the 1860s and the 1920s, speak of the reproductive crisis of the feudal agrarian socioeconomic system.

On the other hand, there is little basis for speaking of the reproductive crisis of the feudal agrarian society - its inability to reproduce its members - as the population decline after 1350 was mainly linked with the Black Death and the subsequent epidemics, and since these events were only linked by relatively few historians with socially-created factors. The epidemics were largely seen as exogenous forces that harrowed an agrarian society which even seemed capable of withstanding the consequences of the high mortality. In that sense, given the general development of research, especially after 1900, one can speak less and less of the reproductive crisis of feudal agrarian society.

IV.6 Socioeconomic and technological agrarian conditions

If one compares the first sentence of the above-mentioned article by Lord Ernle from 1885 with the first sentence of his book from 1912, recalling the relationship between these two works (see note 92 on Chapter I) one gets some hint of the changed climate of opinion that developed in the interval between the two as regards the reasons for the breakdown of the medieval agrarian structures.[72] While Ernle in his earliest work simply made the connection between the population development and socioeconomic change, in his later works he elaborated a theory where population development and the primary natural condition of agriculture - the quality of the soil - were seen as interrelated determinants of the foundation and development of the socioeconomic conditions as well as of the stage of technological development.[73]

In keeping with the consensus among authors of works on agrarian history in the first decades of the twentieth century, Ernle thought that the medieval two-course and three-course systems were sufficient for the age in which they flourished, and that they had arisen under the influence of the population growth of the high middle ages as a safeguard against the exhaustion of limited natural resources.[74] But as Curtler remarked: "Agriculture under feudalism suffered from many of the evils of socialism".[75] The dominant relations of production were an obstacle to the development of agricultural technology.

In accordance with Nasse's general theoretical view of the relationship between socioeconomic factors and the technological nature of agrarian production, Ernle therefore thought, even in his major work from 1912, that research on the agrarian history of the late medieval centuries must necessarily concentrate on the socioeconomic changes that prepared the ground for later technological advances in agriculture.[76] But his views developed. In two articles from 1920 and 1921 this is clearly evident.[77] Contradictions in the technological structures were given ever greater emphasis as explanatory factors.

In these articles Ernle drew attention to a number of technical problems, all of which can be traced back to the purportedly collective form of cultivation and the splitting-up of the open fields, and all of which are seen as having contributed to the inadequate exploitation of the natural resources of agriculture which led to the progressive exhaustion of the arable land in a time of increasing population. The common field system was in itself an obstacle to the improvement of cultivation methods. All the members of the community were subject to fixed rules for the treatment of the soil, so an individual peasant could obstruct the attempts of other members to improve cultivation.[78] Moreover, any form of individual innovation was prevented.[79]

The splitting-up of the arable in the first place led to a waste of tillage land because of the walls and paths that separated the various plots; and secondly to a waste of time because of the distances between the farms and the scattered plots.[80] The permanent separation of tillage and pastures, and the resulting relatively poor manuring of the land, and on the whole the disproportion between tillage and livestock farming, were also problems which according to Ernle had their origins in the collective form of cultivation. So "the remedy for many of these defects was individual occupation".[81]

One can easily register certain changes in agricultural technology between 1300 and 1500: first, inasmuch as manorial land was in certain places withdrawn from common cultivation; secondly, because of the beginnings of the breakdown of the open-field system entailed by the incorporation of new, previously uncultivated arable land and the enclosure of commons. But these measures led to no real rupture of "the agrarian partnership". "They were rather devices to adapt the old system to changing needs, and were extensively practised in the fourteenth and fifteenth centuries".[82]

IV.7 The soil exhaustion theory

While the communal medieval open-field system grew up in Ernle's opinion under the influence of population growth, as a practical method of preserving the fertility of a limited natural resource - the soil - the system was broken down for the same reasons. "So long as population and farming remained stationary, and so long as the virgin soil retained its natural fertility",[83] this agrarian system was reproductive. But "from natural causes the open-field system was breaking down" from the thirteenth century on, and did so with increasing rapidity up until the fifteenth century.[84]

One might expect that the falling population after 1350 would have tempered the exhaustion of the soil that Ernle claimed to be in progress from the thirteenth until the fifteenth century;[85] but this does not seem to have been the case. "The worst feature in the existing system was the inevitable and progressive decline in the productivity of the soil".[86] This necessitated much more than the above-mentioned adjustments in the open-field system. "In 1485-1560 the only remedy for the exhaustion of fertility was the conversion of the worn-out arable land into pasture, and the substitution of existing grassland for the necessary tillage".[87]

Agreeing with Denton, and citing Fitzherbert and Tusser, Ernle observed that several of the useful techniques that had been used to cultivate manorial land in particular in the middle ages fell into disuse in the course of the fourteenth and fifteenth centuries. But in direct contradiction of

Denton, who as we recall explicitly considered the progressive exhaustion of the arable land to be a result of the decline of the manorial system, Ernle claimed that "even without this deterioration in farming practices, the loss of fertility was becoming sufficiently serious".[88]

With his articles from 1920 and 1921 Ernle had well-nigh reached a position of natural determinism. "Nature defies human regulations"[89] - which meant that nature forced changes in the technological form of agrarian production, and this again required changes in the socioeconomic relations. The natural basis - the soil - was presented as the ultimate limiting factor for the continued reproduction of the feudal agrarian relations of production, and thus as the primary cause of their dissolution.

The theory of the exhaustion of the arable land was not Ernle's invention; nor was he alone in modifying the theory from Denton's position, where the exhaustion problem was linked with the breakdown of the manorial system and the erosion of medieval agrarian technology, in order to understand the exhaustion problem as a something derived from the feudal agrarian relations of production.

In an article from 1913, G. Simkhovitch had argued for the same view. Simkhovitch claimed that medieval agriculture's failing production of fodder led to a permanent imbalance between tillage and stock farming, which had the result that a technology familiar in medieval times - natural fertilization - could not be exploited to an extent that would ensure the preservation or improvement of the fertility of the soil. This hypothesis led Simkhovitch to believe that the introduction of grass seed, the various types of clover, and later lucerne from the eighteenth century on, heralded a revolution "that fundamentally changed the basis of agriculture",[90] and that the late medieval enclosures were due to the gradual destruction of the fertility of the soil.

Simkhovitch's hypothesis was taken up in a treatise by Harriet Bradley from 1918,[91] in which she set out to prove that "the enclosure movement is explained not by a change in the price of wool but by the gradual loss of productivity of common-field land".[92] But Bradley further widened the explanatory scope of the theory in her work, analogously with Ernle. It was "the most important of many causes which were at work to undermine the manorial system in the fourteenth century".[93]

IV.8 The area under wheat at Witney, 1209-1349

Bradley argued that the fifteenth-century enclosures were simply a continuation of a process that had begun back in the thirteenth century, and
that this process could be explained independently of factors external to
it, like the development of prices and the Black Death.[94] As far as the
last point is concerned, Bradley demonstrated, on the basis of Rogers'
constructions, that the development of wheat and wool prices ran parallel
until the mid-fifteenth century; then the wool prices took a relative drop
and therefore could not explain the progressive conversion of arable to
pasture.[95]

As for the first point, she tried to establish that the cultivated wheat
area at Witney, one of the Winchester estates, according to a collation of
tables taken from N. S. B. Gras and the previously-mentioned works by
Levett and Ballard, decreased greatly between 1209 and 1349.[96] Her conclusion was undisturbed by the fact that the wheat area sown, as is evident from Bradley's table, also seems to have decreased to less than half
between 1349 and 1397.[97] Bradley concluded: "This withdrawal of land
from cultivation took place without the occurrence of any such calamity
as the Black Death....It affords an indirect proof of the fact that much
land was becoming barren"[98] - a rather bold interpretation of the
material, considering that we are talking about a single manor. True, it
was called representative of the Winchester estates by Bradley, and these
were in turn considered representative for England.[99] Nevertheless, it was
precisely the evident decrease in Witney's wheat-sown areas that formed
the empirical cornerstone of Bradley's attempt to verify the theory of the
exhaustion of the arable land.

This does not mean that she made no attempt to present other
empirical evidence. Beginning with Gras, Levett and Ballard, she tried to
draw up tables to indicate a general drop in yield from the arable of the
Winchester estates. But she did not succeed, although she attempted to
compensate for the lack of proof of a real drop by relating the yield to
price developments and thus modifying the factually quite stable yield
between 1208 and 1397 in favour of the soil exhaustion theory.[100]

And in fact Bradley too thought that the tables for the development of
yield provided a poor basis for saying that the fertility of the soil was
being eroded. "If the fertility of the soil is declining, this is shown by the
gradual withdrawal from cultivation of the less productive land".[101] In
general, she had to admit of the empirical material that "no material in

this field entirely satisfactory for statistical purposes is accessible at the present time".[102]

IV.9 The impoverishment of the peasants

Bradley did think, however, that her "statistical indications of declining productivity are supported by the overwhelming evidence of the poverty of the fourteenth century peasantry - poverty which can be explained only by the barrenness of their land".[103] The alleged deterioration in the fertility of cultivated land in the thirteenth century not only led to enclosures, but also to an extensive impoverishment of the peasants. "The problem which confronted landowners during the Black Death was not so much absolute lack of men on the manors, as a stubborn unwillingness on the part of these men to hold the land".[104] Indeed, "even before the Black Death, it was frequently the case that villein holdings could be filled only by compulsion".[105] "Land holding was regarded as a misfortune in the fourteenth century. The decline in fertility had made it impossible for a villain to support himself and his family and perform the accustomed services and pay the rent for his land".[106]

As the compulsion of villeinage itself did not succeed in keeping the primary producers, who fled in large numbers because of their poverty, the situation had to lead to reductions in the rent burden and to commutation.[107] Bradley saw both commutation in the fourteenth century, and the consolidation and leasing of tenant holdings, as derived from the impoverishment that became the fate of the peasants as the fertility of the soil deteriorated.[108]

With Bradley, then, one can speak not only of the crisis of the manorial system, but also of the crisis of the primary producers, which must make one sceptical about any talk of the fourteenth century as "a golden age of the people". Bradley cited a number of examples of how the primary producers in the fourteenth century had difficulties in meeting their obligations. But these examples, taken at second hand from Levett, Ballard, Page and Davenport among others, can hardly be accepted as firm proof of her claim concerning the relationship between the decreasing fertility of the soil and the general impoverishment of the peasants. Apart from the fact that these examples are taken without further ado to be indications of an overall situation for the English peasants in the thirteenth and fourteenth centuries, other explanations could be given for this apparent impoverishment - and we shall see later that they were. At any rate these examples of poverty do not in themselves support Bradley's statistical indications that the fertility of the soil was deteriorating.

IV.10 The wheat yield at Witney in 1397

Lord Ernle's statistical proof of the increasing exhaustion of the arable from the thirteenth until the fifteenth century is rather weaker than Bradley's. Although he claims that "strong evidence exists",[109] his documentation is restricted to a dubious comparison between Walter of Henley's alleged assessment of an optimal wheat crop in the thirteenth century,[110] and the yield at Witney for 1397.

Thus Ernle tries to verify the soil exhaustion theory on a basis that Bradley found unsatisfactory. Since he was apparently aware that the circumstantial evidence of the Winchester material used by Bradley did not indicate any striking drop in the yield between 1200 and 1400,[111] he attempted the above mentioned rather forced comparison instead - a comparison that indicated that the wheat yield in the period in question fell from ten bushels an acre to six.

In the first place it is unacceptable that Ernle inferred a general situation from this scanty basis. Secondly, if less importantly in this context, it was not Walter of Henley, but an anonymous writer from the thirteenth century who averred that "wheat ought by right to yield to the fifth grain"; which, compared with an estimated sown volume between two and two-and-a-half bushels an acre in the thirteenth century should indeed have yielded between ten and twelve bushels an acre.[112] In the third place, and not insignificantly in this context, this unnamed author thought that five-fold was what wheat "ought by right to yield", writing however that "one cannot be sure of the yield above mentioned".[113]

IV.11 The soil exhaustion theory refuted

It is evident from the above that it was very difficult to present convincing proof for the soil exhaustion theory. This of course does not in itself mean that the theory was wrong, but its weak empirical verification gave rise to criticism of both the theory itself and its empirical foundations.[114]

The first serious attack on the theory was mounted by R. Lennard, who tried to shed light on agrarian productivity in the late middle ages with a better elaborated empirical basis.[115]

Lennard's study was based on secondary source material which he classified chronologically into groups covering the thirteenth, fourteenth and fifteenth centuries respectively. Depending on the nature of the source material, he either calculated the grain yield from simple comparisons between the amount of seed used and the harvested crop, or simply noted the yield in bushels per acre specified in the accounts.[116] He also compared his results with price developments for the respective crops as they appeared in Rogers, so as to assess whether the years included in the study could be regarded as normal, abnormally good or bad harvest years.

If we look first at Lennard's calculations of the grain yield in the thirteenth century, we see that in the four manors covered by the table from 1278 to 1300, there seems to have been an average grain yield for all crops - oats, wheat, barley and peas - below the level the author of *Hosebondrie* thought there should be.[117] Even if the calculations had allowed for a ninth part in "tithes", the average grain yield would have been below the level specified in *Hosebondrie*.

If we look further at the seven manors, and at an unspecified number of Winchester estates for which Lennard tabulated the yield in bushels per acre for eighteen seasons in the thirteenth century, we find that the wheat yield at least was much smaller than Ernle supposed.[118]

Comparing these results with what Lennard arrived at for the fourteenth century, we observe first that the material on grain yield has been considerably expanded. It includes far more manors, only one of which - Forncett - is a duplicate from the table for the thirteenth century. It is also worth noting that only six seasons are represented, five of which are from the 1303-1336 period. All these years, as well as 1388, must be considered normal according to Lennard. The average grain yield for these seasons on the manors in question was higher than the corresponding figure for the thirteenth century.[119]

The calculations for the yield in bushels per acre for the fourteenth century again cover eighteen seasons in the 1305-1397 period. But now, more manors have been included than in the calculation for the thirteenth century. Several of these are duplicates. Here, too, the wheat yield has risen compared with the table for the previous century.[120]

One must concede Lennard that great caution must be exercised in the use of this material, which he did not supplement to any great extent with the few, scattered sources he took from the fifteenth century and which incidentally showed the same tendency towards a moderate rise in the wheat yield at least. Lennard's documentation is a combination of very scattered material. There are data from a number of different manors, distributed over relatively few for each century, and they exhibit so many elements of uncertainty in general that Lennard would not claim on this basis that the yield either rose or fell in this period. He did consider it completely unscholarly to claim on the basis of Ernle's and Bradley's empirical documentation that the fertility of the soil was deteriorating in general in the fourteenth and fifteenth centuries.[121]

Lennard further found the idea of widespread exhaustion of the soil in this period improbable considering that the social division of labour increased with the growth of wool manufacturing, and that wages were rising. He thought these general "facts" of economic history did not accord with a drop in agricultural productivity; and that the theory of the progressive exhaustion of the soil, and Ernle's claim for a wheat yield around ten bushels per acre in the thirteenth century, would require a quite unrealistically high yield in the preceding centuries.

IV.12 The Rothamsted experiments

Lennard also used the so-called Rothamsted experiments in his arguments against Ernle and Bradley. Although these did not address themselves to the study of medieval arable farming methods, but to the introduction of innovative cultivation methods, certain of the results published in A. O. Hall's report from 1905,[122] seem to speak against the formation of general theories of the gradual exhaustion of the arable land by medieval forms of cultivation.[123]

When Hall's report appeared experiments had been done at Rothamsted involving uninterrupted cultivation of a plot without manuring since 1844. As is evident from Fig. 2 on p. 37 of this report, there was a minor drop in the grain yield from this plot during the first decade after 1844; in the next two decades it fell rather more, but then seemed to stagnate, with a slight tendency to rise. However, the biggest fluctuations were apparently connected with climatic conditions. Thus the lowest grain yield can be seen in the figure for the 1872-81 decade, which was typified according to Hall by many seasons of bad weather - and this is confirmed by the fact that the decline corresponded very closely with developments on the plots cultivated with wheat in the same period that were fertilized with either animal manure or artificial fertilizer.

Against this background Hall drew the following conclusion: "All the evidence seems to point to the fact that this plot...has reached a stationary condition, and that the average crop of twelve and a half bushels for the last forty years will in future diminish very slowly, if at all".[124]

A similar experiment was done over the same period of years with barley. This showed that the grain yield from constant cultivation without manuring fell steadily - but it also did this on plots that were artificially fertilized. On the other hand the experiment showed that fertilization with animal manure seemed to stabilize the grain yield.[125]

Another experiment with barley showed that the decline in grain yield was significantly modified by simple rotation between cultivation and fallow periods.[126] But a corresponding experiment with wheat showed that even if the level of grain yield with rotation between cultivation and fallowing was higher on average - 17.1 bushels - than with constant cultivation - 12.7 bushels - the overall yield during the seventy-year course of the experiment was higher with constant cultivation.[127]

The Rothamsted experiments were done on a soil which was described in 1834 by the originator of the experiments, the owner of the Rothamsted estate J. B. Lawes, as heavy loam on a chalk bedrock. The manor is in Hertfordshire and thus lies in the central English area where the open-field system was dominant. Although the experiments, as mentioned before, were not done to study medieval agricultural techniques, and cannot be directly compared with medieval cultivation, they seem to have shed light on some of the ecological aspects of the open-field system. At any rate they appeared to be a strong argument for Lennard against the soil exhaustion theory, although he only referred to the experiments with continuous cultivation of wheat and barley.[128]

IV.13 Grain yield, 1200-1450

When Lennard wrote his article in 1922 he lacked an empirical basis from which to build up statistics for the relationship between seed sown and yield on one or more manors, preferable over a longish, relatively uninterrupted period. Such material was in fact available. As early as 1903

Hubert Hall had translated and published the first of the so-called Winchester Pipe Rolls from 1208-09.[129] Furthermore, both Levett and Gras, whose work Lennard quoted freely, made extensive use of the still unpublished Pipe Rolls from the subsequent period. An edition of this material aimed at clarifying the issue of agricultural productivity in the late middle ages was also in the offing.

When Lord Beveridge was preparing a tabular work on the development of prices in medieval England along with Hubert Hall,[130] he started on a survey which was published in 1927. Among the more than fifty manors covered by the material, Beveridge use a geographical sample of nine manors whose productivity he studied for the period between 1200 and 1449.[131]

In the study, Beveridge was interested in the amount sown per acre, the acre yield and the grain yield - the yield as a function of the amount sown. While the ratio of seed sown to area sown can be read directly from the accounts,[132] he calculated both the yield per acre and the grain yield by simply comparing the total yield for a year with the area or seed volume itemized in the roll for the previous year. Here we will concentrate on the calculation of the grain yield, among other reasons because, as mentioned before, and as pointed out by Beveridge, the calculations of yield per acre are rendered uncertain by the fact that this measure of area varied locally.[133]

The necessary basis for calculating grain yield, that there are accounts for two successive years, means that there are more gaps in the material than one experiences, for example, when calculating the yield as a whole. But apart from a single manor - Esher in Surrey - its is still possible to calculate the grain yield for the manors for between 106 and 143 years. Beveridge calculated the average grain yield between 1200 and 1450 on this basis at 3.89 for wheat, 3.82 for barley and 2.43 for oats.[134]

If one compares these figures with the grain yield development in Beveridge's Table III, one sees that the grain yield for wheat, barley and oats was only below this average in certain periods before 1350, while it was always higher after 1350.[135] But the fluctuations are very small. On one manor, Ecchinswell, there is an upward trend; on another, Meon, there seems to be a falling trend, In general the table gives the impression of stability and stagnation. "There is little sign either of material advance in agricultural methods or of that declining fertility of the soil which some writers have discovered in the later Middle Ages".[136]

Nor is there anything in the table to suggest that there were attempts to avert any failing fertility by using more seed per acre. The amount sown seems to have remained constant around 0.28 of a quarter per acre for wheat, 0.47 for barley and 0.48 for oats throughout the period. Of course the table shows a similar stability for yield in quarters per acre.

The only sign of change appears in Table I, where Beveridge has tabulated statements of the area sown calculated as an average over the two fifty-year periods 1200-1249 and 1400-1449.[137] Apart from the manors where the area sown was relatively small, the table shows a fairly drastic reduction in the cultivated area between the two periods. What is not evident from the table is "that, taking the manors as a whole, the decline of acreage covered by the returns proceeds steadily. There is no sudden jump at the Black Death or at any other point".

This trend raises a question which Beveridge did not take up. Was it due to the leasing out and/or tenanting of the land, or was it because the land was simply no longer cultivated due to the declining fertility of the missing areas?

Notes

1. T. F. Tout, *The Political History of England*, Vol. III, 1905, p. 370.

2. Ibid., pp. 371-72.

3. Tout sympathized with the fourteenth-century potentates' lack of respect for "the laws of political economy". "No one in the Middle Ages believed in letting economic laws work out their natural results. If anything were amiss, it was the duty of the kings and princes to set things right". So the Statutes of Labourers were an organic part of the medieval economy, and "not much more ineffective than most laws at that time". Ibid., pp. 372-73.

4. Ibid., pp. 373-74.

5. Ibid.

6. Ibid., pp. 374-75.

7. Ibid., p. 375.

8. E. Powell, *The Rising in East Anglia*, 1896. pp. 64-65.

9. "The immediate consequence of the rising does not seem to have been any general abandonment by the lords of their disputed rights". According to Cunningham, op. cit., pp. 402-03, the remnants of villeinage persisted well into the sixteenth century. "There can be no doubt that in most regions the old system went on....The best proof of this is that the manorial archives of the next ten years are full of conflicts between landlord and villein precisely similar to those which were rife in the years before the great rising. If we had not the story of Tyler and Ball, Wraw and Litster preserved in the chronicles and the judicial proceedings, we should never have guessed from a mere study of court-rolls that there had been an earth-shaking convulsion in 1381". C. Oman, *The Political History of England*, Vol. IV, 1906, p. 64. Despite the absolute formulation Oman presents no documentation or references showing where documentation can be found.

10. Ibid.

11. "If the class legislation on behalf of the landlords had not intervened, the period following the Black Death would have been a sort of golden age for the peasant". Ibid., p. 27.

12. The villeins protested not so much against the labour rent as the many other different obligations - for example, heriots, compulsory use of the lord's mill, and duties payable for hunting and fishing rights on the uncultivated areas. Ibid., p. 29. One reason for the Revolt was the Statutes of Labourers (ibid., pp. 26-27); but it was the Poll Tax of 1381 which triggered off the rising and united the discontent of the peasants with the smouldering protest in the towns, created by social, economic and ethnic conflicts. Ibid., pp. 26, 30-33.

13. By contrast, the fourteenth century must be considered a turning-point for Johnson's subject. The aim of the book was to give a historical explanation of why landed property in England around 1900 was more centralized and unequally distributed than was the case, for example, in France. Johnson thought, like Seebohm and several others, that the historical explanation of this peculiarly English structure had to be sought as far back as the fourteenth century. A. H. Johnson, *The Disappearance of the Small Landowner*, Oxford 1909.

14. A crucial reason why the plague affected the manors in different ways was that there were three categories of villein at the time: those who paid a fixed money rent, those who rendered labour services "at the will of the lord", and those who were burdened with both rent types. Ibid., p. 24.

15. Ibid., pp. 24-25.

16. Ibid., p. 25.

17. Ibid., p. 27.

18. Ibid., pp. 27-28.

19. Ibid., p. 27.

20. Unlike Oman, Johnson thought with Cunningham that the Statutes of Labourers were a new and alien element in the feudal economy. Ibid., p. 28.

21. Ibid.

22. Ibid., p. 30.

23. Although Johnson thought that commutation had already started before 1350, he considered, with support from Page and Cunningham, that Rogers had exaggerated its extent. Ibid., p. 23.

24. Ibid., pp. 22-23. On the manorial system's inability to exploit "the personal element" in a period of rising population, and the falling value of "boon-days" for estate-holders with rising prices according to Vinogradoff, see Chapter III, pp. 82-83 and note 79 on the same.

25. Pollard's view of the late medieval centuries is closely linked with that of Rogers, and we need not go into it further. The author has a rather curious view of the discipline of history, claiming that "history can never be true to life without imagination"; that "facts...only are secondary considerations"; and that "history is not exact science", because it does not deal with abstractions, only with concrete phenomena. As indicated above, these views do not seem to have had any practical consequences. A. F. Pollard, *Factors in Modern History*, London, 1910, pp. 2, 30.

26. Ibid., p. 51.

27. E. Lipson, *An Introduction to the Economic History of England*, Vol. I, "The Middle Ages", London 1915, p. 85.

28. Cf., among others, F. W. Maitland, op. cit., ibid., p. 86.

29. Ibid., p. 110.

30. "This [manorial] system could only break down when the supply of money became sufficiently great, and its circulation sufficiently rapid....But...we have no ground for assuming that by the middle of Edward III's reign a money economy had replaced natural husbandry to the extent which the complete commutation of personal services would necessarily presuppose...the scarcity of money seems to have blocked that path....On general grounds then, we are led to question the progress of commutation [cf. Page, op. cit., p. 42] but fortunately we have abundant evidence that at the time of the Black Death the system of predial services was still in full operation [cf. Page, op. cit., pp. 45-46]". Ibid., pp. 83-84.

31. "The error as to the disappearance of villeinage probably arose from the fact that, even as early as the thirteenth century, the services exacted from villeins were assessed on the account rolls of the manor in terms of money. The practice was adopted from motives of convenience; it laid the basis for commutation, but it did not imply that the tenant invariably paid a money-equivalent in place of personal service" (cf. *Records of Cardiff*, ed. J. H. Mathews, 1898, Vol. I, p. 279). Ibid., pp. 84-85. However, it can be

seen that Lipson's description of the manorial accounts is not generally applicable from F. W. Maitland, *History of a Cambridgeshire* ..., op. cit., e.g. p. 370; cf. also Cunningham's critique of Rogers, Chap. II, p. 51, n. 49.

32. Lipson, op. cit., p. 82.

33. Ibid., pp. 77-78, 101.

34. This class originated among: 1) cottagers; 2) freed villeins; 3) "The younger sons and brothers of villeins, whose services the lord was unable to utilize on the estate..." Ibid., pp. 81-82. Lipson does not relate the latter resource to a possible population growth.

35. Ibid., pp. 78-79.

36. One could draw up a whole list of reasons for the revolt from Lipson's text: dissatisfaction with the administration of Royal officials; the Statutes of Labourers; the Poll Tax; general social unrest; ideological revolt; and the peasants' discontent because the commutation process stopped just after the Black Death. Lipson did not think like Rogers that the landlords directly attempted to reintroduce the labour rent. But he thought that the reaction to this view probably went to the opposite extreme. At least he considers it both logically and empirically reasonable that "the lord had every motive to check the progress of commutation". Ibid., pp. 87, 106. The Statutes of Labourers were seen as one reason for the Revolt, because these measures "for a time, at any rate when the crisis was most acute... must have served in some degree...to check the rise in wages" (ibid., p. 97), and because "the Statute of Labourers must be regarded as...an unfair exercise of political power in the interests of a single class of the community". Ibid., pp. 98-99. The former, but not the latter view, was shared by Bertha H. Putnam, who in her *The Enforcement of the Statutes of Labourers*, New York 1908, had studied, more systematically and thoroughly than anyone before, whether the legislation was fair and whether it was effective in the 1349-1359 period. Ibid., p. 3. Putnam thought that the legislation was administered zealously, and for that reason among others had a real effect in the period; moreover, that it was necessary insofar as "the situation was plainly a crisis of an unprecedented character..."; and fair, because it did not differ in principle from previous legislative initiatives. Ibid., pp. 222-223.

37. E. Robo, "The Black Death in the Hundred of Farnham" in *English Historical Review*, Vol. XLIV, 1929.

38. A. Ballard, "The Manors of Witney, Brightwell and Downton", in *Oxford Studies in Social and Legal History*, Oxford 1916.

39. A. E. Levett, "The Black Death on the Estates of the See of Winchester", in *Oxford Studies in Social and Legal History*, Oxford 1916.

40. For the earliest and middle period Ballard also used the Domesday Book (11th c.) and the Hundred Rolls (13th c.). Levett supplemented this with a few court rolls, a couple of extents, and with the original Compotus Rolls from which the Pipe Rolls were drawn up.

41. Levett, op. cit., pp. 142, 153-54.

42. In the first place Levett thought that Gray had not been meticulous enough in the search for sources. Secondly, she thought that Gray's method was doubtful. The comparison between the price of labour rent - both rendered and sold - and the total "rent of assize" said nothing in itself about the extent of commutation, since Levett did

not believe that the rents of assize had any relation with the commuted labour rent. Rents of assize rather represented archaic money rents. Ibid., p. 19.

43. Ballard, op. cit., pp. 203, 210, 214.

44. Levett, op. cit., p. 142.

45. Ibid., p. 151.

46. Robo, op. cit., p. 571.

47. Ibid.

48. Levett too saw Rogers as the originator of this interpretation.

49. Ballard, op. cit., p. 216.

50. Robo, op. cit., p. 571.

51. Levett, op. cit., p. 155.

52. Ibid., p. 157.

53. Ibid., p. 156.

54. Ibid., pp. 157-58.

55. Ibid., p. 160.

56. Power considered that two principles applied to assessments of the significance of the Black Death after 1900: "First, that there are no cataclysms in medieval economic history...secondly, that there are no general truths in medieval history; progress was everywhere local and uneven..." E. E. Power, "The Effects of the Black Death on Rural Organisation in England", in *History*, Vol. III, 1918, pp. 112, 116.

57. Ibid., p. 112.

58. Thus, on commutation: "In estates where the demesne was small in proportion to the amount of land in villeinage, and where the lord would not want all the services to which he was entitled, it would obviously be resorted to sooner than in estates with large demesne and fewer villein holdings". On leasing: "From very early times parcels of land reclaimed from the vaste (the so-called assarts) had been added to the customary land of the manor, and these new strips were always let at money rents. Farm servants, the sons of villeins, sometimes men who held strips in villeinage in the common fields, thus became rent-paying tenants for the new lands". Ibid., p. 113. Cf. Vinogradoff, *Villainage in England*, op. cit., pp. 329-33.

59. Helen Robbins, "A Comparison of the Effects of the Black Death on the Economic Organization of France and England", in *The Journal of Political Economy*, Vol. XXXVI, Chicago 1928, p. 477.

60. Ibid., pp. 452, 459.

61. "From 1200 to approximately 1450 the production of silver decreased, apparently because the surface mines available had been exhausted and because abortive attempts to mine by hydraulic pressure had flooded many otherwise productive mines". Ibid., p. 457.

62. Ibid., p. 460.

63. Ibid., pp. 468-69.

64. Ibid., p. 476.

65. Ibid., pp. 478-79.

66. Ibid., p. 472.

67. "...commutation had already gone a good way before the Black Death. That catastrophe, here as in other social fields, shook what was already tottering or decaying;

and it was followed by a very rapid increase in commutation". G. C. Coulton, *The Medieval Village*, Cambridge 1926, p. 137.

68. "The ancient rough uniformity of the size in the villeins' holdings had been changing for some time....This process, which was noticeable at the beginning of the fifteenth century, was much accelerated by the Black Death and consequent events. But the great underlying cause of the breakdown of the manorial system was the increasing exhaustion of the soil". M. E. Seebohm, *The Evolution of the English Farm*, London 1927, p. 186. We may note how Seebohm attributes an accelerating effect to the Black Death. It is also interesting to see him postulating a theory of the increasing exhaustion of the arable up through the late middle ages. In the following section we will look more closely at the way this theory had been posited in the second decade of the twentieth century, and how it was refuted at about the same time, as Seebohm presents the theory without any kind of references or other documentation.

69. Rees was very cautious about assessing the consequences of the Black Death - nor did his article afford much of a basis for doing so - because he thought "the fact that the pestilence coincided with a period of change makes it difficult to determine accurately the direct results of the Black Death". All the same, he attributes psychological importance to the plague as well as significance for the transformation of the agrarian relations of production and state formation. Rees too attributed to Rogers the view that the Black Death was a revolutionary event - an interpretation he considered had already been laid to earth, and even more so by his own contribution. William Rees, "The Black Death in England and Wales, as exhibited in manorial Documents", in *Proceedings of the Royal Society of Medicine*, Vol. 16, 1923, pp. 27, 35-36.

70. Power used A. F. Pollard, op. cit., as an example of how "this view, first stated by Thorold Rogers, still holds the field in many text-books". But the passage he quotes from Pollard seems to confirm that both Pollard and Rogers had a different opinion. Power quotes: "Services had been largely commuted for rents, and the serfs had achieved their emancipation, though some remained in bondage as late as the sixteenth century. Then came the Black Death..." Power, op. cit., p. 109.

71. We have seen inflation explained as a result of the drop in silver production and the accompanying rise in the price of silver. An attempt was made to demonstrate that the coinage had been debased under Edward III in 1351 and before in A. Hughes, C. G. Crump & C. Johnson, "The debasement of the coinage under Edward III", in *Economic Journal* VII, pp. 187, 189, and by W. A. Shaw, *History of Currency*, 1895, p. 46.

72. "The progress of English agriculture was in its infancy determined by the growth of population". B. Ernle in the *Quarterly Review* No. 318, April 1885, Vol. 159, p. 323. "Improvements in the art and science of English agriculture were in its infancy dependent on the exhaustion of the soil". Lord Ernle, *English Farming Past and Present*, London 1912, p. 1.

73. However, this theory is also visible in the article from 1885, where he dealt with the constitution of the socioeconomic and technological structures of medieval agriculture.

74. "So long as land was abundant, and the people few and migratory, no rotation of crops was needed. Fresh land could be ploughed each year. It was only when numbers had increased and settlements became permanent, that farmers were driven to devise

methods of cultivation which restored or maintained the fertility of their holdings". *English Farming...*, op. cit., p. 1.

75. W. H. R. Curtler, *A Short History of English Agriculture*, Oxford 1909, p. 47.

76. Ernle, *English Farming...*, op. cit., p. 34.

77. Lord Ernle, "The Enclosures of Open-Field Farms", in *The Journal of the Ministry of Agriculture*, December 1920 (I), January 1921 (II).

78. "Agriculture...was unprogressive. No improved methods or increased resources were offered to farmers, which could only be introduced on open-fields with the consent of a timid and ignorant body of partners, any one of whom could refuse to have them adopted on the farm. The system fostered stagnation, and starved enterprise..." Ernle, "The Enclosures...", op. cit. (I), p. 836.

79. "All the occupiers were bound to rigid customary rules, compelled to treat all kinds of soil alike, unable to differentiate in their cultivation, bound to the unvarying triennial succession, obliged to keep exact time with one another in sowing and reaping their crops. Each man was at the mercy of his neighbours. The idleness of one might destroy the industry of twenty". Ibid., p. 837.

80. Ibid., pp. 836-37.

81. Ibid., p. 837.

82. Ibid., pp. 838-39.

83. Ibid., p. 836.

84. Ibid., (II), p. 900.

85. Ibid., p. 899.

86. Ibid., (I), p. 838.

87. Ibid., (II), p. 900.

88. "One is that of marling..." Ibid., (I), p. 839.

89. Ibid., (II), p. 899.

90. Vladimir G. Simkhovitch, "Hay and History", in *Political Science Quarterly*, Vol. XXVIII, 1913, p. 393.

91. Harriet Bradley, *The Enclosures in England - An Economic Reconstruction*, New York 1918.

92. Ibid., p. 13.

93. Ibid., p. 72.

94. "The process of withdrawing land from cultivation began independently of the scarcity of labor caused by the Black Death and independently of any change in the price of wool..." Ibid., p. 22.

95. Ibid., pp. 24-26, 42.

96. Ibid., p. 55. See also N. S. B. Gras, *The Evolution of the English Corn Market*, New York 1915, Appendix A; and Levett & Ballard, op. cit., p. 190.

97. According to Table V, ibid., p. 55, the area sown with wheat decreased on the manor of Witney between 1209 and 1277 from 417 to 180 acres; between 1277 and 1349 from 180 to 128 acres; while the decline between 1349 and 1397 was no less than from 128 to 51.5 acres.

98. Ibid., p. 56.

99. Ibid., p. 52.

100. Ibid., Table III, p. 53; Table V, p. 55. For example, the demonstrably relatively high grain yield in 1300 was modified with the claim that it was abnormally high. This

was justified by a reference to Rogers, who had said that the price of wheat in that year was 17% below the average.

101. Ibid., p. 55.
102. Ibid., p. 51.
103. Ibid., p. 56.
104. Ibid., p. 60.
105. Ibid., p. 57.
106. Ibid., pp. 58-59.
107. Ibid., p. 66.
108. Ibid., pp. 56-57.
109. Ernle, "The Enclosures...", (I), op. cit., p. 838.
110. As mentioned before, Ernle based his accounts to a great extent on technical agricultural works. Between the publication of the first edition of *English Farming...* in 1888 and the second in 1912, Cunningham and Lamond had in 1890 published certain works from the thirteenth century which have been attributed to Walter of Henley. Ernle incorporated material from these works in the second edition.
111. Table III in Bradley, op. cit., p. 53, drawn up on the basis of Gras's Appendix A, shows a largely stagnating grain yield between 1208 and 1397.
112. Ernle apparently confused Walter of Henley's *Le Dite de Hosebondrie* with the anonymous work *Hosebondrie*. Both works were, as stated in note 110, published together in 1890.
113. *Hosebondrie*, author unknown, ed. E. Lamond, 1890, pp. 66-67, 70-71.
114. The dubious empirical basis of the soil exhaustion theory was also what its early critics objected to in Denton's original postulation of it. See Gibbins, *Industry in England*, New York 1897, p. 181; and Hasbach, op. cit., p. 31.
115. R. Lennard, "The Alleged Exhaustion of the Soil in medieval England", in *The Economic Journal*, Vol. XXXII, 1922.
116. The latter approach is somewhat dubious, among other reasons because of local differences in the exact sizes of acres and bushels, and because one can nowhere in the material be certain that the area sown was the same from year to year. See also Walter of Henley, op. cit., pp. 66-67.
117. Wheat - 5; oats - 4; barley - 8; beans and peas - 8. Ibid., pp. 70-71.
118. By adding a ninth as "tithe" one arrives at the following average figures: "Wheat, between 6 1/4 and 6 1/2 bushels per acre; Barley, between 11 1/4 and 11 1/2 bushels; Oats, between 9 1/4 and 9 1/2 bushels". Lennard, op. cit., p. 20.
119. Wheat rose from 3 17/20-fold to 4 5/8; oats 2 13/14 to 2 37/40; barley from 3 15/28 to 3 27/28; while the yield for peas fell slightly from 4 1/6 to 4 1/8. Ibid., pp. 16-17, 21-22.
120. From about 6 to 7 1/2-7 3/4 bushels per acre for wheat; barley from about 11 to between 15 1/4 and 15 1/2; while oats exhibited a small drop of 3/4 of a bushel per acre. Ibid., pp. 18, 20, 23-24.
121. Ibid., pp. 25-26.
122. A. D. Hall, *The Book of the Rothamsted Experiments*, London 1905.
123. Surprisingly enough, they do not correspond either with the contents of the quotation from E. J. Russell's book, *The Fertility of the Soil*, Cambridge 1913, used by Bradley - this although Russell took part in the Rothamsted experiments. Russell was

quoted by Bradley for the view that continuous cultivation of wheat, as "practiced under modern conditions in new countries", leads in the long term to the exhaustion of the soil. Bradley, op. cit., pp. 46-47.

124. Hall, op. cit., p. 38.

125. Ibid., Table XXIX, p. 72; Table XXX, p. 73; Fig. 10, p. 74.

126. Ibid., Table XXXI, pp. 75-76.

127. Ibid., Table XXIV, p. 62-63.

128. Lennard, op. cit., p. 27.

129. Hubert Hall (ed.), *Pipe Roll of Bishopric of Winchester, 1208-09*, 1903.

130. Volume I of this work, *Prices and Wages in England - From the Twelfth to the Nineteenth Century*, covering the period from c. 1550 - c. 1830, appeared in 1939. Volume II, which was to cover the medieval centuries from the twelfth century on, has never been published.

131. Lord Beveridge, "The Yield and Price of Corn in the Middle Ages", in *Economic History*, 1927. Quoted here from a reprint of the article in *Essays in Economic History*, Vol. I, ed. E. M. Carus-Wilson, 1954, Table 1, p. 15.

132. The account for each year shows the amount of corn produced; the amount of corn preserved from previous years; and the amount of corn purchased. Moreover, certain of the accounts show tithes and payments in kind of corn to servants. These lists are supplemented by notes on seed and area sown the next year.

133. Ibid., p. 23.

134. Ibid., Table II, p. 16.

135. Ibid., Table III, p. 18.

136. Ibid., p. 19.

137. Ibid., Table I, p. 15.

Chapter V

The Medieval Crisis
(1930-1950)

V.1 New horizons

In a historiographical work from 1969 the Czech historian Frantisek Graus distinguishes between "die Vorgeschichte der modernen Krisendiskussion" and the modern discussion of the late medieval crisis after 1945.[1] In a wide-ranging European perspective he sees the prehistory of the discussion as something developed in opposition to a dominant idea that the late medieval ages formed a linking stage in a the continuous progress of the European societies from the high middle ages to early modern times.

He sees the beginnings of this opposition in Ranke's general description of the fourteenth and fifteenth centuries as "eine Epoche der "allgemeinen Auflösung"" and the French historians G. C. Guibals' and H. Denifles' account of these centuries as "une époque de transition, de crise, de révolution" under the influence of, among other things, the devastation caused by the Hundred Years' War; traces it through Thorold Rogers' and the German Lamprecht's talk of occupational crises, to the Austrian Grund's pointing-out of the problem of the deserted farms of the middle ages in 1901 and Pirenne's division of the development of trade into phases, where the late middle ages appear as a period of recession.

Graus further acknowledges that as early as 1931, Marc Bloch had spoken of the great crisis of the fourteenth and fifteenth centuries, and gives several examples of similar interpretations of the period by European historians of the latter half of the nineteenth century and the first decades of the twentieth.[2] Given this background, it can be difficult on the face of it to understand his classification into periods.

But in the prehistory of the crisis debate Graus sees only a tendency to take up individual, isolated manifestations of crisis,[3] and thinks that these do nothing to delimit precisely the phenomenon of crisis.[4] After 1945, on the other hand, he thinks that the discussion of crisis had reached a stage where attempts were being made to identify the nature and causes of the crisis.

Our survey of English historical research in this field clearly disproves these views. English historiography, which Graus saw as one of the cornerstones of crisis research, although he himself made little use of it,[5] could even before 1900 boast several contributions to the description of the nature of the crisis as well as the pinpointing of its causes. Moreover, one can hardly say that English research on the late medieval centuries before and shortly after 1900 was typified by any general view of history as a continuum. The dominant reading of the late medieval centuries saw it as a series of social upheavals, characterizing the period as a watershed during which the feudal social order was transmuted into a higher form.

Graus does not document his view convincingly. In his confusion he must reduce it to a tendency, saying in fact that "den eigentlichen Wendepunkt in der Diskussion...stellt jedoch erst das Buch von W. Abel [1935] dar". The epoch-making element that Graus sees in this book is Abel's development of the concept of *Agrarkrise* into a stringently delimited "quasi technischen Begriff der Wirtschaftgeschichte".[6] Graus touches here on the less rigid chronology of the crisis debate found in one of his earlier works,[7] while at the same time recalling the more or less general assumption that crisis research had its breakthrough in the 1930s - with Marc Bloch in France, Abel in Germany and M. M. Postan in England.[8]

This reading can to some extent be defended in the case of English historiography, if one compares research in the period after 1930 with that of the immediately preceding decades. But it cannot if one goes further back, as is clearly evident from the article - generally considered pioneering - by M. M. Postan from 1938,[9] where he tried to reconcile the conflicting interpretations of two of the past's clearest advocates of conditions of crisis in the late medieval centuries.

V.2 The agrarian crisis

Postan takes his point of departure in the controversy among economic historians over Rogers' and Denton's conflicting views, and argues that Denton's notion of crisis, which unlike Rogers' concerned the country's general political and economic development, was not incompatible with Rogers' opinion that the conditions of the working classes improved in the course of the fourteenth and fifteenth centuries. Since "a relative decline in the total volume of national wealth is fully compatible with rising standard of life of the labouring classes".[10] Here we see Postan's concept of crisis linked with indications of the expansion or contraction of the social reproduction process.

Postan argued that there was a general agricultural crisis in the fifteenth century from a theoretical basis that accorded with Rogers' definition of the crisis of the manorial system. But he did not consider the reorganization of manorial production and the leasing-out of the demesne to be the only reasons for the drop in agricultural production which constituted the crisis for him. He asserted that this was not a simple reorganization, and thus not just a crisis of the manorial system, since "more land was withdrawn from the demesne than was let out to tenants. In other words there was a net contraction of the area under cultivation", and there was a consequent absolute drop in rent incomes.[11] The halt to the colonization waves of the twelfth and thirteenth centuries also influenced the drop in rents.[12]

But was this increase in fallow land not compensated by an increase in sheep farming? Postan indirectly attacked this assumption, often repeated since Seebohm, by pointing out that both wool exports and the cloth industry that had flourished since the mid-fourteenth century stagnated or even tended to decline in the course of the fifteenth century.[13] This rejection of the old assumption was later supported by Eileen Power, who wrote in a monograph on the medieval wool trade that "it is difficult to find signs of that wholesale substitution of pasture for arable farming, which according to textbooks happened after the Black Death". On the contrary, she thought, "one is indeed forced to the irresistible conclusion that there was a serious drop in sheep farming during this period".[14]

According to Postan agricultural production was declining, not only because the manorial system was breaking down, but particularly because land was being withdrawn from cultivation, and because fallowing was not being compensated by an increase in sheep farming. To this Postan added a remark that in N. S. B. Gras' study of the English corn market[15] there were signs of a contraction in the corn trade up to and in the fifteenth century. The reason he gave for this was that the letting-out of the demesne in itself led to a fall in the supply of corn to the market, because "the peasant holding represented a more self-sufficient economy" than that of the manors, which had in the thirteenth century been "capitalist concerns".[16]

140

One can raise a number of objections to this interpretation. First, a more limited corn supply did not necessarily indicate a drop in agricultural production. Secondly, Gras did not in fact speak of a contraction of the English corn market in the period. What he said was that with the advance of the money rent and the leasing system there was a strengthening and expansion of the local and territorial markets.[17] Thirdly, it is hard to see why the disintegration of the manorial system should simply have led to a restriction of the market economy. The money rent and the leasing system required by their very nature exchange relations, while these were only a possibility whose development was historically extraordinarily probable under the manorial system. These were factors that Postan had demonstrated his familiarity with in an earlier work.[18] The breakdown of the manorial system, the desertion of arable land without subsequent use for sheep farming, and a restricted corn market all indicate a drop in agricultural production. On this basis Postan thought, despite acknowledging the improvement in the conditions of the peasants and agricultural labourers after the Black Death, that agriculture as a whole encountered a crisis that began in the fourteenth century, perhaps even before 1350, and which persisted until the end of the fifteenth century.[19] But as we have already seen suggested, he spoke not only of an agrarian crisis, but of a general socioeconomic crisis, which also affected the urban trades.

The crises in the two economic sectors ran their course according to Postan relatively independently of each other, although the agrarian crisis in a country so dominated by this sector of production cannot have failed to affect the other industries. Thus the purported "diminution of buying and selling in the countryside [was] in part responsible for the decline of the corporate towns".[20] But the crisis in the urban industries would have developed anyway.

We will not go further here into the indications of crisis in the urban trades presented by Postan. Instead, let us look at his attempts to provide possible explanations of the agrarian crisis. For he makes no claim in this work to pinpoint "the real causes of the agricultural depression [which] still await investigation". He does suggest that the most important reason must be sought in the development of population.[21]

Insofar as one can really speak of "Revisions in Economic History" in this article, one should direct attention to Postan's concept of crisis. The crisis is defined as "a relative decline in the total volume of national wealth", more particularly as "the decline of agricultural production", "the decline of English trading centres" and "the decline of wool trade and cloth export".

Given this, it is interesting to see how Postan's proposed explanation of the crisis falls back on a simplistic reading of Rogers. The falling population after the Black Death led to a more abundant supply of land, larger holdings for the peasants and better conditions of tenure, sometimes at a lower rent without labour obligations. For the wage-earners, wages rose, while prices either stagnated or fell. "The golden age of the English agricultural labourer" arrived.

"The real sufferers from the agricultural depression were therefore the landlords....In short, in the countryside the main burden of economic changes was borne by the upper ranks of society".[22] These familiar views inevitably suggest that we are looking at the crisis of the manorial system and the landlords, not a general agrarian crisis.

V.3   Trade and forms of rent

A few years before in the above-mentioned article on the chronology of the labour rent, Postan had dealt with the rise and fall of manorial production. Having studied about ten surveys from the twelfth century, he demonstrated in the article how the money rent seems to have been very common in this century, and that this apparently reflects a general movement towards commutation compared with earlier conditions.[23]

In the whole southern, southeastern and southwestern area of England, ("the more progressive parts"), he then found evidence that there was a regression towards the labour rent and a strengthening of the manorial system in the thirteenth century. Unfortunately Postan did not reveal his sources, but he claimed it was a general tendency on over three quarters of the approximately 800 manors he had studied[24] - a tendency confirmed by a number of earlier, more local, studies whose acquaintance we made in the preceding chapters.[25]

"Thus the typical sequence is from labour services to partial or complete commutation, and then back again to partial or complete return to labour services".[26] The typical sequence is however not universal, since there are examples of commutation progressing steadily from the twelfth century on, and of commutation not having started at all at such an early stage in certain places. But according to Postan the tendency was so typical in the specified areas of the country that one can say "the cycle is...complete, though it is destined to be broken again in the late four-teenth and fifteenth centuries, when a new, and this time a final, wave of commutation lays for ever the ghost of manorial order".[27]

This survey of the fluctuations of the manorial system from the twelfth to the fourteenth century, as far as Postan could see, drove a stake through the familiar notion of the disintegrating effect of trade and the money economy on the manorial system, and thus also through the theories based on such a chain of cause and effect. It showed that "the expansion of markets and the growth of production is as likely to lead to the increase of labour services as to their decline".[28]

Postan later took up the critique of the independent significance of the money economy for socioeconomic development for more elaborate treat-ment. Here he attacked many previous historians for using "the rise of money economy" as a formula, a *deus ex machina* invoked for want of better explanations;[29] and he considered that the only reasonable definition of this formula in terms of economic history was "the relative volume of money payments".

In this sense the development of the money economy was not a continuous historical process that could be linked with a particular socioeconomic pattern of change. He argued, for example, that the historical growth of money transactions in the middle and late middle ages must be judged on the basis of class-specific approaches to the problem. Thus the commutation of the fifteenth century, from the peasants' point of view, was a growth in the money economy compared with the thirteenth century - which for the landlords, on the other hand, was a century typified by "the rise of money economy".[30]

"The rise of money economy" thus indicates various economic processes, one of which is "easy to identify and dangerous to miss" - that is, "an overall increase in agricultural and industrial production, but also a greater emphasis on production for sale and the spread of more or less capitalistic agriculture on large estates, and the growth of towns, markets and mercantile classes".[31] An expanding money economy can be related in a particular way to large-scale agricultural production, which in the feudal context concentrates the focus on the labour rent. But this is not the same as denying the relationship of the exchange economy with the opposite process.

This is because - and here we come to the real point of Postan's criti-cism - it was not the money economy in itself, but economic development in general, and in particular the formation of prices established by the exchange economy, that he linked with the socioeconomic transformation process. Periods of rising population, production and prices - like the thirteenth century - are associated with arbitrary labour rent and the expansion of manorial production; while periods of declining population and production, and of falling or stagnating prices, are associated with the emergence of fixed money and kind dues - like the twelfth and fifteenth centuries.

It must be remarked of this paradigm - which was not without its conflicts with Postan's earlier works, for example his earlier observation of an incipient definition of labour services in the thirteenth century - that he succeeded in penetrating the "black box" that "the rise of money economy" really had been for many economic historians; especially as he managed to present the socioeconomic changes of the late middle ages in the light of the landlords' objective, economically rational interests in situations with rising or falling prices and population.[32]

V.4 A theory of population in the making

Postan continued throughout the 1940s to ask what the factors were that triggered off the agrarian fluctuations he had helped to map out. But it becomes ever more evident from his work that the development of the population seemed to him to furnish the definitive explanation of these fluctuations - a connection also made in a 1942 work where he tried to determine the social effects of the Hundred Years' War.

Besides thinking that this war had certain social consequences as regards the redistribution of the wealth of the mercantile and agrarian ruling classes, certain economic consequences in the form of a drastic decline in wool production, and a possible negative effect on agriculture's investment in technical equipment, he concluded: "It would be useless to attempt to explain the agrarian revolution by the effects of the war....In the machinery of social change the War was not so much the mainspring as a make-weight....What caused it [the agricultural depression] we do not yet know, but all the indications point either to the falling prices, or the declining population, or to both....What caused the decline of population (if there was one) in the early fourteenth century it is impossible to say". But the Hundred Years' War cannot have been of great importance in that respect, nor for the beginning of the decline in prices, since prices had turned around even "before Edward decided to invade France".[33]

We may note here that Postan cited the developments in prices and population as possible causes of the agrarian crisis, and again we see more than a hint of a connection between these factors. The beginnings of the population decline in the first half of the fourteenth century, for which Postan had no direct documentation and therefore made no firm claims, seem to be indicated in his interpretation by the turnaround in prices before the beginning of the Hundred Years' War.

Postan's contribution to our field was, as we have seen, concentrated in the 1930s on the description of the fluctuations in the agrarian sector in the middle and late middle ages. He presented evidence and arguments indicating how the English manorial system had developed historically towards its final decay in the crisis that he considered had led to a general crisis of society, without offering a definitive answer as to the general causes of this process.

In the early 1940s he was approaching an answer to this question through his criticism and refutation of the established theories of the overall socioeconomic significance of the money economy and the Hundred Years' War. But he did not get as far as an attempt at a final answer to the question, let alone the establishment of a general theory of the causes of the process he had described.

Judged in their entirety, Postan's works are in the first place original, as we have already noted, by virtue of his concept of crisis. His definition of the late medieval crisis as "a relative decline in the total volume of national wealth", including "the decline of agricultural production". Secondly, they were original because of his competent analysis of certain possible connections between the rise of the money economy and the developmental variables in the agrarian structures of production. Finally, one glimpses the outline of a "neo-Malthusian" body of theory aimed at the understanding of the late medieval agrarian crisis to the extent that the developments of prices and population before 1350 were seen as interrelated.

V.5 Forms of rent in the thirteenth century

In 1931 E. A. Kosminsky had drawn attention to the so-called Hundred Rolls from 1279-80.[34] These fragments of a nation-wide survey, along with the source type recommended by Gray - the *inquisitiones post mortem* - were used by Kosminsky in a 1935 work to clarify the relationship between money and labour rents in the thirteenth century.[35]

Just as Gray's choice of sources drove a stake through Page's chronology of the commutation process, Kosminsky disturbed Postan's account of "the chronology of labour services" in one important respect - two years before it was published, in fact. However, this was not a matter of direct disagreement between Postan and Kosminsky. It will be recalled that Postan accounted explicitly for the parts of England his survey covered. This demarcation left open the question of the dominance of the labour rent in the country as a whole in the thirteenth century.

Here Kosminsky conceded that the labour rent had actually been growing in the thirteenth century; but this "feudal reaction", which he thought, like Rogers, had made its impact after 1350 and had been an important stimulus to the 1381 Revolt, took place, then as in the thirteenth century, precisely on the big, well consolidated, wealthy, mostly ecclesiastical estates.[36] This was the picture at least in those areas of central England covered by the Hundred Rolls from 1279-80.[37] Within this area Kosminsky's study suggests that the labour rent was dominant in the thirteenth century and persisted longest on this type of estate.[38]

But Kosminsky did not think this type of estate was the constitutive cell for agricultural production in thirteenth century England. It was an economically important part, the basis of the life of the wealthier classes; but Kosminsky attached at least as much importance to the smaller manors, for which his study showed quite a different picture. On the smaller and medium-sized holdings the money rent was dominant, both among villeins and free tenants in the same century.[39]

Against the previous advocates of the dominance of the labour rent and the manorial system in the thirteenth and the first half of the fourteenth century - Page, Vinogradoff and others - Gray had argued that their documentation was not representative, as it was mainly taken from the big ecclesiastical estates in the southeastern area of England. Kosminsky repeated this objection,[40] but how representative was his own survey?

Kosminsky studied 650 villages in the Hundred Rolls, looking first at whether the land was associated with a typical manorial structure - that is, demesnes with dependent villein land. Sixty per cent of the territory studied was of this type. The rest was mostly free tenement - thirty per cent - while demesne without villein land amounted to only nine per cent. "Thus the manor was not altogether a myth in Midland England...[but] at the same time we are no longer justified in regarding non-manorial tenements as a mere exception".[41]

The thirty per cent of freely-tenanted land more or less corresponds to the percentage of villein land - 34% - but it should be noted that this category also includes small manors under 500 acres. On these the tenements burdened with labour rent made up only 27%, while on manors of over 500 acres it made up a whole 47%. On the biggest manors of over a thousand acres, the area subject to villein labour rent reached an average 52% of the total area of the manors.

The opposite is true of the distribution of money rents and wage labour. These "non-manorial elements" only accounted for 25-27% on the large manors. Moreover, if we compare the same structures on ecclesiastical and lay estates, we find, with Kosminsky, that "on the whole, ecclesiastical manors are characterised by a higher percentage of land in villeinage than those belonging to lay owners".[42]

Then Kosminsky began comparing the itemizations of money and labour dues rendered. This comparison presents many problems which would mean too much of a digression here; it should be mentioned however that the author explains why he thinks that the itemizations of labour dues must in general be considered exaggerated, while not all the actually existing money dues could be expected to appear in the material.[43] Because of the nature of the sources this part of the study also had to be restricted to an eastern and a western region of the area covered by the Hundred Rolls.

The very rough result shown by this comparison was that labour rent was predominant in the eastern region, especially in the three northern hundreds of Huntingdonshire. In Warwickshire, in the western region, money rent was calculated as constituting 70% of total rents. The distribution over large and small manors showed that "on the whole, labour dues are most strongly represented on large, and especially on ecclesiastical, estates. This fact stands out most plainly in the Eastern group, and it is therefore quite clear why the importance of labour dues was exaggerated in the classical theory, which was chiefly based on material drawn from large, and above all from ecclesiastical estates".[44]

An exemplary cross-check of certain *inquisitiones post mortem* against selected manorial accounts, and the very varied quality of the source type as regards terminology and consistency[45] leaves Kosminsky rather sceptical about the usefulness of these sources. All the same, he is tempted to use them, but only with great caution, as they cover, as we recall, a very wide geographical area".[46]

The overall impression of his study of the *inquisitiones post mortem* is "the predominance of money rent in thirteenth century England". The money rent was apparently predominant in all the counties it was possible to study. Even in the parts of the country where the labour rent made up a solid proportion of rents, it accounted for a relatively small part of total rent payment. In eastern England, for example, labour rents made up only 40% of total rent payments; in southern England only 24%; and in western England and the Midlands between 21% and 23%.[47]

Kosminsky's study of 650 villages in the Hundred Rolls, and the more than 400 manors where there had been *inquisitiones post mortem* during the thirteenth century, indicated that the money rent was just as important a factor in the thirteenth-century agrarian economy as the labour rent - if not, as Kosminsky thought, a more important one. Similarly, the assumption, common since Rogers' day, that there had been considerable wage labour in agriculture as early as the thirteenth century, seemed to find support in this survey.[48]

It was further supported by the regional differentiation in the chronology of the commutation process demonstrated in principle by Gray. Gray's break with the rough generalizations of former times was further refined by Kosminsky, who added another important dimension to the regional perspective; that is, the distinction between different types of manor - large, medium-sized and ecclesiastical.

In the areas of the Midlands covered by the Hundred Rolls, Kosminsky found, besides the big ecclesiastical estates, many small and medium-sized estates where the manorial system and labour rent had more or less disintegrated as early as the thirteenth century. "In general it may be said that labour services were maintained on...large estates which had succeeded in establishing and maintaining their hold over an unfree peasantry". On the smaller manors, on the other hand, the exchange economy led to a reinforcement of the money rent. "The development of exchange in the peasant economy, whether it served the local market directly, or more distant markets through merchant middlemen, led to the development of money rent. The development of exchange in the lord's economy, on the other hand, led to the growth of labour service, though the process is complicated by the fact that he could also develop production for exchange through a combination of both systems".[49]

The exchange economy led to an underpinning of the stratification of the feudal estate-holders. The great centrally-placed manors intensified the labour rent, while the smaller ones and those more peripheral to the commercial routes and centres increased rent pressure on the peasants by means of the money rent. Kosminsky thus agreed with Postan that the exchange economy could and did lead to a reinforcement of the feudal manorial system, although he saw the opposite process making more of an impact in the thirteenth century than Postan did.

V.6 Rising rents in the thirteenth century

The theory of rising rents in the last centuries of the high middle ages had already been provided with evidence in Levett's study of the Winchester estates. Although she had claimed that the rise in entry fines she had observed on certain of the manors up through the thirteenth century should probably be connected with the relative drop in the rest of the rent burden, this bare observation, along with Levett's point that widows sometimes refused to take over the tenancies of their deceased husbands, gave the impression of a generally rising rent pressure.[50]

The theory was supported by Kosminsky, who saw the increasing number of appeals to royal courts against increased rent burdens as proof of it.[51] Kosminsky took these appeal cases as evidence of generally rising rents. Unlike Postan, he thought that there was evidence here of an increasing demand for both labour services and money payments from the peasants.[52]

For Postan had claimed that the expansion of the manorial system in the thirteenth century only led to heavier labour burdens for the peasants. This happened in the first place because the expansion of the demesne "was unaccompanied by any considerable increase in the amount of land subject to villein labour";[53] and secondly because of "the absence of a large reservoir of free landless labour"; and thirdly because "the legal and political safeguards of the liberal state" did not exist.[54]

The deterioration in the conditions of the peasants thus consisted, according to Postan, not only of a simple regression to labour rents, but also of a relative rise in the rent burden for the individual peasant, since the total extent of tenanted land did not expand with the demesne.[55] On the Winchester estates more labour burdens were simply loaded on the peasants; while on the Glastonbury Abbey estates, the tenanted land was further split up without a corresponding reduction in the labour services. Moreover, the more fixed definition of the formerly arbitrary labour obligations that began in the thirteenth century sometimes conflicted with the interests of the peasants - at least when, as on the estates of the Ely bishopric and some of the Worcester ones, it is said to have led to increased labour burdens.[56].

V.7 Summing-up

While Postan - although still only by suggestion - related rising rents to population growth, Kosminksy saw them in the light of the expanding exchange economy. For Kosminsky, trade was a determinant of the development of the feudal rent types and the manorial system in the thirteenth century. But inasmuch as he realized that trade could be said to have both a consolidating and a disintegrative effect on the manorial system, this explanation cancels itself out. Nevertheless it was his only explanation of the development of the rent types he described, and of his view that rents were rising. When we confront these authors with one another, we thus not only find differences of opinion as to the extent and chronology of the feudal rent types, but explicitly different approaches to the understanding of the dynamics of feudal agrarian production and its incipient decline.

Kosminsky's revision of the prevalent view of the dominance of the manorial system in the thirteenth century, which we can incidentally already see signs of in Maitland's work from before 1900,[57] is theoretically and methodologically built up from the interpretation of the nature of feudal rents developed by Karl Marx in his account of the *Genesis der Kapitalistischen Grundrente.*[58] Both Kosminsky's description of the three feudal rent types - labour rent, product rent and money rent - and the interrelations of each with the development of trade and the money economy can be matched with their models in Marx; nor does Kosminsky conceal this.[59]

Kosminsky used this theoretical and methodological point of departure to underpin an interpretation of the economic dynamics of feudal agriculture within a commercialization model. In his paradigm, money circulation and trade cycles stimulated manorial production and peasant production within the framework of "a complex class war within the manor".[60]

By contrast, we have seen how the embryonic demographic theory developed through Postan's criticism of the commercial theory. We have also seen that the development of the demographic theory with respect to the population growth of the high middle ages and its possible negative effect on the reproductive capability of the manorial system can be traced back to certain earlier historians - especially Paul Vinogradoff.

If we glance at a couple of striking general accounts of English economic history from the 1940s,[61] we can see how these theoretical approaches to the understanding of our field dominated the decade. In J. H. Clapham's *Concise Economic History*, the main emphasis is on a demographic explanation of general socioeconomic development from the thirteenth until the fifteenth century; it underscores the importance of the population expansion of the high middle ages, but is far from rejecting the idea that the population decline after the Black Death marked "a watershed in social and economic history".[62]

In *The Cambridge Economic History*, population development and the accompanying changes in agrarian reproductive structures also assume a very prominent place; but in particular Hans Nabholz, in his chapter on "Medieval Agrarian Society in Transition", tried to break with this trend towards monocausal explanation. Population development and the related internal colonization "made the first break in the earlier framework" in the period up to just before 1300.[63]

But "the steady progress of a money economy proved far more destructive to old relationships than active clearing was", since trade led to a split in agrarian production, expressed as an intensification of the labour rent and its conversion to money rent or leasing.[64] And this process of dissolution "could not fail to be affected by the pestilence".[65]

It is notable that the authors of these general works did not use the demographic or commercial theory to prove the existence of a crisis, but to document the transformation process they thought was beginning as early as the thirteenth century and was in full swing by 1300. It is further interesting to observe how the significance of the Black Death for this transformation process, despite the chronology adopted, was once again taken into consideration. Finally, it should be mentioned that the theory of the exhaustion of the arable in the late medieval centuries was once again rejected.[66]

The criticism of the soil exhaustion theory had found new nourishment in the 1930s. R. Lennard, who as we recall was involved in the original attack on the theory, followed up on the criticism along with the American M. K. Bennett. The latter argued for the possibility of a rise in productivity in arable farming between 1200 and 1450,[67] and claimed that Beveridge had underestimated the average yield per acre of the Winchester lands.[68] Lennard thought that all previous research had been based on a crucially false premiss that had led to a general undervaluation of medieval productivity in arable farming.[69]

The two authors mentioned first and foremost consolidated the refutation of the soil exhaustion theory we dealt with earlier; but they were not alone in this consolidation, since the years immediately after the publication of Beveridge's results in 1927 saw the appearance of other empirical analyses which agreed on the most general issues with his, Lennard's and Bennett's rejections of the theory.[70] After this there were few arguments for declining productivity and the exhaustion of the soil in the late medieval centuries - at least not in connection with the manorial system.[71]

V.8 A neo-Marxist synthesis

In Maurice Dobb's book *Studies in the Development of Capitalism* from 1946, we find a bold attempt at a synthesis of the commercial theory and the beginnings of a demographic theory we have seen develop. In his very general reflections on the decline of feudalism, Dobb started by identifying the internal structure of the feudal mode of production. He believed that the crises from the fourteenth century on should basically be understood in the light of the contradictions inherent in these structures.

It was first and foremost the inefficiency of feudalism as a mode of production, paired with the ruling class's growing need for earnings, that were responsible for the decay of the system, because this growing need caused increasing exploitation of the primary producers.[72] This general contradiction was aggravated during the high middle ages by two factors: the development of trade and the development of population.

Dobb argued for the inefficiency of the production system in a way that is not unfamiliar to us. We can recognize it from Rogers' writings and for that matter also from Walter of Henley. The unwillingness of the peasants and their lack of incentive to work on the lord's fields, combined with the age's low and static technological level, meant that there was only a small margin within which to increase the surplus production that was the only source of income for the ruling class.[73]

Nevertheless, constant attempts were made to increase it; first, because trade stimulated the increase in the lords' demand through the attraction of the exotic commodities it made accessible; secondly, because the lords' households were constantly being expanded; and thirdly because of the wars that could almost be called an integral feature of the feudal order.[74] The result was a growing pressure on the primary producers which not only led to "an exhaustion of the hen that laid the golden eggs for the lord", but also a mass flight of producers which came to drain the system of its "essential life-blood and to provoke the series of crises" in the fourteenth and fifteenth centuries.[75]

The population increase before 1300, however, compensated somewhat for these losses. "This, it is true, would have served to provide more labour to support the system and to furnish additional feudal revenue. But except in areas where the increase in numbers was accompanied by an increase in cultivable land available to the peasants (which would in return have required a sufficient increase in draught animals and instruments in the hands of the cultivators), the eventual result was bound to be an increase in the peasants' burden owing to the increased pressure on the available land".[76]

This view corresponded excellently with Postan's opinion that villeinage was intensified in the thirteenth century, and with Kosminsky's view of a general rise in rents. But Dobb went further. Without having any convincing documentation, he connected the rising overexploitation of the peasants in the thirteenth century with what he claimed was the beginnings of a population decline from about 1300 - a negative trend whose "immediate effect was to precipitate what may be called a crisis of feudal economy in the fourteenth century".[77]

The reason for considering Dobb's book in this context is not that it, like the other literature, addresses itself to the study of the English middle and late middle ages, or that it contains radically new approaches or empirical data in our field; as is evident from the title of the book, it is a study of the development of capitalism, and Dobb's reflections on feudalism and its decline, given the main subject of the book, primarily have a genetic significance for the breakthrough of capitalism in the European context. The book was written as a generalizing account of this historical development on the basis of source material and interpretations of material gathered by others. Given this, it was natural enough that the

various elements in Dobb's account of our problem area were all more
or less recognizable from the literature we have already dealt with.

For this reason, and because Dobb's method, which he is incidentally
sceptical about himself,[78] has certain theoretical advantages, the book has
been used here. For Dobb's contribution, as we have already seen, was
to assign priorities to the influence of various elements on what he called
the decline of feudalism. Instead of attributing equal importance to the
various factors, or providing rigorous empirical documentation for the
dominance of a single factor, Dobb ordered the factors available to him
in a theoretical system which reduced the cause of the crisis of feudalism
to certain general contradictions in the agrarian reproductive structures,
so that on this basis the importance of the other factors for the whole
process could be explained interactively.

Dobb's debt to Postan is striking, and the bold demographic reflections
he presents make it all the more necessary to look more closely at their
documentation, such as it appeared between 1930 and 1950. Dobb spoke
of the beginnings of a population decline at the start of the fourteenth
century, and directly related this negative trend with the disparity between
the preceding population growth and the land resources available to the
peasants. We saw that these two relationships were touched on by Postan
- the first, however, only as a hypothesis. So there are also grounds to
concern ourselves with research on the medieval colonization process.

Finally, we have noted that the plague epidemics regained some of their
former explanatory credibility after 1930. One can thus say that between
1930 and 1950 the study of the demographic factor took two directions.
First, the population growth of the high middle ages was again more or
less directly connected with the late medieval agrarian crisis or transformation process. Secondly, the classical theory of the crucial influence on
this process of the Black Death and subsequent epidemics was taken up
again, after having been increasingly rejected in the first three decades of
the twentieth century - cf. p. 151.

V.9 The epidemiological theory

In a lecture to the Cambridge Historical Society,[79] John Saltmarsh ad-
vocated a view which once more put the population decline after 1350 at
the centre of attention. Saltmarsh's theory was built up on the foundation
of Postan's chronology of the crisis,[80] Creighton's observations of the long-

term effects of the plague over several centuries (see Chapter III, p. 70), and the refutation of the immediate consequences of the Black Death we saw presented in Chapter IV by Levett, Ballard and Robo (see pp. 111-114).

The plague - that is, not only the Black Death of 1348/49 - was thus in the last analysis seen as a triggering factor for the crisis of the fifteenth century.[81] "The Black Death of 1349, standing alone, will not fill the bill; but a permanent infection of England by a new and fatal disease will fill it very well".[82] This "cumulative effect may well explain more in English history than the Black Death itself".[83]

The connection between the crisis and the population decline caused by the plague appeared in Saltmarsh's work as a relatively unverified hypothesis. He was aware of this, and claimed no more than that it was one of many possible explanations of the crisis, which for him seemed an indisputable fact.[84] He could not applaud Postan's and Rogers' claims for a more even distribution of the social product in the fifteenth century. But he did think that "everything seems to indicate a decline in total wealth, and in the whole scale of economic life in England".[85]

In the period around the publication of Creighton's and Gasquet's works, epidemiology had gained scientific knowledge of the nature of the plague, but this progress was not, as we recall, reflected in these authors.[86] Nor was the new knowledge incorporated into research immediately after this on the social and economic history of the late middle ages. One might therefore say that it was high time the breakthroughs in epidemiological research were used to clarify the field.

Although his approach centred on the epidemiological factor, Saltmarsh did not update his material with the latest research results either. Thus he failed to distinguish between pneumonic and bubonic plague - which Creighton at least did - and associated the spread of the plague far too closely with the black rat's invasion of Europe.[87] In 1941 epidemiological research had come much further. It was now known, for example, that the pneumonic plague, whose clinical symptoms were often described in late medieval sources, could be spread from human to human through the air. But Saltmarsh did break new ground when he associated the subsidence of the plague with the increasing use of stone in construction and with better hygiene, as these measures give the black rat poor reproductive conditions. For it is true that the association of this rat species with human dwellings, combined with its flea's willingness to bite humans, are crucial to its ability to spread bubonic plague. By contrast, the hardier outdoor life of the brown rat and the habitat of its fleas in the rats' nests do not constitute a similar risk for the spread of the plague bacteria to human beings. It would be going too far, however, to say that the dying-out of the plague epidemics in the seventeenth century was related to the brown rat's invasion of England.[88]

The plague played a central role in Saltmarsh's understanding of the late medieval crisis. Its consequence, population decline, was seen as the reason why the manifestations of crisis - falling rents, the desertion of marginal land, and a "decline in total wealth" - developed. The plague epidemics affected population development, which led to the manifestations of crisis. But as a corollary the population development also seems to have been a determinant of the socioeconomic consequences of the epidemics.

For the population growth of the high middle ages was indicated as a possible explanation of why the Black Death of 1348/49, despite the high mortality of approximately a third, had no great social and economic effects.[89] Saltmarsh, who also rejected the idea that the 1381 Revolt and any climatic changes at the beginning of the fourteenth century had anything but negligible significance for the development of the crisis,[90] also stressed in his account the increasing need for more accurate demographic research - not least when, in contrast with his view of population development before 1300, and regardless of his rejection of the soil exhaustion theory,[91] he cited examples of how marginal sandy land in Norfolk had already been withdrawn from cultivation in the thirteenth century.[92]

V.10 The population growth of the middle ages

England has long-standing traditions of demographic research, going back long before this discipline attracted serious interest with the publication of Malthus' legendary *Essay on the Principle of Population* in 1798.[93] Attempts to calculate the population development of the middle and late middle ages also have a long prehistory, stretching back to the seventeenth century.[94]

All the same we have observed how estimates of population development diverged greatly, even in the twentieth century. By and large there was only a general consensus that the population figures up through the middle and late middle ages could be illustrated by an S-curve. There were several reasons for this. Demographic research on the period had been quite sporadic; the relevant sources had only been partially investigated in this discontinuous process - and as a rule without comparisons among them.[95] It is perhaps therefore not so surprising to see reservations in the literature of the 1930s and 1940s about making specific estimates of fluctuations in population figures.[96]

Just before 1950, however, J. C. Russell's very comprehensive in-depth analysis of the British population in the middle ages appeared.[97] This work was distinctive in that it used a far wider and more cohesive body of source material than had been seen hitherto in demographic assessments of the period. Russell made use of data from Domesday, surveys, *inquisitiones post mortem*, the Poll Taxes, the registries of the bishoprics, monasterial surveys and to a lesser extent secular court rolls.

It would lead us too far to describe Russell's method in detail, but it seems appropriate to list some of its features. The Domesday Survey and the Poll Taxes were the foundations of his calculations. He considered these sources the most complete ones, and that they afforded a basis for the most accurate calculations. He placed special emphasis on the Poll Taxes,[98] because this form of taxation, as the name suggests, was applied to everyone - with few exceptions - over the age of fourteen.[99] On the basis of the Poll tax of 1377 he calculated the population of England to have been a good two million - 2,232,737.[100]

Russell took the view that the aim of the Domesday Survey was to obtain an overview of the available agricultural labour in the eleventh century. He therefore assumed that everyone, including craftsmen and cottagers, was covered by the main category of the survey - "landholders".[101] This meant that the total of the survey, after due subtraction of clerics and town-dwellers,[102] was equal to the total number of rural households. The total rural population was then calculated by multiplying this total by the number of children, servants and spouses, about whom Domesday provides no information. For this reason Russell used a multiplication factor of 3.5 for his calculation.[103]

Disregarding possible problems of a more technical nature in Domesday - missing areas in various counties, etc. - Russell only had one problem left in calculating the total rural population - that certain counties were missing and some were included in the total for others. He tried to solve the problem by analogy with the method he used with the Poll Tax. He assumed that the land/population ratio in northern England, where the problems were, had been fairly uniform, and then calculated the population figures in the missing counties on the basis of those that had been included.[104]

Russell arrived at the conclusion that the rural population amounted to about a million - 987,805; added the urban population to this - 111,971 - and the number of clerics - 5,440,[105] - which gave him a population in the eleventh century of about a million: 1,105,216.[106] So we see that Russell arrived at substantially lower figures than we have previously encountered, especially for the population in the eleventh century. On the other hand, his figures suggest greater population growth between 1086 and 1377 than we have seen before.

V.11 Population losses after 1348

This result does not of course mean that between the termini of the period one can speak of a continuously rising population. As we know, the Black Death intervened. This raised two questions for Russell. First, how big was the population loss between 1348 and 1377, and what had the population been just before the Black Death? Secondly, had there been a continuous population growth between 1086 and 1348, or can the population development between these two years also be drawn as an S-curve? Russell attacked these problems partly through a discussion of birth and death factors, partly through an attempt at a numerical assessment of the population development. We must here content ourselves with looking at the latter approach. Let us look first at the calculation of the population development after 1348.[107]

Russell divided the period between 1348 and 1430 into three phases because of the available source material. He used *inquisitiones post mortem* to clarify population development during the first plague epidemics in 1348/50 and 1360/61, while for the 1360-90 period he primarily used the Poll Taxes, comparing them with data from various surveys. For the next period, up to and after 1430, the documentation was weaker. The surveys were fewer, and Russell mainly had to rely on indications of population development based on calculations of so-called "successions of generations", to the extent that there were lists of heirs in the *inquisitiones post mortem*.

Russell arrived at a death rate of 1.67 between 1348 and 1377, which, given his calculation of the population in 1377 as 2,232,373 individuals, meant that the population before the first plague epidemic must have been about 3,757,500, and that the epidemics between 1348 and 1377 wiped out 40% of the English population.[108] "The decline from 1377 to about 1400 was probably about 5 per cent, bringing the population of England down to around 2,100,000....During the last phase from 1400 to 1430...the population probably held about even";[109] that is, at about the lowest level before population figures again began to develop in a positive direction.[110]

V.12 Population developments, 1300-1348

To answer the important question of the demographic trend between 1086 and 1348, Russell compared data from Domesday and the Poll Tax with information from surveys from the period between the twelfth and the fourteenth centuries. The latter source type is commonest, we know, in the second half of the thirteenth century, among other reasons because the English Crown carried out more general surveys of the taxation basis of the manors,[111] but it is also extant from previous centuries, covering the holdings of various monasteries.

Russell drew up a comparison of the information in the surveys with Domesday in so-called "outcome groups" whose categories reflected the comparative rates of population growth between Domesday and each survey. By comparing these categories, divided into twenty-year periods between 1250 and 1348, he arrived at an average growth rate of 1.22 for the whole period, and further concluded that this ratio had a falling tendency towards the end of the period.[112]

If one sets this result beside a similar comparison of Domesday data with surveys from before 1240, one finds a growth ratio that is rather lower on average. On the basis of this "outcome group method" Russell therefore concluded that the population growth levelled out in the course of the twelfth century. But "perhaps in the thirteenth century another period of rapid growth occurred which our evidence would seem to show had been retarded just before the plague".[113]

It is difficult to understand this admittedly cautious conclusion when the comparison of surveys before 1250 with the Poll Tax figures indicates a continuously falling population between 1250 and 1349.[114] On the other hand Russell has grounds to think he is "justified in feeling that an S curve would illustrate the development [from 1086 to 1348] better than a compound interest curve". But when did the decrease in population growth, or the population decline, begin?

Russell's constructions for the "succession of generations" suggest that ever-fewer children survived their parents from the early fourteenth century on;[115] but his survey of mortality and its factors before 1348 told a different story, for which reason "the plotting of the curve as yet would be largely hypothetical".[116] For in accordance with the population development indicated by the comparison of the surveys and the Poll Tax figures, he concluded as regards "general aspects of mortality" that "from the middle of the thirteenth century conditions of mortality seemed to be getting worse even though the Black Death had not yet appeared".[117]

Russell rejected the idea that climatic changes around 1300 could have had any influence on the rising mortality.[118] The reason for the rising mortality from the mid-thirteenth century on "would seem to be that population had begun to push against the means of subsistence in the island, causing a crowding which was unfavourable to the mortality rate there".[119] He went so far as to say that the most densely populated areas of the country before the Black Death were actually overpopulated.[120] But he also expressed views that went against this interpretation.

In a comparison between certain ecclesiastical sources and *inquisitiones post mortem* he had noticed that mortality was highest among the oldest age groups during all the four plague outbreaks he dealt with.[121] Russell made much of this observation, although the high mortality in the older age groups, which the sources indeed indicate,[122] might well reflect the general fact that mortality is normally highest in the oldest population groups.[123]

The rejection of the social and economic significance of the Black Death which we saw making serious impact in the first decades of the present century becomes understandable for him in this light. The high excess mortality among the old in the period after 1350 paved the way for an accelerated succession of generations, that is the transfer of holdings to younger generations. This must have meant according to Russell that the average marrying age dropped and the birth rate accordingly rose.[124]

The interrelationship between available holdings, marrying age and birth rate had been pointed out before by G. C. Homans, who thought that there was plenty of proof that people, where the open-field system was prevalent, did not marry before inheriting a holding. The average marrying age must therefore have been relatively high in the thirteenth century. "And lateness of marriage may have had some effect on the number of children born....The working of the rule, no land no marriage...secured a stable adaption of society to its economic conditions. Despite the logic of Malthus...[it] limited the number of persons who pressed on the land for subsistence".[125]

"The ideal was that a living must precede marriage", thought Russell. "This restricted population to some extent..."[126] - but apparently no more than that a latent overpopulation situation had probably developed before the Black Death took its toll. However, no crisis - demographic or social - seems to have developed until after 1348.

"Three characteristics seem to be typical of growing populations which develop within sufficient resources: potential supply of men, prosperity, and an aggressive spirit, self-confident and self-reliant. All of these were typical of particularly the phase of English history from 1199 to 1348".[127] The crisis only came in connection with the subsequent population decline, not just as a consequence of the first epidemic's reduction of the population by 20%.[128] Like Saltmarsh, Russell thought that it was the cumulative effect of successive plague epidemics that set the population figures back by up to 50% between 1348 and 1400,[129] and meant that "in the fifteenth century depression seems to have extended to nearly all classes in England".[130]

V.13 Colonization

The colonization issue became a major theme in German and French medieval research at an early stage. In the English context, too, it was hardly virgin ground in the 1930s and 1940s. It was rather the case that research in this field was developed and gradually incorporated into the body of theory on the late medieval crisis or transformation.

In what follows we will therefore be looking at the phenomenon on the basis of the available literature in the light of three main issues: the quantitative extent of colonization and its regional distribution - among other reasons because this was often used as non-statistical evidence of population growth;[131] the connection of colonization with manorial culti-vation and peasant farming, since it was claimed that the fairly clear con-nection of colonization with the expansion of the manorial system was a contributory factor in the deterioration of the peasants' reproductive conditions in the period up to 1300; and finally, the consequences for agricultural technology that followed with colonization, with emphasis on the question of the relationship between cattle farming and tillage, since it had been asserted that colonization aggravated the problems inherent in an agrarian monoculture.

Research on medieval colonization in England[132] had in the 1930s and 1940s progressed far enough to identify the main chronological features of the movement. There was thus a general consensus that the Anglo-Saxon and Scandinavian colonizations had been superseded in the eleventh century by widespread desertions due to the Norman Conquest. In the next century the Cistercian movement and the Benedictines were very active. They cultivated new areas in the southern and central areas and established extensive sheep farming in the wildest and most deserted areas of the North.[133] But colonization was not just a matter for the clerics, especially after about 1200.[134]

One of the consequences of the desertions that followed on the Norman Conquest was the expansion of the forests[135] - a process said to have reached its peak under Henry II (1154-1189), and in the light of which the disafforestation process of the subsequent centuries, mainly in southern and central England, must be seen as a struggle between the Crown on the one hand and the barons and clergy on the other.[136]

One cannot conclude from the Crown's progressive ceding of its hunting preserves that this in itself "marked the economic progress of the country-side".[137] That the King yielded up hunting areas did not in all cases lead to agricultural exploitation of those areas, as indicated by the fact that "a great deal of private afforestation went on in the twelfth and even in the thirteenth century".[138] For this reason, and because of the regional disparities mentioned, one should not use the disafforestation process as evidence of the spread of colonization without some caution.

T. A. M. Bishop's study of the colonization process in central and eastern Yorkshire,[139] and H. C. Darby's of the Fens of Cambridgeshire, Norfolk and Lincolnshire[140] only remedied this problem to some extent. They both dealt with the colonization problem in narrow terms of the structure of production and agricultural technology, contributing little to the assessment of the amount of colonization in these areas. Primarily, they simply documented that colonization happened, and that it was probably at its height in the thirteenth century.

On the available, rather slender, evidence, we can claim that in large parts of England there was land reclamation for agricultural purposes. Colonization was strongest in the thirteenth century, then waned, and it must to some extent be seen in the context of the desertions that followed on the Norman invasion. Thus to some undetermined extent it was a matter of the reclamation of land which had fallen into disuse as arable in the eleventh century. It is very difficult to assess the regional importance and national spread of colonization.

V.14 Colonization, manorial production and peasant farming

From Lord Ernles's reflections on the earliest enclosures one could get the impression that colonization in the thirteenth century was simply a matter of parts of the common being drawn into manorial cultivation either as pastures or arable areas. Postan reinforced this idea, or at least the notion that the reclamation of new land took place through an expansion of the demesnes. But is the idea tenable?

As early as Vinogradoff one can find evidence that certain of these newly-reclaimed lands on the manorial demesnes, the so-called forelands, were plots placed at the disposal of the servants of the manor instead of, or as supplements to, other payments; or which were let out to the growing group of small farmers.[141] Vinogradoff also had documentation that the expansion of the cultivated area in the thirteenth century did not only take place through an expansion of the demesne, although the lord's right to the uncultivated areas was enacted in 1235 in the Statute of Merton, provided the lord showed consideration for the needs of the community for pasturage.

"Documents show...that the spread of the area under cultivation was effected in different ways; sometimes by a single settler with help from his lord, and sometimes by the entire village, or at any rate by a large group of peasants who club together for the purpose".[142] The community also itself reclaimed new land as a supplement to the open-field system,[143] but sometimes only after a single colonist had cleared the land.

Individual clearances were in time incorporated into the common in the above-mentioned areas of Yorkshire studied by Bishop, judging from the fact that personal names and terms like "toft" and "croft" often featured in place-names for collective field systems in the thirteenth century.[144] But more than this, it was Bishop's thesis that parts of the open-field system

163

that existed in the thirteenth century but not in the twelfth in the area he studied had their origins in clearances by individual colonists as a result of the population expansion.[145]

The open-field system and the collective type of production among the Yorkshire yeomen must thus have arisen through the parcelling-out of individual colonists' land among kin and immigrants; and after this system had formed, the reclamation of new land continued. There arose a symbiosis between "open-fields" and "assarts", which had the effect that the open-field system could maintain its structure and the individual tenancy its extent despite population growth and the liberal rules for the transfer and sale of land to which these peasants were subject.[146]

In the first instance the reclaimed land was drawn in as individual supplements to the open-field system (cf. note 143).[147] Then it was directly incorporated as an extension of the field complex. The addition of assarts improved the reproductive basis on the individual farm and laid the foundations of a population growth that could be absorbed by parcelling out individual tenancies through inheritance, since the inherited plot combined with the assarts incorporated into the open-field complex could return the individual farm to a size compatible with a single farming unit.[148] The open-field system among the Yorkshire yeomen, combined with continuous expansion of the cultivated area, therefore constituted in the thirteenth century, according to Bishop, "a power of growth and reproduction".[149]

But on the basis of the available evidence it is hard to say whether colonization in general matched population development and countered the overpopulation problems and increasing exploitation claimed to have been prevalent up to 1300. What was found in colonization and the population increase was an explanatory model that illustrated parts of the socioeconomic transformation which had long been thought to have begun as early as the thirteenth century, and which it was generally agreed had been hastened by the Black Death.[150]

Thus Clapham saw the social stratification of the free peasants, the development of a class of landless rural labourers and peasants with expanding land holdings, in the light of colonization and the division and subdivision entailed by the population growth.[151] But these and other familiar incipient changes were assigned no great importance.[152] The thirteenth century was generally considered as "the one in which the social and economic order of the Middle Ages was most prosperous and least challenged".[153]

V.15 Colonization and agrarian technology

The expansion of the arable "was a comparatively silent revolution but its cumulative effect was to change the face of the countryside and to provide for an increasing population". Yet the uncultivated area was not only a reservoir for colonization, but "played an integral part in manorial economy": and "common rights...were vital elements in the tenements of a village"; while "medieval husbandry was conducted on the principle of maintaining a balance between agriculture and pastoral pursuits. Progressive encroachment upon the waste must therefore, sooner or later, reach a limit that could not but disturb the balance of rural economy".[154]

More or less the same line of thought as Darby's can be found in Charles Parain, who thought that pasturage facilities in the uncultivated areas in themselves stimulated population growth, which in turn increased the need for colonization and thus brought about a decrease in the necessary uncultivated areas, as well as a tendency to overexploit them.[155] "The rearing of animals would have to face increasing difficulties, which would react dangerously on the whole agricultural system".[156]

This problem of agrarian technology had, as we recall, already been pointed out by E. Nasse in the 1860s and brought up by other agrarian historians before and after the turn of the century. For example, it will be remembered how the drop in cattle-breeding was part of Ernle's argument for a decreasing grain yield in the late middle ages. We saw above how the intensification of the agrarian monoculture was related by Darby and Parain to the colonization process. But the study of the colonization of the Fenland areas published by Darby in 1940 must nevertheless, seen in *isolation*, be regarded as a refutation of this thesis.

Just as Vinogradoff had demonstrated that assarts in the form of "enclosed plots of...meadows, pastures and wood" could benefit animal breeding, Darby's study documented that the increasing exploitation of the Fens throughout the middle and late middle ages primarily took place for purposes of livestock farming. Insofar as the Fens were drained,[157] this was done mainly to extend meadow areas. Surveys from the fourteenth century indicate a relative scarcity of arable in proportion to meadows in these areas, compared with the adjacent areas. Darby thought that this disproportion was due to an absolute expansion of the meadowlands between the eleventh and fourteenth centuries, stating that "meadowland constituted a most important element in the economy of the Fenland during the later Middle Ages".[158]

So on the one hand there was the reclamation of meadowland for cattle farming, and on the other the leaving of the marshes untouched insofar as it suited the same ends.[159] "At least as early as the twelfth century the grazing of animals had become one of the most important fenland occupations".[160] In the course of the twelfth and thirteenth centuries the Fens were more meticulously subdivided: "...the practice of enclosing was frequent". And this subdivision, the enclosures, the disputes among monasteries and other religious houses about this, as well as "the practice of leasing out pieces of fen for an annual rent" in the fourteenth and fifteenth centuries, indicated for Darby "the vital position occupied by pasture rights in the economy of the fenland".[161]

The theory of the increasing disproportion between cattle farming and tillage and its fatal effect on agriculture's maintenance of an ecological/technological balance was not verified by Darby's study, and can hardly be said to have been by Bishop's either. At any rate it is hard to imagine this consequence of colonization in the northern English area, if only because the density of population there, despite a stronger tendency to rise than in the southern and Midland areas after the eleventh century according to Russell, was still relatively low at the end of the fourteenth century.[162] This is of course not to say that the "monocultural" theory of the increasing disproportion between cattle farming and tillage can be *generally* rejected with Darby's and Bishop's studies, among other reasons because these were concerned with particular localities where other factors prevailed than in the typical arable farming areas.

In this respect it is a problem that the colonization issue, besides having been dealt with peripherally in various treatises, inasmuch as assarts are frequently mentioned in surveys, manorial accounts and other source types, and as the studies of the forest problem mentioned here dealt with the colonization issue in the broad geographical perspective, has only been thoroughly researched in these very few areas: central and eastern Yorkshire and the Fens. Given this, it does not seem unreasonable when Nellie Neilson remarks that "English movements of approvements and colonization fill a smaller and less conspicuous place in social life than similar movements in the continent".[163]

One reason for the apparently smaller amount of colonization in England than on the Continent may have been "the existence in England of great royal forests which included a very considerable portion of the uncultivated land".[164] The probability of this thesis is supported by the fact that the royal forest system, according to Bazeley, did not break down in earnest until after Edward III's accession to the throne, that is about half a century after colonization, according to Clapham, had ceased.[165] It is also supported by Bazeley's finding that "a great deal of private afforestation" took place in the twelfth and thirteenth centuries. Finally, Bazeley's map of royal forests in the thirteenth century confirms Neilson's supposition.[166]

The possibly relatively modest colonization in England, along with the rising population, had consequences for agrarian technology, if we are to believe C. S. and C. S. Orwin. It meant that the three-course system gained ground in the thirteenth century. "For it is clear that the two-field system of farming was expensive of land, only one-half of it being applied to productive use each year. When the village community was small, and land was plentiful, this might not matter, but as population increased, and the need for greater food production grew, the wisdom of a more economical use of land must have suggested itself, if such could be devised".[167]

Notes

1. Frantisek Graus, "Das Spätmittelalter als Krisenzeit" in *Mediævalia Bohemica*, Supplementum 1, 1969.

2. Ibid., pp. 8-15.

3. Ibid., p. 15.

4. Ibid., p. 40.

5. Graus's study, which includes a bibliography of about 200 titles, focuses on German, French and Czech historical research, with some discussion of English and Scandinavian work.

6. Ibid., pp. 12-13; W. Abel, *Agrarkrisen und Agrarkonjunktur in Mitteleuropa vom 13. bis zum 19. Jahrhundert*, Berlin 1935. Abel later disclaimed the credit given to him by Graus for an epoch-making contribution, referring to M. M. Postan's work in the 1930s, to which we shall shortly be turning. See W. Abel, *Strukturen und Krisen der spätmittelalterlichen Wirtschaft*, Stuttgart 1980.

7. Frantisek Graus, "Die erste Krise des Feudalismus", in *Zeitschrift für Geschichtswissenschaft*, Vol. III, 1955.

8. Ingela Kyrre, "Teorier om Førkapitalistisk Samfundsdynamik", in *Kritiske Historikere*, pp. 59-60.

9. M. M. Postan, "Revisions in Economic History - The Fifteenth Century", in *The Economic History Review*, Vol. IX, 1938-39.

10. Ibid., p. 161.

11. Ibid.

12. Ibid., p. 162.

13. Ibid., pp. 163-64. See *Studies in English Trade in the Fifteenth Century*, ed. Eileen Power & M. M. Postan, London 1933, especially articles by H. L. Gray, "The English Foreign Trade from 1446 to 1482", pp. 1-39, and Power's "The Wool Trade in the Fifteenth Century", pp. 39-91.

14. Eileen Power, *The Wool Trade in English Medieval History*, London 1941, p. 35. Power thought that epidemics among the sheep, and more particularly taxation and the Company of the Staple's near-monopoly caused a drop in exports of approximately 35% between 1310 and 1448, despite the fact that wool manufactures stagnated. Power thought she could document wool exports below the 1310 level even after the start of the enclosure movement at the end of the fifteenth century. Ibid., pp. 36-37.

15. N. S. B. Gras, *The Evolution of the English Corn Market*, New York 1915.

16. Postan, "Revisions...", op. cit., pp. 162-63.

17. Gras, op. cit., pp. 30-31.

18. In the article "The Chronology of Labour Services" from 1937 in *Transactions of the Royal Historical Society*, 4th Series, Vol. XX, Postan argued that the labour and money rents were complementary seen in relation to the development of the market and the exchange economy.

19. It should be noted that Postan, in view of the peasants' position after the Black Death, concluded that the crisis was not quite as serious in the first and probably also the second decade of the fifteenth century. Postan, "Revisions...", op. cit., p. 162.

20. Ibid., p. 163.

21. Ibid., p 166.
22. Ibid.
23. These surveys show the extent of land tenanted against money and/or labour rent. But it is more interesting that Postan has found in this material a varied body of evidence that there had been commutation recently - either itemizations of changes in terms of tenure, notes on concessions of land against money rent, or references to the parcelling-out of the demesne against money rent. Postan, "The Chronology...", op. cit., pp. 176-84.
24. "Out of about eight hundred manors investigated, over six hundred and fifty show clear signs of an extension or intensification of demesne cultivation". Ibid., p. 186.
25. Maitland, *The History of a Cambridge Manor*; Nielson, *Economic Conditions of the Manors of Ramsay Abbey*; Feiling, *An Essex Manor in the Fourteenth Century*, et al.
26. Postan, "The Chronology...", op. cit., p. 189.
27. Ibid., pp. 189-191.
28. Ibid., p. 193.
29. M. M. Postan, "The Rise of a Money Economy", in *The Economic History Review*, Vol. XIV, 1944, pp. 123-24.
30. Ibid., pp. 127, 131.
31. Ibid., pp. 127-28.
32. Ibid., pp. 128-29, 132-33.
33. M. M. Postan, "Some Social Consequences of the Hundred Years' War", in *The Economic History Review*, Vol. XII, 1942, pp. 1-12.
34. E. A. Kosminsky, "The Hundred Rolls of 1279-80 as Sources for English Agrarian History", in *The Economic History Review*, Vol. III, 1931-32.
35. These are of course *inquisitiones post mortem* from an earlier period than those Gray used: to be precise, from the 1216-72 and parts of the 1272-1307 period. The Hundred Rolls from 1279-80 cover only a couple of English counties, and these only partially: that is, southern Cambridgeshire, Huntingdonshire, parts of Bedfordshire and Buckinghamshire, the bulk of Oxfordshire and two of the Warwickshire hundreds. E. A. Kosminsky, "Services and Money Rents in the Thirteenth Century", in *The Economic History Review*, Vol. 5, 1934-35. pp. 29-30. A more exhaustive discussion of these sources can be found in the above-mentioned article - see note 34 - and in E. A. Kosminsky, *Studies in the History of the English Village*, Moscow 1935.
36. Kosminsky, "Services...", op. cit., p. 45.
37. Ibid., p. 40.
38. Ibid., p. 42.
39. Ibid., p. 40.
40. Ibid., p. 25.
41. Ibid., p. 32.
42. Ibid., p. 34.
43. "...the labour services are all, or almost all, contained in the documents, the money rents are given only in parts. As a rule we get only that part of the rent which is described as "redditus assisae": the "auxilium", which may often be arbitrary and irregular, is not always given, nor are some other rents.... Furthermore, money payments might be partially substituted for labour services in the annual practice of the estate (venditio operum) which need not be reflected in our sources..." (Ibid., p. 35) and

further, "the money payments are not always sufficiently differentiated from labour services". In Kosminsky's opinion this is because the source's specification of the various labour services, like its rigid classification of the land into "virgates", first and foremost expresses a methodological practice. "The study of villein holdings and villein services leads us to the conclusion that the Hundred Rolls have reproduced the forms of manorial bookkeeping rather than living facts....The virgates...are a more or less artificial unit for the allocation of manorial services. The various forms of the labour services ...represent...a method by which services and money-payments were reckoned". Therefore, "...the Hundred Rolls exaggerate the part played by labour services...which is made particularly clear by a comparison with the Inquisitiones Post Mortem". Kosminsky, "The Hundred Rolls...", op. cit., pp. 30, 35.

44. Kosminsky, "Services...", op. cit., p. 37.

45. He is in fact faced by the same problems in the study of the Hundred Rolls. The terminology is inconsistent, although the source can be considered a standardized and simplified version of the thirteenth century survey. The Hundred Rolls are in principle built up analogously with these, although the village and not the manor is the structuring unit. Kosminsky thus finds a number of traps. For example, it is hard to distinguish between freeman and bondsman, between tenancy, inherited tenure and tenancy "at the lord's will", etc. See also note 43. Kosminsky, "The Hundred Rolls...", op. cit., pp. 30-32.

46. Thus Kosminsky does not think, for example, that one can assume like Gray that there was no obligatory labour rent on a manor if it was not explicitly evident from the source. Gray, "Commutation of Villain Services...", op. cit., p. 630. Kosminsky claims that often, in the *inquisitiones post mortem*, where only money rents, not labour services, are explicitly mentioned, we encounter a monetary assessment of a number of different rents, including labour services. In consequence, he has not used any source where labour services are not directly itemized or are said not to exist. Kosminsky, "Services...", op. cit., p. 39.

47. Ibid., pp. 39-40.

48. Kosminsky arrives at this supposition through an implicit comparison of the amount of demesne and available peasants with labour duties on the smaller and medium-sized manors, as well as indications of the relatively large percentage of cotter farms itemized in the Hundred Rolls. Ibid., p. 42. If we must be content with this indirect evidence it is because "only those who hold land are included in the survey" (ibid., p. 36), as "the inquest was to define the land property of every land-owner in England, whoever he might be and whatever the conditions of his tenure". Ibid., p. 26. Nevertheless, Kosminsky thinks that "a very considerable part was played by hired labour recruited from the large reserves of the small-holding cotters" in the cultivation of demesne. Ibid., p. 38.

49. Ibid., p. 43.

50. A. E. Levett, "The Black Death...", op. cit., pp. 49-51.

51. "...there are sufficient grounds for regarding the thirteenth century, and perhaps even the twelfth century, as a period of the general growth of feudal rent. In view of the fact that complaints by peasants against increases of rent could reach the royal courts only in exceptional circumstances, the abundance of such complaints in the records of the royal courts for our period is very significant". E. A. Kosminsky, "Services...", op. cit., p. 41.

52. "...labour services increased side by side with the money rents". Ibid., p. 42.

53. Postan, "The Chronology...", op. cit., p. 186.

54. Ibid., pp. 192-193.

55. Ibid., p. 173.

56. Ibid., pp. 187-89.

57. Maitland, as early as 1894, questioned generalized views of the manorial system. *The History of a Cambridgeshire Manor*, op. cit., p. 417.

58. Karl Marx, *Das Kapital*, Vol. III, Chapter 47, MEW, Bd. 25, Berlin 1977.

59. Kosminsky, "The Hundred Rolls...", op. cit., p. 19.

60. Ibid., p. 44.

61. J. H. Clapham & Eileen Power (eds.), *The Cambridge Economic History of Europe*, Vol. I, Cambridge 1942; J. H. Clapham, *A Concise Economic History of Britain*, Cambridge 1949. The work appeared posthumously, and was written in the years immediately preceding Clapham's death in 1946.

62. Ibid., p. 116.

63. *Cambridge Eco. Hist.*, op. cit., pp. 500-501. See also Clapham, *A Concise...*, op. cit., p. 77.

64. *Cambridge Eco. Hist.*, op. cit., pp. 503, 510-11.

65. Ibid., p. 512.

66. Ibid., p. 527; Clapham, *A Concise...*, op. cit., p. 84.

67. From eight bushels an acre in 1200 to 8.5 in 1450 because 1. "...the level of agricultural technique may at the beginning have been as low as it could be"; 2."Famines...seem to have been less common after the Black Death than before"; 3. "...the population tended to become distributed less in the country and more in towns"; 4. "...enclosures probably made some progress". M. K. Bennett, "British Wheat Yield per acre for Seven Centuries", in *Economic History*, Vol. III, 1935, p. 22. One could of course raise several objections to this non-statistical evidence, but we may refrain from doing so because of Bennett's very cautious use of it.

68. In the first place, because the yield per acre on Russian peasant farms with three-course rotation and more or less the same amount of rainfall as in England between 1883 and 1900 was 11.5 bushels as against Winchester's 7.5 (according to Beveridge). Secondly, because the Rothamsted experiments had shown that successive wheat cultivation without manuring stabilized at a yield of 12 bushels per acre. Ibid., p. 21. Incidentally, Bennett did not think that Winchester can have been representative, even though he accepts Beveridge's calculation of Winchester yields. Productivity at Winchester should not be used inductively, as it is said to lie about ten per cent below the level of the rest of England in the twentieth century. Ibid., pp. 17, 19-21.

69. Previous research had been "ready to assume that a simple comparison of the amount of grain sown with the amount threshed and garnered is all that is required" to calculate grain yield. Thus grain yield according to Lennart cannot be calculated in most cases, as tithes were normally assessed and deducted from the total crop - even while the harvested sheaves were still in the fields. This means that in statistical calculations a ninth must be added to the total crop before comparing it with either the amount or area sown. R. Lennard, "Statistics of Corn Yields in Medieval England", in *Economic History*, Vol. III, 1936, pp. 173-74, 182-183.

70. See for example H. W. Sauders, *Introduction to the Obedientary and Manor Rolls of Norwich Cathedral*, 1930, pp. 57, 60; N. S. B. & E. C. Gras, *The Economic and Social History of an English Village*, 1920, pp. 338-72; and F. M. Page, *The Estates of Crowland Abbey*, 1934, pp. 329-30.

71. The situation might conceivably have been different with the cultivation of the peasants' own land. H. S. Bennett, who does not otherwise think that there is any basis for the soil exhaustion theory, thus argued that the peasants' generally poor opportunities for cattle farming, combined with the *jus faldae* rights of the lord, must have influenced their arable productivity. H. S. Bennett, *Life of the English Manor*, Cambridge 1937, p. 78.

72. Maurice Dobb, *Studies in the Development of Capitalism*, London 1947, p. 42.

73. Ibid., pp. 42-45.

74. Ibid., pp. 44-45.

75. Ibid., p. 46.

76. Ibid., p. 47.

77. Ibid., p. 48.

78. Ibid., p. vii.

79. John Saltmarsh, "Plague and Economic Decline in England in the Later Middle Ages", published in *The Cambridge Historical Journal*, Vol. VII, 1941.

80. The description of the general course of the crisis is in complete agreement with those of Postan and other eminent European historians. See, for example, Henri Pirenne, *Economic and Social History of Medieval Europe*, London 1935, p. 193. This entails that even though Saltmarsh acknowledges that medieval civilization had peaked around 1300, the effect of the crisis was thought to have culminated in the middle of that century. Saltmarsh, op. cit., pp. 23, 40.

81. But the plague alone cannot explain the impact of the crisis. The Hundred Years' War and its consequences, and political factors abroad influencing the decline of English trade must also be taken into account; along with just as chronic civil strife (the Wars of the Roses), the Hanseatic offensive against English Trade in Scandinavia and the Baltic, the loss of the French possessions, high taxes, piracy and raids on the English coasts, and Scottish invasions. Ibid., p. 27.

82. Ibid., p. 30.

83. Ibid., p. 41.

84. Ibid.

85. Nor were these questions clarified empirically. The issue of the social distribution of the social product was allowed to hang unanswered except by a loose claim, and the documentation of the social crisis was limited to an enumeration of familiar positions - rent drops and the desertion of the arable land. Ibid., pp. 23-24.

86. The bacteriological science recently developed by Pasteur and Koch formed the knowledge base that allowed a pupil of Pasteur, the Frenchman Dr. Yersin, and a pupil of Koch, the Japanese Kitasato, to identify the plague germ during the outbreak in Hong Kong in 1894. As the germ was found not only in plague victims but also in rat organs, suspicion was directed at this animal as a bearer of the infection. But it was only in 1908 that a number of experiments conducted by the Plague Research Commission under the leadership of Sir Charles Marti established: 1) that the true bearer of the plague is the rat flea; 2) that the plague cannot be spread continuously except through the agency

172

of this parasite. L. Fabian Hirst, *The Conquest of Plague*, Oxford 1953, pp. 106-07, 172-74.

87. The black rat has apparently been known in Europe since antiquity, and in England at least since the Crusades. M. A. C. Hinton, *Rats and Mice as Enemies of Mankind*, 3rd ed., British Museum of Natural History, Economic Series, No. 9, London 1931.

88. Saltmarsh, op. cit., p. 33.

89. Ibid., p. 37.

90. Ibid., p. 28.

91. Ibid.

92. Ibid., p. 24.

93. See for example John Graunt, *Natural and Political Observations upon the Bills of Mortality*, London 1662. David Ricardo wrote in 1817 that he was pleased to have the occasion to express his admiration for Malthus' *Essay on Population*, and that the attacks of opponents to this great work had only served to show that its justified reputation would spread with the cultivation of the science of which it was such an ornament (*Principper for den politiske ¢konomi og beskatningen*, Copenhagen 1978, p. 307). Later, Karl Marx wrote that the great attention aroused by this work was exclusively due to partisan interests to which Malthus was attentive. Marx thought that it was reactionary hack-work, plagiarized from Defoe, Sir James Stewart, Townsend, Franklin, Wallace and others, which had no interest in itself. *Das Kapital*, Vol. I, op. cit., p. 644.

94. See for example Matthew Hale, *The Primitive Origination of Mankind*, London 1677, where information is sought about population development in the as yet unpublished Domesday Book and village lists from early fourteenth-century Gloucestershire. Attempts to calculate population development in the whole country on the basis of the "enrollment of the poll tax of 1377" published by Sir John Topham in 1785 can be found in George Chalmers, *An estimate of comparative strength of Britain during the present and four preceding reigns*, London 1772, and in David MacPherson, *Annals of Commerce*, Vol. I, London 1805. The latter's calculation of the population in 1377 formed the basis of both Seebohm's and Rogers' assessments.

95. We do however have demographic studies from the first decades of the twentieth century. The Domesday Book, whose data had been studied at the end of the nineteenth century by Maitland, and the 1377 Poll Tax figures were analysed by A. H. Inmann in *Domesday and Feudal Statistics*, London 1900. Inmann corrected MacPherson's estimate to 3,069,000 in 1377 and Maitland's to 1,800,000 in 1086. Ibid., pp. 115, 119-22. The clerical sources were also used again to discover Black Death mortalities in regional studies. A. Hamilton, "The Registers of John Gynewell, Bishop of Lincoln, for the years 1349-50", in *The Archaeological Journal*, Vol. 68, 1911; and "Pestilences of the Fourteenth Century in the Diocese of York", ibid., Vol. 71, 1914; P. G. Mode, *The Influence of the Black Death upon English Monasteries*, Chicago 1916; J. Lunn, see G. G. Coulton in *Medieval Panorama*, New York 1938.

96. At least one attempt was made, though: 2.5 million in 1086 and five million in 1349 were "the figures commonly accepted", thought H. S. Bennett in *Life on the English Manor*, op. cit., p. 239.

97. Josiah Cox Russell, *British Medieval Population*, Albuquerque, New Mexico, 1948.

98. "The poll taxes of 1377 and 1379 can enable us to secure a more accurate picture of the population and society of England than at any time until the first census of 1801". Ibid., p. 119.

99. Ibid., p. 121.

100. Apart from children under the age of fourteen, this taxation did not apply to mendicant monks and other paupers. One must also allow for the fact that the counties of Cheshire and the diocese of Durham do not feature in the lists from 1377. Russell compensates for these shortcomings by calculating the Durham and Cheshire populations in terms of the population and area of the surrounding counties, and by adding 50% to the total population to compensate for the under-fourteens. This is a lot, considering that Russell thought that under-fourteens made up a third of the population. Ibid., p. 143. Moreover, the number of mendicant monks is estimated to have been 2,500 and other paupers as having made up 5% of the total population. Ibid., p. 146.

101. Ibid., pp. 37-38.

102. Information on priests and churches from certain counties provided Russell with the opportunity to calculate deductively a general ratio of priests to population, and thus to assess the number of priests in the eleventh century. According to Table 3.2 the ratio was 1:258, which, over the total agrarian population, gives about 3,800 priests. This figure is increased to 4,500 to allow for the urban population and "unbeneficed clerks". Ibid., p. 43-45. Russell estimated the urban population to have made up a ninth of the rural population in the eleventh century, but was somewhat uncertain of this estimate. Ibid., p. 54.

103. Russell thought that this factor, which was "based upon the actual number of persons taxed house by house in 1377..." (ibid., p. 61), could be used not only to calculate the "Domesday population" (ibid., pp. 38-39), but also for the thirteenth century (ibid., p. 69). Here Russell deliberately broke not only with Maitland's view of the medieval extended family; but also with G. C. Homans, who has claimed that a household could consist of the following total members in several houses: husband, wife and children, servants, possibly subtenants, grandparents and unmarried sisters and brothers. G. C. Homans, *English Villagers of the Thirteenth Century*, Cambridge, Massachusetts, 1942, pp. 209-11. A more developed criticism of Russell's 3.5 factor came later in J. Krause's article, "The Medieval Household: Large or Small?", in *The Economic History Review*, 2nd Series, Vol. IX, No. 3, 1957, pp. 420-32.

104. Ibid., pp. 52-54.

105. To the above-mentioned 4,500 priests (cf. note 102) were added 940 monks and nuns. Cf. Russell's "Clerical Population of Medieval England", in *Tradition*, No. 2, 1944, pp. 179, 185.

106. Russell, "British Medieval...", op. cit., p. 54.

107. In this connection there is criticism of a number of the source types we have seen used before to shed light on the problem: chronicles, court rolls, and frank-pledge lists. Ibid., pp. 227-229.

108. Ibid., pp. 262-63.

109. Ibid., p. 269.

110. The slight rise in the population in the latter half of the fifteenth century developed into a rapid rise in the middle of the next century, then flattened out again according to Russell in the seventeenth century. Ibid., pp. 270-72.

111. These surveys can be divided into two main types. One had to do with royal tax-gathering. The best-known and most comprehensive of this type were the *rotuli hundredorum* or Hundred Rolls. The other type was drawn up for the internal administration of the manors. These are known from the twelfth century, primarily from the monasterial estates, but in the course of the thirteenth century they also became common in the administration of the lay manors. Both types involve certain problems as sources for demographic studies, especially in view of Russell's comparative method. See ibid., pp. 56-70. For example, there are a number of identification problems with manors and villages when one compares Domesday and surveys.

112. Table 10.8. Ibid., p. 250.

113. Ibid., p. 259.

114. Table 10.11. Ibid., p. 258.

115. Table 10.2 and Fig. 10.1. Ibid., pp. 240-43.

116. Ibid., p. 260.

117. Ibid., p. 232.

118. This does not preclude the possibility that natural conditions might have been important on the Continent. Cf. Fritz Curschman, *Hungersnöte im Mittelalter*, Leipzig 1900; and H. S. Lucas, *The Great European Famine of 1315, 1316 and 1317*, 1930. But "England certainly suffered less from the forces of nature than did the continent; she was of a singularly equable climate". "The climate of the British Isles has not changed greatly in historical times". Cf. C. E. Britton, *Meteorological Chronology to 1450*, London 1937.

119. Russell, "British Medieval...", op. cit., p. 232.

120. In Table 11.8, Russell listed the ratios of acres to persons and persons to square miles for 1086 and 1377 distributed over various counties. By dividing the acres/persons ratio for 1377 by 1.66 (the death rate between 1348 and 1377) Russell found the acres/persons ratio for each county just before 1348. In twenty of the Table's thirty-nine counties this ratio was less than 8.6 acres per person, which Russell considered acceptable. The premiss for this judgement was that thirty acres was "a good allotment of land", and that each farm should have been able to feed 3.5 people; and that these thirty acres did not include the share in the necessary common woods and pastures. Ibid., pp. 312-14.

121. 1348-50, Table 9.13; 1360-61, Table 9.14; 1369, Table 9.15; and 1375, Table 9.16. Ibid., pp. 216-18.

122. Ibid.

123. Table 9.2, ibid., p. 200. See also Tables 9.3-9.5, pp. 202-03. The situation is reasonable clear from these tables of age-specific mortality from the Black Death. In general, Russell builds on a very slender body of material. Table 9.13 includes information on a total of 505 persons; Table 9.2 information on 187 persons for the years 1280-82; 309 persons for 1310-12; and 351 persons for 1340-42.

124. Ibid., pp. 230, 370.

125. Homans, "English Villagers...", op. cit., pp. 142, 158-59.

126. Russell, "British Medieval...", op. cit., pp. 162-64.

127. Ibid., p. 378.

128. Ibid., p. 367.

129. Ibid.

130. Ibid., p. 381.

131. See for example Clapham, "A Concise Economic...", op. cit., p. 77.

132. Usually called "assarting" in English. This term is taken from contemporary sources and is derived from the French word *essart*, which means clearance land or clearance.

133. See A. M. Cooke, "The Settlement of the Cistercians in England", in *English Historical Review*, 1893; F. M. Stenton, *Facsimiles of Early Charters from Northamptonshire Collections*, 1930; the same, *Documents Illustrative of the Social and Economic History of the Danelaw*, 1920.

134. H. Lindkvist, *Middle English Place-Names of Scandinavian Origin*, Upsala 1912.

135. The term "forest" has had several meanings - woods or privatized hunting area. The medieval meaning primarily applied to areas where the King's game was protected by special legislation, or a separate administrative unit within such an area. F. H. Baring, "The Makings of the New Forest", in *English Historical Review*, Vol. XVI, London 1901, pp. 427-38.

136. Margaret Ley Bazeley, "The Extent of the English Forest in the Thirteenth Century", in *Transactions of the Royal Historical Society*, 4th Series, Vol. IV, London 1921, pp. 146, 151, 152-56.

137. As H. Darby does in *An Historical Geography of England before A. D. 1800*, Cambridge 1936, p. 178.

138. Bazeley, op. cit., p. 142.

139. T. A. M. Bishop, "Assarting and the Growth of the Open Fields", in *The Economic History Review*, Vol. VI, 1935.

140. H. C. Darby, *The Medieval Fenland*, Cambridge 1940.

141. Vinogradoff, "The Growth of...", op. cit., pp. 330-31.

142. Vinogradoff, "Villainage...", op. cit., pp. 332-33. See also Richard Koebner in *The Cambridge Economic History*, op. cit., p. 78.

143. H. S. Bennett, "Life on the English...", op. cit., pp. 51-52.

144. Bishop, "Assarting...", op. cit., pp. 22-23.

145. Ibid.

146. Ibid., pp. 26-27.

147. Ibid., pp. 24-25.

148. Bishop documents this practice with a single charter from 1250. Ibid., p. 28.

149. Ibid., p. 29.

150. "It is now generally appreciated that the Black Death merely hastened the change that it was formerly thought to have originated". R. A. Pelham, in *An Historical Geography...*, ed. H. C. Darby, op. cit., p. 230.

151. Clapham, *A Concise Economic...*, op. cit., pp. 111, 115.

152. "The period 1250-1350 was the most prosperous century of medieval agriculture....Whatever views may be taken of the significance of the Black Death, the fact was that in the following century a growing individualism undermined the conservatism of manorial life". Darby, *An Historical Geography...*, op. cit., p. 238.

153. Homans, "English Villagers...", op. cit., p. 4.

154. H. C. Darby in *An Historical Geography...*, op. cit., pp. 188-89.

155. Charles Parain, in *The Cambridge Economic History*, op. cit., p. 162.

156. Ibid., p. 168.

157. Darby thought there were "certainly no signs of any large-scale project of draining". There was "a continuous piecemeal encroachment upon the edges of the fens, and this

was a movement that seems to have been very general". H. C. Darby, *The Medieval Fenland*, op. cit., p. 43. See also p. 52.

158. Ibid., p. 61.

159. An important reason for this was that "the Fenlands were more valuable in their natural condition than if they had been converted permanently into winter ground [i.e. drained]. Common rights were no incidental appurtenance in manorial economy. They represented an important and profitable way of exploiting certain types of terrain....Winter floods made the pasture richer, and the damage done by summer floods was not too critical". Ibid., pp. 52, 67.

160. Ibid., p. 66.

161. Ibid., pp. 74-81.

162. Russell, "British Medieval...", op. cit., Table 11.8, p. 313.

163. Nellie Neilson, in *The Cambridge Economic History*, op. cit., p. 458.

164. Ibid., p. 459.

165. Clapham, *A Concise Economic...*, op. cit., pp. 79-80.

166. Bazeley, "The Extent of...", op. cit. See the map between pp. 140 and 141.

167. C. S. & C. S. Orwin, *The Open Fields*, Oxford 1938, pp. 50-51. It should incidentally be noted that the Orwins found a rather wider spread of the open-field system than Gray. Thus they thought that the system had also existed in Northumberland, Durham, Norfolk, Suffolk, Essex and Kent. Ibid., pp. 60-61. See also the map, p. 65. This disagreement may be due to the fact that the map in question "is a diagram of the geographical spread of the Open Fields, not a measure of their extent", suggests Evert Barger, who further states that Gray had already in 1915 "determined once and for all the main pattern of medieval English agriculture". E. Barger, "The Present Position of Studies in English Field-Systems", in *The English Historical Review*, Vol. LIII, 1938, pp. 385-86.

Chapter VI

Theories of Late Medieval Crisis
(1950-1960)

VI.1 The five theories

Clearly, apart from the recapitulation of the "epidemiological" theory we have encountered, the theories that crystallized between 1930 and 1950 derived the extensive changes in agriculture, with increasing conviction, from factors endogenous to society. What was still a moot point was whether the late medieval centuries exhibited a restructuring of the agrarian relations of production, or a gradual change to quasi-capitalist forms of organization, although the period had earlier been described (cf. Chapter IV, pp. 117-118) as a watershed between feudal and capitalist forms of organization. So on this basis one cannot speak of the reproductive crisis of feudal society.

Similarly derived from the theoretical developments of the 1930s and 1940s was the idea that the reduction in population could now partly be explained in the light of socially-created conditions. In this respect one can to a certain extent speak of the reproductive crisis of feudal society, since a deterioration in the conditions of reproduction and a failure of the population to reproduce are in fact major elements in these theories - elements which found some support in Russell's demographic studies, but not in research on the English colonization process, which however did not directly refute them either.

The historians of this period, however, were far from agreeing on whether the agrarian transformation process entailed crisis conditions at all, and even more so on the exact nature of the crisis, if crisis there was. Some argued for a crisis of the manorial system, others for a general crisis of society. Beyond this, as the assumption of the unique significance of the Black Death was increasingly abandoned, views differed on the chronology of the crisis. While some tried to place its impact as far back as the thirteenth century, others argued that its effect came much later in the period after the Black Death.

In these decades, as before, the disagreements were tied to the development of various bodies of theory concerning late medieval crisis or transformation. In the years between 1930 and 1950 these theories developed to such a level that we should now be able to identify the five principal modern theoretical schools of English research. In the period before 1930, relatively untouched by research between then and 1950, we found a "political" theory of crisis, in which the development of state power, military activity and fiscal factors were offered as determinants. The even earlier "neoclassical" theory survived despite serious attacks on its essential features, especially during the formation of the "neo-Marxist" theory in the 1930s and 1940s; while the "epidemiological" theory was recapitulated. Finally, we saw how a "neo-Malthusian" theory began to take form in the years up to 1950. In the following, we will deal with the further fate of these five main theoretical schools, and their importance for research on our subject matter.

VI.2 The "neo-Malthusian" theory

In a report to the Ninth International History Congress in Paris in 1950, Postan presented a hypothetical explanation of the causes of the late medieval crisis which must be described as neo-Malthusian, and which he considered at the time "tentative in the extreme; a mere guess which may well turn out to be untrue".[1] Yet some of this uncertainty seemed to vanish during the drawing-up of his contribution to *The Cambridge Economic History* two years later. Now he considered that "this particular argument emerged from evidence purely descriptive and from arguments largely empirical".[2]

Postan thought he could relatively safely assert that "the commercial and industrial depression of the later Middle Ages must be accounted for by decline in numbers [of population]".[3] But although he still believed that population development in itself "in some ways...was more fundamental than any other economic changes...it would be difficult to treat the population trends as the sole or final cause..." For "what caused the fall and rise in population?"

Theoretically, he could conceive that biological factors had an influence: perhaps, for example, changes in fertility. But he would not accept the Black Death as the manifestation of a biological factor. He further rejected the climatic factor which some Scandinavian scholars, among others, had suggested,[4] although he admitted that the plentiful rainfall in

179

the first two decades of the fourteenth century influenced the famine years of the second decade. But he did not think that the climatic factor could be assigned any crucial long-term demographic importance.

According to Postan, for the worsened climatic conditions at the beginning of the fourteenth century to have had a serious effect on population development, we must imagine that the population was already at "the Malthusian level of subsistence". We must imagine a situation where the stimulation of population figures by the colonization process had gradually provoked a situation where "the marginal character of marginal lands was bound to assert itself, and the honeymoon of high yields was succeeded by long years of reckoning, when the poorer lands, no longer new, punished the men who tilled them with failing crops and with murrain of sheep and cattle. In these conditions a fortuitous combination of adverse events, such as the succession of bad seasons in the second decade of the fourteenth century, was sufficient to reverse the entire trend of agricultural production and send the population figures tumbling down".[5]

It is important to keep in mind that Postan proffered this hypothesis much more under the influence of Continental research and criticism of other theories than on the basis of empirical documentation from English studies. He thus agreed with the German W. Abel, the Norwegian J. Schreiner, the Belgian van Werverke and the Frenchman M. Bloch on the chronology of the crisis process, and agreed with the first three of these on the reasons for the crisis.[6] And he used, for example, Abel's and Schreiner's topographical studies as evidence for the population decline before 1350 in his contribution to *The Cambridge Economic History* - in general, dubious evidence of these rather crucial factors for the neo-Malthusian theory.[7]

Postan's interpretation of the reasons for the late medieval crisis were based on two critical premisses: first, that it could be proved that the population was already developing negatively before 1350; and secondly, that it could be established with reasonable certainty that this demographic downturn was due to relative overpopulation, where the marginal lands that had been incorporated to cater for the rising pressure of population emerged as a natural barrier to the continued reproduction of society. He himself contributed actively to the effort to obtain documentation of the purely demographic problem in the course of the 1950s. Yet he made no direct attempt of his own to verify the "marginal land theory". But some light was shed on this theory by the continued research on the medieval colonization process and the problem of the deserted farms

180

which became a widespread theme of English research in the same decade, and which also shed light on the demographic issue.

VI.3 Economic evidence

In the course of the 1950s the following broad consensus on the principal features of English economic history from the late middle ages until the nineteenth century was worked out among supporters of the neo-Malthusian theory: "Rising population, rising prices, rising agricultural profits, low real incomes for the mass of the population...this might stand for a description of the thirteenth century, the sixteenth, and the period 1750-1815. Falling or stationary population with depressed agricultural profits but higher mass incomes might be said to be characteristic of the intervening periods".[8] Postan's two attempts to assess late medieval population development in the same period were made within the same terms of reference, which, as we have seen, he was already influenced by in the 1930s.

His first attempt was published in 1950.[9] In this he was primarily looking for so-called economic evidence of the population development - that is, indications based on fluctuations in prices, wages and rents. As a secondary approach he used indirect evidence of the deserted farms problem as a demographic indicator. Starting from Rogers' calculations of developments in wages, his own collection of data on wage development at Glastonbury and Peterborough Abbeys in the Duchy of Lancaster, and many small manors scattered all over England, and not least Beveridge's estimates of wheat prices and wage development on the Winchester estates,[10] Postan thus took the rising wages in this material after 1300 as an indication of falling population.[11]

He concluded that the rising wages could be taken as a direct expression of falling population; first, because monetary factors in the form of the coinage debasement known from the period were not directly related to the actual development of rising wages; secondly, because rising wages were accompanied by rising prices;[12] and thirdly, because nothing suggested that agricultural labour could be absorbed by other industries in the fourteenth century. This was partly because "a falling curve would fit into facts of English industry between 1350 and 1470 much better than a rising one", and partly because "wages of artisans also rose in proportion and at a pace little different from wages of agricultural labour".[13]

181

It should be noted that even if one accepts these premisses, and on the whole the general relationship claimed between wage and population developments, Postan's interpretation involved a very narrow reading of the tables of wage development he had drawn up. In some cases they show only very moderate rises in the period before 1350.[14] And the documentation was hardly bolstered up by his use of another economic indicator: "the value of land", or rather the development of rents.

Postan thought that the "value of land" and its fluctuations were best illustrated by the rent that came closest to what he called "economic rent", that is payments in free tenancies. But the material used, taken from Davenport among others, has no indications at all of a drop in rents before 1350. The same thing was true when he examined the development of entry fines at Glastonbury Abbey. Here there was a rise from less than a pound per virgate in the early thirteenth century to about twelve pounds in 1345, and a corresponding drop in the period between 1350 and 1450.

The entry fines and their development, according to Postan, can be taken as an expression of the fluctuations in the naturally invariable customary rent. Fixed rents could thus vary with rises and drops in the entry fines. Commutation was another way the terms for customary land could be changed - that, is eased. So Postan took commutation as a sign of falling land value, a situation he took to be prevalent in the second half of the fourteenth century and in the fifteenth. So there was nothing here either to indicate falling "land value" and population figures before 1350.[15]

Nor was there much to suggest a rise in due, but unpaid rents. - *decasus redite* - in the manorial accounts until after 1350. For Postan this meant that there were few signs of deserted tenant farms in the first half of the fourteenth century, nor any sign either in this respect of the beginnings of a population decline.

In fact Postan provided only one example of *decasus redite* from the period before 1350. But he claimed that these items in the manorial accounts were very numerous after 1350; and further that "in the earlier period [before 1350] the decasus redite could sometimes be a sign of rising demand for land". Thus the example he cites from the accounts of the manor of Rolleston in 1314 was explained by the fact that "a number of tenements had been drawn into the demesne and its park". It was only after 1350 that the itemizations of *decasus redite* showed that "the landlord was prepared to let...on lower terms".[16]

Given this, it seems risky when Postan ventures to conclude that "the data of wages and those of rents and vacancies agree in placing the beginning of the decline somewhere in the first twenty years of the fourteenth century". But the author is apparently no more convinced than that he finds it necessary to add: "What makes the data all the more certain is the continental evidence, for there too the turning point in employment and occupation of land comes in the first quarter of the fourteenth century".[17]

VI.4 Heriots

So one turns expectantly to Postan's and J. Titow's 1959 study of "Heriots and Prices" on some Winchester estates in the period between 1245 and 1350.[18] The two authors set out here to study the development of heriots[19] in order to outline population developments in the period before the Black Death. To begin with, it must be remarked that the material is relatively slender, since the study only covers five of the many estates of the Winchester complex, and since there were only about 1700 villeins on these estates in 1321.[20] Moreover, it must be added that the material does not really become dense and continuous until about 1270.

Because of the lack of available information both on overall population and the total number of villein tenants, Postan's and Titow's statistical constructions only cover the number of heriots paid, which, all other things being equal, ought to be the same as the number of deceased villein tenants year by year.[21] This approach seemed plausible to the authors, as they claimed to have a background for saying that the total number of villein tenant farmers throughout the period was stable. It was thus thought that there was no splitting-up of the individual holdings, nor were there apparently any absolute expansions of the number of villein holdings through colonization.[22]

Two general points emerged from the statistical account of the number of heriots in the period (see Table I and Fig. I). In the first place one saw a considerable rise from a point in the 1290s, and secondly the level from 1300 to 1310, and again from 1325 to 1347 was by and large the same, but interrupted by a drastic rise in the famine years in the second decade of the fourteenth century. The authors did not exclude the possibility that the general rise from the last decade of the thirteenth century on may have been connected with the development of better administration, and thus more diligent collecting of heriots, but the sources showed no direct traces of this.[23] Postan and Titow therefore concluded that from about 1300 the number of heriots on these Winchester estates rose to a level which, apart from the famine year in the second decade of the fourteenth century, was relatively stable.

On this basis it was difficult to come to any definite conclusions on population development before the Black Death. At any rate the authors were not prompted to see it as evidence of the beginning of a population decline. On the contrary, they rather argued that the rising mortality among villein tenants could have been due to a new age mix within this group caused by ever-increasing pressure of population combined with a relatively stable number of tenements. This could have forced the age up among "the reservoir of would-be successors to dead tenants", so that the death rate among the villein tenants accelerated.[24] This is a line of thought we also saw pursued by Russell.

However, it was not Postan's and Titow's main conclusion. Through a detailed analysis of the fluctuations in heriots year by year, especially in view of the striking rise in the second decade of the fourteenth century, they connected the rising mortality primarily with the rash of epidemics and years typified by bad harvests. They further thought that by means of a systematic classification into "money heriots" and proper "animal heriots" (the first type being particularly associated with cottagers without livestock), they could establish the points where the peasants had been most vulnerable to bad harvests and accompanying famine. On the other hand they concluded that the epidemics took their toll across social and economic differences among the villeins.

Their conclusion was therefore that, apart from the years when the epidemics were rife, "all the...spectacular rises in mortality coincide with or follow bad harvest..."[25]. So "if the secular trend was rising it was at least in part because the proportion of smallholders in village population was also rising. The catastrophic failures of crops in the first half of the fourteenth century may of course have been due to unprecedented rainfalls, or to the progressive impoverishment of the soil, or to the spread of cultivation to marginal lands, or to all these factors in combination. But the sharp response of mortalities to crop failures undoubtedly reflected the high and rising number of tenants whose holdings were too small to enable them to sustain themselves in years of bad harvest".[26]

This is a surprising conclusion, as an important premise for the method used to assess mortality was precisely that in the period it was supposedly impossible to find signs on the manors in question that the total number of villein tenants developed significantly in either a positive or negative direction.[27] If we accept the conclusion, this has a crucial retrospective effect on the assessment of the observable rise in heriots as an expression of a corresponding rise in mortality. And as a corollary, if we accept Titow's and Postan's assurances that the total number of villeins in the period was on the whole stable, how can this claim be reconciled with the fact that the percentage of smallholders was rising, and further that precisely this, because of the smallness and poorness of the holdings was seen as a sign that tillage in the first half of the fourteenth century had been driven to a point where the land could feed its tillers only in the good years?

It can only be done provided either part of the cultivated area was withdrawn from cultivation because the soil had lost its fertility, or part of the peasants' land was annexed to the demesne or converted to park or pasture. There is no documentation for either of these two preconditions in this work, but we recall how Postan presented an example in a work mentioned above of how a *decasus redite* itemization in a manorial account from 1314 was due to the fact that part of the villeins' lands were absorbed "into the demesne and its park". No further documentation of the two premises can be found.

So Postan was not successful either in firmly verifying, by means of the heriots, his old hypothesis that the population development had already turned around at the beginning of the fourteenth century. On the contrary, there emerged from his study a new hypothesis about the consequences of the overpopulation he saw as a result of the population growth of the high middle ages. Besides the marginal lands theory he asserted that the percentage of smallholders was rising, and that both factors in combination with climatic and epidemiological circumstances gradually turned the population development in a negative direction. As we have seen, neither of these claims was convincingly documented. In particular, the marginal lands theory remained pure theory. In the following we will look at two of these three hypotheses in the light of the 1950s research on the colonization process and the deserted farms problem.

VI.5 Colonization

It is an important premiss of the marginal lands theory that, like A. R. Lewis, one sees the period before and around 1350 as characterized by "the closing of the Medieval Frontier".[28] But the possibility certainly cannot be excluded that there was still after this point a potential for what Lewis called external colonization. Thus W. G. Hoskins, in a 1955 work, thought that "despite the great colonization movement of the twelfth and thirteenth centuries, England still contained many wild and secret places ..." when the Black Death hit the country and "put an end to the land-hunger of the thirteenth century..." The pressure of population subsided and effected "a retreat from the marginal soils".[29]

Nevertheless, Hoskins thought in a later work that "the open-field villages had reached their frontiers by the closing years of the thirteenth century". But this did not mean that "they had cleared and ploughed up all their waste, for the waste - yielding as it did common pasture for animals and firing for villagers - was an essential part of the peasant economy, and it was necessary to strike a balance between the claims of arable, pasture and woodland".[30] This is a conclusion that clearly conflicts with the view that what Lewis called internal colonization in particular had to result in a problem for agricultural technology, as Nasse and Ernle had suggested, but which Darby was the first to relate to colonization - see Chapter V, p. 165 - and which Rodney Hilton saw in the 1950s as "the principal defect of medieval agriculture" for the peasants, - that is, the disproportion between tillage and animal farming. And R. Trow-Smith was able to support Hilton in this view. Despite partially conflicting sources he observed that livestock breeding was greatest in the areas where tillage was least "on the villein holding".[32]

On the present basis it is hard to accept that colonization had reached its absolute limits when it ceased. But the question is whether it was possible to exploit any free resources; whether they were tied because of their geographical location and possible status as "forests" or the like. In addition, the expansion of demesnes in the thirteenth century probably stifled the peasant's possibilities for expansion.

On the other hand it is easier to accept that internal colonization approached its absolute level in the first half of the fourteenth century in the most densely populated, mainly tillage-orientated, parts of the country; and that it was perhaps in these areas that the disproportion between tillage and cattle breeding developed that was seen by some authors as a basic defect in medieval agriculture - among other reasons because this monocultural form of production could lead to the exhaustion of the arable.

VI.6 The late medieval desertions

Since Saltmarsh had pointed out that marginal sandy soils had been with-drawn from cultivation in the thirteenth century - see Chapter V, p. 156 - Hoskins and in particular Maurice Beresford had begun a search for vanished medieval villages in Leicestershire,[33] Buckinghamshire, Yorkshire and Warwickshire. This work had been begun with Canon Foster's pioneering study of Lincolnshire from 1924.[34]

In the course of the 1950s and 1960s these efforts were followed by a large number of local studies, among which Beresford's well-known work *The Lost Villages of England* from 1954 stands out by dealing with more than a single county.[35] The development of this research has been de-scribed by Beresford in the anthology *Deserted Medieval Villages* from 1971, where his co-editor John G. Hurst contributes, among other things, an ample select bibliography of the subject.[36] Since Beresford sketches out in the first chapter a line of research that saw its peak in the 1950s, and since his chapter may thus serve as a general assessment of the subject in the present context, we will give this work some consideration in the following, although it may seem to be a digression from our chronological approach. But let us first look at the conclusions Beresford had reached in 1954 as regards the desertions before 1348.

Beresford thought that in general one would find traces of lost villages throughout the middle ages, whether the reason was war, fire, natural dis-asters, the movements of the Cistercians as they abandoned villages, or the exhaustion of the natural resources of the village referred to by Pos-tan in the marginal lands theory.[37] But because of the nature of the sources Beresford found it difficult to date these desertions more accur-ately, before the state in 1485 began to collect information on definitive desertions. Neither manorial accounts nor archaeological excavations pro-vided him with reliable information for dating the early desertions. So, apart from the Domesday Book, he decided that he would mainly have to look at taxation rolls,[38] and on this basis he obtained a general impression that the bulk of the desertions must have taken place after 1400.[39]

This impression does not differ in its main features from his account of the problem in 1971, after two decades of intensive research. Here, too, he thinks that the desertion of the villages is a phenomenon one can observe throughout the middle ages, and that "vacant holdings and uncultivated acres are a recurring feature of manorial surveys of the early fourteenth century". But "they seem to be especially frequent on the high, dry chalklands of Lincolnshire, and the East Riding of Yorkshire, and on the sands of the Breckland in Norfolk and Suffolk". It would therefore be logical and convenient to follow Postan's hypothesis that the early desertions from about 1300 "were on marginal soil, reluctantly colonised in a period of population expansion, but soon disappointing the over-optimism of their settlers by poor yields". But Beresford does not think this theory can explain early desertions in the middle of fertile counties like Oxfordshire in a time of expanding population and general scarcity of land.[40]

Research on the lost villages of England neither confirms nor directly refutes the marginal lands theory. Inasmuch as it mainly sought and could find vanished villages, this research primarily shed light on what we have above, with Lewis, called external colonization, and the subsequent external desertion. We do not have the same picture of a putative internal desertion process. The development in the size of medieval villages is a more obscure area which provides us with no opportunity of assessing the probability of the marginal lands theory.

But this much is clear: the external desertion process extended throughout the middle ages; neither the thirteenth nor the fourteenth century can be seen as its peak. And neither in the continental research mentioned above nor in the English work is the serious escalation of the desertion process dated to any period before at least 1400. If we continue to take Beresford as representative of English research, which we can do with some justice, we must observe that in the 1971 anthology he does not deviate in the essentials from a dating where he thinks that "the assembled general evidence...points to a period between 1440 and 1520 as that in which the main flood of depopulation took place".[41]

As part of the background for this dating in Beresford's studies of his own and others' primary research, there is plentiful evidence that the Black Death and subsequent epidemics had a long-term, not an immediate, effect on the desertion process.[42] This suggests that the presumably high mortality in the second half of the fourteenth century did not result in desertions in the first instance, as the population surplus from before 1350 was enough to compensate for the immediate losses - a line of thought we have encountered several times, and which clearly conflicts with the notion of a striking loss of population before 1350.

The crucial difference between Beresford's accounts from 1954 and 1971 was due to the fact that local studies of the desertion process had in the meantime reduced the number of areas where it could be said with certainty that they avoided village desertion. It thus appears that only areas where the soil was eminently suited to tillage, and areas where the monoculture was less in evidence, were fairly immune.[43] But there is a fair amount of agreement between these two works as far as the indication of the causes of the village desertions is concerned.

Without going into detail, one can sum up with two main reasons. Disregarding various possible causes mentioned before, Beresford's explanation of the desertions relates them either to the marginal lands theory or the transition from arable to cattle and sheep farming,[44] which he associates with the long-term demographic effect of the Black Death and the subsequent epidemics. Quantitatively, though, the marginal lands theory only has a tangential importance which he relates exclusively to certain externally colonized areas in Norfolk, Suffolk, Lincolnshire and the East Riding in the first half of the fourteenth century.

"The main force of depopulation was not felt until after c. 1450. The incentive was the demand for wool for the expanding English cloth industry, while the post-plague population had not recovered enough to increase the demand for corn". And further, "pastoral farming was also tempting since it used a smaller labour force at a time when the bargaining power of labour was still high".[45]

The research on the deserted English villages did not directly support the hypothesis of a falling population before 1350. At least, any population losses in the decades before the Black Death were not manifested as a measurably rising number of abandoned settlements. Moreover, this research suggests that the marginal lands theory can only be verified for peripheral areas of the country before 1350, and that large parts of the areas colonized during the period of expansion before 1300, for example the Fens, in fact proved highly resistant for geomorphological reasons at the peak of the desertion process after 1450. Finally, we saw that there was no sure evidence that the colonization process had reached its absolute limits around 1300.

These factors, combined with the relatively limited available knowledge of agrarian productivity, that is the development of yields, make it difficult to accept Postan's general view that "when population rose agriculture expanded under conditions which economists would recognize as those of steeply diminishing returns, and agricultural prices were bound to rise".[46] And by contrast "when population fell, some marginal lands would in all probability be abandoned and food would be produced on better land. Relative to the amount of land and labour engaged in food production and relative to the demand for food, supplies would then be more plentiful and therefore cheaper".[47]

VI.7 The "neoclassical monetary" theory

The immediate connection between prices and wages development and demographic fluctuations which Postan had already pointed out in the 1930s, and which later led him to seek evidence of the allegedly falling population before 1350 in rising wages and falling rents, was thus further clarified by the marginal lands theory. But for Postan, the late medieval changes in the structure of production, too, might justify the development of prices and wages; or this development could be regarded as evidence of the changes in the structure of production to the extent that he thought the breakdown of the manorial system entailed "the decline of commercial production".

He saw a shift from a developed exchange economy based on a quasi-capitalist form of manorial production towards a higher degree of subsistence economy, along with the progressive breakdown of the manorial system and an increasing supply of land structured by various forms of lease. The peasant farms with their mainly self-reproductive economy, according to Postan, "multiplied at the expense of large units of commercial agriculture...", which led to a drop in demand and the prices of agricultural produce, since more people now held enough land to reproduce themselves directly. Wages also rose, as there were fewer people, and they were less dependent on wage labour.[48]

For Postan not only thought that thirteenth-century demesne production was quasi-capitalist because it was primarily market-oriented. He also thought he had demonstrated that the bulk of the work volume on the manors was done by wage labourers - *famuli* - not by tenant farmers subject to labour rent, at least not at the end of the thirteenth century.[49] But

190

these elaborations of a formal, logical explanation of the price and wage fluctuations from 1300 on did not ward off the attack from "neoclassical" quarters that would almost inevitably come, not least as a consequence of the hard attacks Postan had mounted against the advocates of this school of thought since the 1940s.

The assault on Postan's "neo-Malthusian" interpretation of the causes of the crisis of the fourteenth and fifteenth centuries came from two "neoclassicists", neither of whom would acknowledge his economic documentation of the alleged population decline before 1350. The first was the Norwegian Johan Schreiner, whose acquaintance we have already made.

Schreiner agreed with Postan as far as the development of wages before and after the Black Death was concerned, and inasmuch as he too saw the striking rise in wages after 1350 in the light of agrarian structural reorganization: wages rose because the supply of tenant land rose.[50] He also agreed with Postan that the population drop had a crucial influence on this process after 1350, since he thought that two factors determined economic development in the late middle ages - falling prices and falling population.[51]

Schreiner therefore sought the explanation for the agrarian restructuring process - that is, the manorial system's incipient breakdown, the increasing availability of tenant land, and the subsequent rising wages - before 1350 in the drop in the price of corn. And as we have seen he did not acknowledge that the population declined before the Black Death; so he found another explanation of this price drop in certain monetary considerations, the starting-point of which he called a "technological obstacle" in the silver-mining industry.

Schreiner thought, like so many other European historians before him, that the apparent exhaustion of the European silver mines in the fourteenth century "was occasioned by the fact that mines had gradually become so deep that the miners were everywhere stopped by underground water". This "technological obstacle" was allegedly not overcome until the end of the fifteenth century, when "a successful solution of the problem "appeared through" the invention of more powerful drainage engines". But in the meantime this technological problem had led to a decrease in the supply of silver and a consequent rise in the silver price; and since "silver coinage was the main medium of local trade and everyday exchange, it would seem natural to seek the explanation for the falling prices in the fourteenth and fifteenth centuries in these changed monetary conditions".[52]

This theory explained for Schreiner why the manorial system was already on the decline in the decades before the Black Death, and here he was also in agreement with Postan. A falling supply of silver and rising silver prices led to "the fall in grain prices, which reduced markedly the profit from demesne cultivation", which led, for the wage-dependent peasants and labourers to "possibilities of a social "upgrading" before the great epidemics...the chance to take over land on favourable terms. And this again led to a general increase in the level of wages".[53]

We will not here go into the polemics Schreiner directed at Postan's rejection of the monetary theory. But we will not neglect to point out a serious inconsistency in Schreiner's interpretation of the theory. As mentioned before, he thought that the development of corn prices throughout the fourteenth and fifteenth centuries could be explained in the light of the rising silver price - at least after 1320 when the price drop happened, with a couple of decades' delay after the decline in silver production.[54] This means that the theory should also be able to explain the temporarily positive trend in prices that appeared just after the Black Death.

Faced with this fact Schreiner claimed that the population losses modified the falling supply of silver and thus the price rises so much that the result was rising corn prices. With the knowledge that existed in the 1950s of the general demographic trend after 1350, one would expect that a rise in corn prices would have been maintained at least until about 1400. But statistical information about the development of the price of corn shows otherwise. Schreiner knew about this, but apparently not about the demographic trend. For he persisted in saying that after 1369 "as population increased once more the consequences of the lack of silver were increasingly felt".[55]

The American W. C. Robinson did not think either, as suggested before, "that population was levelling off before the Black Death".[56] Robinson's view of the population development came from a clear misinterpretation of Russell, who, as we recall, argued that "an S curve would illustrate the development [from 1086 to 1348] better than a compound interest curve". In a response to Robinson's article, Postan pointed out, indirectly, but correctly, that Robinson distorted Russell's choice between the two hypotheses that had been drawn up.[57]

Apart from this misinterpretation, Robinson justified his view of population development exclusively in terms of the following theoretical paradigm: in the first place he thought that "births and deaths are related in

192

a primitive pre-modern society; the more births, the more deaths..."; secondly, he thought that fertility in stable social and political conditions was
dependent on and could be limited by "institutional checks" and "periodic
catastrophe". With these demographic regulation factors Robinson was
dissociating himself from Postan's Malthusian arguments: first, because the
so-called "institutional checks" were seen as a normal, factually existing
social birth control; secondly because "periodic catastrophes" were not
seen as directly derived from social conditions.[58] "We argue that they
were in every way fortuitous "acts of God" uncontrollable by men, given
the state of science and the social and economic organization of the
time".[59]

The continued slight rise in population before 1350 did not result in
overpopulation, because of society's internal, normal "institutional checks"
on the birth rate. And after this period the population fell because of the
extra-societal, abnormal "periodic catastrophes", primarily the Black Death
and the epidemics that followed. So Robinson attributed a certain amount
of importance to the Black Death, the Hundred Years' War and the
internecine strife of the fourteenth and fifteenth centuries; but the decisive
factor, especially of course for the social and economic changes before
1350, was "changes in money supply" and the price fluctuations derived
from them.[60]

Postan's objections to the monetary theory can be summed up in two
main points: first, he could not understand why the price movements in
the fourteenth century were not uniform, when they were traced to the
same cause - the rising price of silver due to its failing supply; secondly,
he doubted the claim for a diminishing money supply in the late middle
ages.[61] Schreiner and Robinson agreed in principle when countering the
first objection. They thought that price formation depended on the
relationship between supply and demand of the individual commodities,
but that the elasticity of prices varied for the different products - so the
movements of the silver price must also have had different effects on
different goods.[62] However, their arguments against Postan's second
objection differed somewhat.

Schreiner restricted himself to conceding that the volume of money
developed in the thirteenth century, and that England must have possessed considerable reserves of silver around 1300. He therefore found it
logical, as mentioned before, that the drop in the price of corn came
some time after the "technological obstacle" in the silver mining industry.
Robinson, on the other hand, insisted on a formal logical explanation of

193

price development, still assuming that the money supply and rate of circulation were restricted, while price development and the physical volume of commodity exchanges were in principle unlimited.[63]

This explanation establishes neither a real logical nor an empirical probability that the means of exchange decreased in the fourteenth century. In principle, Robinson stuck to the points that towards the end of the thirteenth century the silver mines yielded less; that the price rises of the same century led to an accumulation of the means of exchange; and that exchanges in this and the following century could only be mediated by means of payment in precious metals, so the volume of these in circulation was considered to determine almost totally the rate of circulation.

Unlike Schreiner, Robinson did not argue that the supply of silver took a direct drop in the fourteenth and fifteenth centuries; but that the supply of silver and the volume of money had reached their extreme limit in the fourteenth century. And a stagnating supply of the means of payment in combination with a rising population and a general development of the volume of trade "is the explanation for whatever downward price movements took place. Population increase may thus have actually caused price decreases ultimately, not increases".[64]

It should be obvious that this interpretation could not provide a general explanation of the development of corn prices in the late middle ages, since it conflicted with the available empirical knowledge of both population development and the fluctuations in corn prices. Like Schreiner's monetary explanation, it clashes with the empirical facts, at least when we look at the period after 1369, when population declined concurrently with a drop in corn prices.

VI.8 Summing-up

Both the "neoclassicists" and "neo-Malthusians" related the causes of the social and economic change that was seen as a crisis situation to phenomena whose emergence dated back to long before the Black Death hit England. For this reason, the discussion between them concentrated on the preceding half-century.

The "neoclassicists" had abandoned the simple reflections on the connection between supply/demand relations and price formation which had been prevalent among the earliest representatives of this school of thought. This was in the first place because their picture of price development was not uniform, as the price of various commodities fluctuated irregularly. Thus the prices of different agricultural products - especially when one compares plant and animal products - did not develop in the same way, and there was also a striking difference in price development between agricultural goods and those from the urban industries. Secondly, they were not convinced by the evidence for a declining population before 1350 available to demographic research.

It therefore seemed natural to follow the line of neoclassical theory which had developed earlier in Continental research on the economic history of the late middle ages, but which had not previously gained a foothold in English research (but see Chapter III, pp. 79-80 and Chapter IV, pp. 110, 117, note 71). The monetary factor was incorporated, and was thought capable of explaining the price fluctuations of the fourteenth and fifteenth centuries in the light of a declining supply of silver. At the same time it was claimed that there was a direct connection between price development and the socioeconomic changes that typified the period.

Apart from the above-mentioned clash with the empirical facts for the period after 1369, the most common objection that can be raised to these monetary variants of the neoclassical school has to do with whether price development had any crucial importance at all for social and economic development in the late middle ages. Postan did not regard price developments as the cause of either crisis or fundamental changes in a society where the great majority of the producers only had a marginal connection with the market.

He acknowledged that demesne production, and first and foremost the great landlords, were under the influence of market and price developments. But "peasant producers were far too self-sufficient to order their production and to regulate their marketable surpluses in direct response to commercial considerations; and in the later Middle Ages this element of self-sufficiency increased rather than diminished, or, what is the same thing, the proportion of agricultural producers sensitive to prices was smaller than before. There is thus even less reason for seeking in grain prices the main cause for the falling trends of the later Middle Ages than there is for seeking in them an explanation of the rising trends in the earlier centuries".[65]

The decline of the manorial system, according to Postan, entailed on the one hand commutation and a falling fixed rent pressure, and on the other that the number of *famuli* who were an integral, important part of quasi-capitalist demesne production, diminished. Purely theoretically, it is conceivable that the falling rent pressure after 1350 resulted in a contraction of the market, and that "the golden age of the English labourers" came because higher percentages of society's agricultural produce were withdrawn from the market for immediate consumption. But it is hard to see why more people should have been made more independent of the market, since commutation and the increasing spread of leaseholding must logically have required sale on a market. It is further doubtful whether the thirteenth century's *famuli* were in fact dependent on a market and can thus be considered as a species of medieval wage labourers. There is much to suggest that their wages mainly consisted of various payments in kind.

The neo-Malthusian theory is predicated on the theoretical premiss that around 1300 England was overpopulated, given its stage of technological development. The following two empirical minimum requirements must therefore have been fulfilled if the theory is to be accepted as a possible explanation of the late medieval crisis. There must be documentation, in the first place, establishing the probability that the population figures before the Black Death really did fall continuously; and secondly, this fall had to be connected with a serious decline in agricultural productivity, or the exhaustion of the natural agricultural resources.

It should be clear that no such probability could be firmly established at the end of the 1950s - not even though Postan, in his response to Robinson, referring to Beveridge's and Bennett's studies of the Winchester yields, pointed out that the stagnating grain yield from the lands of this manorial complex, considering the attempts that must have been made to improve the yield, suggested a relative decline in yields.[66]

Demographic research's problems with the more accurate determination of population development in the first half of the fourteenth century,[67] and the neoclassicists' and neo-Malthusians' attacks on one another led to a lack of clarity as to the causes of the changes in England's agrarian socioeconomic structures, which were now regarded by most scholars as crisis conditions. This lack of clarity was precisely expressed in May McKisack's brief but excellent 1959 description of the process of upheaval between 1307 and 1399. Although her account is influenced by the neo-Malthusian theory; and although she directly relates the possible overpopulation before 1350 to the fact that "the age of high farming was well over", she still does not think that research in economic history has yet been able "to offer any fuller satisfactory explanation of these pheno-mena".[68]

McKisack was right in her assessment, inasmuch as Postan at this point had got no further with his arguments than pointing out that the demo-graphic checks were connected with technological problems in English agricultural production around 1300; and inasmuch as Maurice Dobb had only given a theoretical outline of another approach to the understanding of the decline of the manorial system, which also included the assumption that there had been a falling or stagnating population development before 1350.

Dobb's theoretical deliberations were related, as we recall, to both Postan's and Kosminsky's early works, and we have called Dobb's work a "neo-Marxist" synthesis. In the following we will examine in more detail how this theory developed in relation to the other principal theoretical directions, in order, among other things, to judge whether McKisack was right in her assessment of the research front immediately before 1960.[69]

VI.9   The "neo-Marxist" theory

The neo-Marxist contributions to the debate on the social and economic crisis of the late middle ages in England were few. They came from Kos-minsky and R. H. Hilton, whose efforts from about 1950 until 1960 took the form of a discussion of some social and economic consequences of the late medieval upheavals - determining their nature and causes.

Both Kosminsky and Hilton accepted the assumption that England was overpopulated around 1300. But, like Dobb, they thought that this was a relative overpopulation - one that could not simply be explained by Pos-tan's pointing-out that the rising population in the thirteenth century gradually reached its reproductive limit, because agricultural technology stagnated, or at least could no longer keep up with the rising population; nor by the idea that the natural resources, given existing technology, were exhausted. Hilton and Kosminsky claimed like Dobb that the overpopu-lation problem had to be understood in the light of an increasing exploi-tation of the peasants which took place under the influence of population growth and the increasing unproductive consumption of the ruling class.[70]

Behind the growing exploitation of the primary producers was an ob-struction to productivity due to the fact that the landlords' investments in productivity-promoting technology were too small compared with their

197

growing unproductive consumption. Despite colonization, better fertilization of the soil and various other productivity measures in the manorial
system, the rising unproductive consumption was "largely the fruit of increased exploitation and oppression as well as of rationalised estate administration. There were no basic improvements in the techniques of
production".[71] Contradictions grew up between productivity and unproductive consumption which were seen by Hilton as "les causes primitives
de la crise de la société feudale".[72]

The overpopulation was a sign of the lords' exhaustion of the peasant
economy. This threat to the physical existence of the peasants led, according to Hilton, to malnutrition, which in turn was followed by an increased
sensitivity to the influence of climatic and epidemic factors on the reproductive capacity of the population. Seen in this light, "the Black Death
...was not a visitation from heaven but the consequence of oppression and
poverty".[73] "Probably, even given the level of productive forces then prevailing, England could have easily supported a much larger population, if
the feudal lords, the feudal church and the feudal state had not sucked
the labouring classes dry..."[74]

If we are to believe Hilton, the growing exploitation of the primary producers throughout the thirteenth and fourteenth centuries resulted on the
one hand in a halt to the development in population even before the
Black Death, and on the other in increasing social unrest and conflicts
between lord and peasant, culminating in the Peasants' Revolt of 1381. By
contrast, Kosminsky doubted that the population decline could have begun
so early, because "we have no reliable evidence on the chronology of that
decline, when it began and when it ended, how strong and how significant
it was".[75] He concluded that the significance of the demographic factor
had been exaggerated,[76] but did not deny that it may have had some
influence, perhaps because his criticism of the documentation of the
postulated population decline before 1350 was restricted to a number of
objections to Postan's version of the development of real wages for
agricultural labourers on the Winchester estates and his use of this as a
demographic indicator.[77]

The central issue for Kosminsky, though, was that the socioeconomic upheavals of the late middle ages could be explained in another way - the
way Hilton had pointed out (see above). The increasing exploitation
according to Kosminsky led perhaps to an incipient population decline
before 1350, certainly to increased social conflict, and for him this was

198

the essential explanatory factor. For the peasants did not simply passively accept the increasing exploitation.

"English peasant communities in the thirteenth and fourteenth centuries had been capable of great resistance, even successful resistance, to attacks on their conditions by the landlords".[78] Because of "the fact that the peasants were organized in an organic community and that they were in effective possession of their own means of subsistence",[79] Hilton saw the medieval peasant as relatively resistant to overexploitation; but this did not mean that rents and thus the lord's appropriation of surplus labour were fixed. "It was in fact a shifting compromise between peasant resistance based on the mutual solidarity produced by common interests and a common routine of agriculture on the one hand, and the lord's claims on the other, more or less urgent as they might be, and backed up by more or less political and military power".[80]

And yet this explanatory factor is only very tentatively and theoretically developed in Kosminsky's and Hilton's works from the 1950s. The resistance of the peasants "found its clearest expression in the revolt of Wat Tyler",[81] and this "great revolt...was at once a symptom and a cause of the collapse of a decaying order of society..."[82] So the demands of the peasants in the course of the fifteenth century "could be fulfilled only because they corresponded to the whole course of social development of feudal England".[83] "The revolt was no sudden or chance uprising but the outcome of a number of complicated and inter-related antagonisms".[84]

The resistance of the peasants to increasing exploitation and the dominant relations of production, which Hilton had taken upon himself to demonstrate had gathered strength in the course of the thirteenth century in the article "Peasant Movements in England before 1381"; and which saw a temporary culmination in the Revolt of 1381, was thus seen as at once a cause and a consequence of the breakdown of the feudal order. The protests of the peasants against being legally bound under the yoke of villeinage, their refusals to pay the higher rents and their opportunities of making the labour rent in particular unprofitable by deliberate wastage or direct sabotage provoked a situation where "serf production had begun to hamper the productive forces of medieval society".[85]

This resistance hastened the "commutation process", led to the parcelling-out of the demesne and created a class of "kulak peasants" who increasingly became a competitive factor in ailing manorial production. "By lowering prices and by raising the wages paid to peasants, the well-to-do peasants could make demesne economy unprofitable for the feudal lords".[86]

Interpreting and formulating the problem in sharp contours, one can say with Kosminsky and Hilton that demesne production was broken down from within by the obstruction of the work process by peasants with labour service obligations, by the resistance of the bigger farmers with money rent obligations to the payment of higher rents - or any rent at all - and by their rebellion against all measures which restricted their opportunities for accumulation, and in particular restrictions imposed on them as regards the consolidation and extension of their holdings.[87] From without, it was broken down to the extent that this resistance created conditions favourable to the growth of a class of big farmers whose market-oriented production and use of wage labour placed manorial production in a situation of direct competition which helped to weaken it.

The competition was intensified by the Black Death; and the subsequent labour legislation aggravated the above-mentioned social conflict between peasants and landlords. The Statutes of Labourers united and strengthened what had otherwise become a highly heterogeneous class of primary producers against the estate-holding class. This was not so much because all the parties were opponents of the attempts of the state and the landlords to freeze wage development - this must have aroused most resistance from groups that were more or less dependent on wage incomes, and some sympathy from the bigger farmers who used wage labour. What united them was the struggle against villeinage and the attempts of labour legislation to hamper the mobility of labour.[88]

For the bigger farmers, the right to import outside labour to their farms was of vital importance. For the more or less wage-dependent smaller farmers their personal free mobility was of crucial importance to the improvement of their conditions of life through higher wages and/or better terms of tenancy. The Statutes of Labourers were a manifestation of the attempt of the landlords to block these possibilities by means of state power. In this sense Kosminsky and Hilton, like Rogers, saw the 1381 Revolt partly as the result of the reaction of the landlords.

VI.10 Rising rents in the thirteenth century

We have seen that it was a central postulate of the neo-Marxist theory of the late medieval crisis that exploitation of the primary producers increased in the century before the Black Death; and that growing exploitation was registrable as a rising rent burden. There are therefore good grounds to examine whether this claim could be documented in the 1950s.

Two fundamentally different types of documentation had been presented: one direct, and one indirect. The direct documentation was based on manorial rolls, surveys and similar types of source, insofar as at least two of these types of sources could be found for a period and comparisons could be made of rent development in them. The indirect method was based on the interpretation of various legal sources and other accounts of cases concerning rent obligations and social and economic obligations in general.

Hilton and Kosminsky mainly used the latter, indirect method; but we have encountered the direct method in connection with Postan's claim that the expansion of the manorial system in the thirteenth century not only led to an extension, but also to an intensification of the labour rent.

Among Postan's sources for his view were two detailed studies which have been considered before in our survey: Maitland's 1894 study of the history of the Cambridgeshire manor of Wilburton, and Feiling's 1911 analysis of the Essex manor of Hutton.

For Maitland, comparing two surveys from the fourteenth century, it looked as though "in 1277 the bishop was exacting from the Wilburton tenants a greater amount of "week work" than he exacted in 1221"; and, further, that the same tendency could be traced in extents from other estates of the Bishop of Ely - Lyndon, Stretham and Thriplow. So on the face of it, it seems that the volume of labour rent per tenant was in fact rising. But was this rise necessarily a sign of an overall rising rent burden?

It emerges from Maitland's description of these surveys that the money dues from customary tenants and cottagers were fixed between 1221 and 1277. The only possible sign of a rising rent burden over and above those already mentioned was that "freeholders'" payments of heriots were mentioned in 1277 but not in 1221.[89] Considering the general development of prices in the period, this means, all other things being equal, that the money rent must have been dropping relatively.

Whether this fall was wholly or partially compensated by the above-mentioned expansion of "week work" depends on several things - among others, whether the relative strengthening of the labour rent was not an advantage for the peasants in a period of rising population and perhaps expanding households, which may have made it less of a burden to divert labour to the cultivation of the lords' land.

Nor can the rise in both money and labour rents that Feiling thought he could prove by comparing a rental from 1283 with two more elaborate extents from 1312 necessarily be taken as proof of a rising rent burden. For the comparison shows considerable expansion of the peasant's fields, so that several peasants in 1312 had two or more holdings of the size typical in 1283. We must therefore suppose that the allegedly rising overall rent, in view of this and the general rise in prices in the period, was at least not an equal burden to all the customary tenants on the Hutton estate.[90]

Vinogradoff had already found examples of peasants complaining over the imposition of labour obligations with which they had not previously been burdened because of their freeman status. There appear to be signs in these examples of an expansion of the labour rent, although Vinogradoff would not exclude the possibility that the complaints might have been exaggerated.[91] But in themselves they do not tell us that the overall rent burden had become heavier.

This sort or reservation far from affected Hilton's interpretation of a large body of legal cases taken from various types of source. He took the peasants' complaints over increased and new labour obligations at their face value, and saw them as expressions of the peasants' individual and sometimes collective resistance to growing labour dues. Since these cases were rarely decided in favour of the peasants, and since Hilton also claimed with no further documentation that the money rent rose, he could conclude that the overall rent burden must have been rising.[92]

It is probably not surprising, as Hilton himself remarks, that the resistance of the peasants seems to have increased from the beginning of the thirteenth century on, since it is precisely from this period on that sources become plentiful. On the other hand, he also sees a connection between the increasing documentation and the intensification of exploitation.[93] Like Postan, he thought that the documentation of the peasants' obligations was connected with better definitions of them, making the terms on which they held their land stricter than before, when obligations had been more diffuse, depending on custom.[94]

It is probably indisputable that labour obligations grew with the advance of the manorial system in the thirteenth century, whether it was a matter of imposing such burdens on formerly "free" peasants or increasing the labour burdens to which they were already subjected. But the many complaints over labour dues, and in particular the many cases of peasants refusing to work[95] perhaps tell us as much about the problems confronting the landlords in their attempts to develop and consolidate manorial cultivation.

In reality, they perhaps say just as much about the reasons why "demesne production for the market depending on the labour of unfree villeins began to reveal itself". So "by the second half of the thirteenth century the practice of "venditio operum" was becoming more widespread".[96] Another reason may have been that not every expansion of the labour rent meant a net profit for the landlord.

VI.11 Increasing exploitation in the thirteenth century

It was thus, as we recall (see Chapter IV p. 109), not an uncommon practice on several large manorial complexes across the country that the peasants, when rendering the so-called "boon-days" at the time of the most labour-intensive summer and autumn jobs on the estates, were either fed at work or could take home various goods. Vinogradoff concluded that such labour obligations gradually became unprofitable because of price developments, and that it was therefore "no wonder that such "boon-work" has to be given up or to be commuted for money".[97] Seen in this light, this kind of labour service may have had a directly supportive effect on the reproductive capacity of the peasants.

Even if we accept that both the labour and the money rent rose in the course of the thirteenth century, this does not necessarily mean that the reproductive conditions of the peasants deteriorated or that exploitation of them increased. A formal, perhaps measurable, rise in rents is not in itself the same as a real increase in the exploitation of the peasants. It has been mentioned that the rising population and the subsequent rise in the number of household members on the individual farm, all other things being equal, could have compensated for the burden of the labour services imposed. We also saw above how labour services could be associated with counter-payments from the lord which directly strengthened the reproduction of the peasants, especially in a period of rising prices and population. In addition, the increase in money payments, the documentation of which we found to be doubtful,[98] must have been counteracted by the upward development in prices. The latter point is of great importance in a general assessment of the rent burden on the peasants, if one inclines towards Kosminsky's reading of the relative spread of the feudal rent forms in thirteenth-century England.

For it emerges from this, as we have seen, that product rent was negligible all over the country except in the northernmost regions, whereas money rent is considered to have been generally dominant, even though it is not disputed that the labour rent was gaining ground. Thus in the eastern areas, where Kosminsky found the labour rent to be most widespread, it accounted for 39% of total rent earnings; it dropped to 23-24% in the central and southern counties and dropped further towards the west, while it was almost negligibly small in Yorkshire, Northumbria and Kent.[99] The necessary relations of the peasants with the market thus seem in general to have been very strong - so strong that a calculation based on 32 Winchester estates shows that on these typically market-oriented estates, "on the whole the peasants...put more produce on the market than came from the demesne".[100]

If we accept this calculation, the rent burden on the peasants, and thus on their level of reproduction, must have been extremely sensitive to market fluctuations; both because the rising corn prices progressively devalued the money dues, thus bolstering up the peasants' reproductive capacity; and inasmuch as this devaluation, and the interest of the landlords in obtaining increasing percentages of the market fluctuation gains, triggered off a continuing struggle between lord and peasant over the form and size of the rent.

This is not the place to determine who benefited from this struggle. But given the available evidence, it seems doubtful whether one can confirm Kosminsky's and Hilton's far too firm view that this conflict resulted in great impoverishment of the English peasants up to and in the fourteenth century; even though one of the results of the struggle was perhaps that manorial production was increasingly the job of more or less landless agricultural wage-labourers whose wages exhibited a declining tendency.

For there was some disagreement on the importance of these *famuli* for manorial production around 1300. Kosminsky and Hilton, who acknowledged their existence (and Kosminsky in particular related their significance to the incipient commutation process in the thirteenth century)[101] do not seem to have assigned the same importance for the feudal manorial system to these "free labourers" as Postan did (cf. note 49).[102]

Moreover, the *famulus* of the thirteenth century cannot be directly compared with the wage-labourer under capitalism. In the first place, he was not entirely "free" of possessing the means of subsistence. Thus some of the *famuli* on the manors had land in "base serjeanty" and were paid for their work - in this way and others, inasmuch as certain jobs, for example, the ploughing of their allotments, were taken care of by the lord. In the second place neither these nor the completely landless labourers were free in the sense that they could look for work anywhere they liked. Thirdly, they received very substantial portions of their wages in the form of subsistence goods.[103]

It is therefore difficult to concur in the classic assumption that the conditions of the agricultural labourers deteriorated as a result of falling money wages and rising corn prices in the thirteenth century.[104] Nor is there much to suggest that the population development and the contraction of colonization necessarily had a negative effect on the reproductive conditions of the rural labourers.

Postan distinguished between two classes of rural workers in the twelfth and thirteenth centuries: the "famulus in serjeanty" or agricultural labourer with the disposal of a small plot of land; and the "stipendiary famuli" or absolutely landless labourers. His thesis was that the first class was typical of a situation with expanding manorial production and extensive colonization, while the second grew numerous at the expense of the first as colonization reached its limits.[105] If one accepts this thesis, one could perhaps be tempted to conclude that the conditions of the agricultural labourers deteriorated in the course of the thirteenth century; but this need not necessarily have been the case. For Postan documented how the payment of subsistence goods was inversely proportional to the ownership of land in this century. "It is in this respect that holders in serjeanty may have differed most from stipendiary labourers..." The less land an agricultural labourer owned, the more payments he received in subsistence goods, and vice versa.[106]

VI.12 The crisis of feudalism

The nature of the social and economic changes that took place in English society in the fourteenth and fifteenth centuries justifies us in speaking, along with Kosminsky and Hilton, of "une époque de crise de la société anglaise"[107] or of "a crisis of feudalism".[108] The crisis led not only to upheavals in the agrarian structures of production, but also to constitutional and political turbulence manifested as "une guerre sans succès, une banquerotte nationale et finalement, la guerre civile".[109] Hilton's and Kosminsky's definition of crisis was thus not only concerned with the breakdown of the manorial system. It was not only the crisis of the manorial system.

Hilton described the crisis both as the reproductive crisis of the feudal social system and as the crisis of feudal society, since he saw the crisis of the social system in relation to the growing problems of reproducing the population. For Kosminsky on the other hand, the crisis was first and foremost a crisis of the social system. They agreed, as we saw in an earlier section, on the most general causes of the crisis. They both sought the cause of the crisis in the generation of crisis symptoms that they saw as derived from the antagonisms in the agrarian relations of production, when they reached their zenith in the thirteenth century. They saw the crisis as something derived from factors endogenous in the basic relations of production of the feudal social system.

In general, the crisis was seen as a national phenomenon, but Kosminsky thought that it was most marked in the eastern areas of the country, where the manorial system and the labour rent were predominant. This followed from his view of the nature of the changes in the relations of production that constituted the crisis, and his assessment of the national spread of the manorial system. For there were slight differences between Kosminksy's and Hilton's views of the extent to which the crisis involved a true rupture in the feudal agrarian relations of production.

Kosminsky thought that the crisis was primarily "reflected in a fall in the level and extent of feudal rent, and in the establishment of the dominance of its money form, which is already in decomposition". Commutation, when more or less accomplished, led only to a reorganization and shaking-up of the feudal relations of production. "Feudal landed property and the right to feudal rent remained in the hands of the feudal lords, and the liberation of the peasants from feudal dependence was far from complete; consequently it was still possible to increase rents and to evict the peasants when conditions were favourable". And furthermore, "the feudal state was preserved; this made it possible for feudal lords to gain some compensation for their losses in manorial rent by raising the centralised rent". But "the very "victory" of the peasantry bore within itself the seeds of future defeat. The differentiation of the peasantry was gathering momentum. The petty bourgeois economy of peasant commodity producers was a bridge to the capitalist economy..."[110]

206

Precisely "the stratification of the peasantry was one of the most important developments in the English countryside in the fourteenth and fifteenth century", wrote Hilton.[111] While Kosminsky emphasized the commutation of the feudal labour rent as a crucial factor in all his studies, Hilton had directed his attention ever since his earliest works - for example, the above-quoted one from 1947 - to the transmutation of the feudal money rent to the lease, and of the medieval village community to a polarized social and economic stratification of the primary producers. When, for example, the peasants were successfully driven from their land in the fifteenth century, this should not according to Hilton be seen as a sign that "feudal landed property...remained in the hands of the lords", but rather in the light of the fact that this polarization had destroyed the resistive force that the village community had constituted.

This difference between Hilton and Kosminsky means that in the former's assessment of the outcome of the crisis there was stronger emphasis on the signs of the negation of the feudal relations of production - although Kosminsky also saw this. Although Hilton, like Postan, thought that market-oriented agricultural production, with the fall of the manorial system, was restricted in the fourteenth century,[112] and although on the whole he assigned the market economy less importance than Kosminsky for the generation of the crisis,[113] his general view was that ""farm" or leasehold rents, competitive and adjustable according to market conditions, as distinct from assize rents",[114] took on increasing importance during the crisis, as one of its causes, and as a result of it.

The development of leasehold, which should be seen in the context of the progressive abandonment of the manorial system and the devaluation of assize rents which the rising corn prices must have entailed, can justifiably be regarded as a qualitative break with the feudal relations of production, even if leasehold had in fact existed as far back as the high middle ages. This form of holding changed the relationship between lord and peasant into a more or less purely economic relationship. The customary fixing of the nature and extent of ownership and payment was replaced by contractual provisions whose variability was adapted to market fluctuations.

Moreover, leasehold was accompanied by a break with the customarily determined village structures, which prevailed as long as the feudal relations of production were maintained. Thus leasehold not only created a general situation of competition between landlord and peasant over the fixing of contractual obligations, but also led to competition among the peasants of the village; and this prepared the way for increased social and economic inequality among them.

In his study of certain manors in Leicestershire belonging to Leicester Abbey and Owston Abbey, Hilton concluded that this development was so advanced that "leasehold rents...formed the greater part of the rent income" as early as 1348, and in the fourteenth century he found "a definitely landless class" of agricultural labourers.[115] This polarization was reflected in the Mile End programme from 1381. The peasant of the rich and expanding class was "as revolutionary a figure as the poor or landless peasant oppressed by the statute of labourers..."[116] Hilton saw two classes in embryo around 1400 - capitalist petty bourgeois peasants and a rural proletariat -[117] and the breakdown of the village community.

Well into the 1950s, the fifteenth century has to be regarded as a lacuna in research on England's economic history. The main reason for this, and the reason why this century may always remain more obscure than the fourteenth and especially the thirteenth, is that the fall of the manorial system spread a blanket of darkness over the rich opportunities we have of studying the earlier centuries. The leasing out of the demesne meant that production fell into the hands of people who were not very good at documenting the results of their efforts. Therefore we know very little, for example, about the significance of the cessation of manorial production in terms of agricultural technology - whether it meant progress or a step backward.[118]

Postan and many other historians before him saw the decline of the manorial system as a step backwards. Kosminsky took a diametrically opposite view. He thought in the first place, as we have seen, that the manorial system was flawed with internal antagonisms which hampered the development of productivity; and in the second place that the manorial system also restricted the development of productivity on the peasants' own holdings. With the fall of the manorial system in the fourteenth and fifteenth centuries these restrictions disintegrated: productivity increased and larger volumes of corn were offered on the market.

Kosminsky's documentation for this state of affairs was his reading of the development of corn prices, based on Adam Smith's theory of the value of labour. The drop in corn prices in the fifteenth century indicated rising agricultural productivity because the falling price reflected the fact that the working time invested in the volume of corn offered was falling.[119] Surprisingly enough, Kosminsky did not apply this theory in his assessment of the development of prices in the thirteenth century.

Although Kosminsky consistently spoke of the crisis of the fourteenth and fifteenth centuries, his use of the above-mentioned theory implies an earlier dating of the crisis. The application of the theory to the economic history of the thirteenth century would inevitably have led him to speak of a decline in productivity. And despite the manifestations of crisis - "the collapse of the demesne economy, possibly the slowing down of population growth and even its temporary decline, the political disturbances, the baronial wars and so on, the temporary fall, or at any rate the slowing down of the growth of productive forces..." - he did in fact think that "on the whole this crisis bore a progressive character, since as a result of it a number of obstacles were removed which the feudal order had placed in the way of further development of the productive forces of English society".[120] These reflections prepared the way for a revision of the history of the late medieval centuries.

Notes

1. M. M. Postan, "Moyen Age", Rapports I, *IXe Congrès International Des Sciences Historiques*, Paris 1950, pp. 236, 241.

2. M. M. Postan, "Trade and Industry in the Middle Ages", in *The Cambridge Economic History*, Vol. II, 1952, p. 214.

3. Ibid., p. 216.

4. Axel Steensberg, "Archaeological Dating of the Climatic Changes in North Europe about A.D. 1300", in *Natura*, Vol. 168, 1951.

5. Postan, "Moyen Age", op. cit., pp. 234-35.

6. Bloch's variant of the monetary theory, briefly, says that the overall European balance of trade with Near and Far Eastern trading partners in the period from the last centuries of the Roman Empire had drained Europe's gold reserves. Marc Bloch, "Le Problème de l'Or au Moyen Age", in *Les Annales d'Histoire Économique et Sociale*, 1933. This situation may have been reversed in the late medieval centuries, so that gold now flowed in the opposite direction, with inflationary consequences for European price development. Cf. M. Lombard, "L'Or Musulman du VIIe au XIe Siècle", in *Annales*, No. 2, 1947.

7. Wilhelm Abel, *Die Wüstungen des Ausgehenden Mittelalters*, 1943, pp. 1-13, 25-30; cf. Postan, "Trade and Industry...", op. cit., p. 214. In this work, Abel did not directly relate the desertion problem before 1350 to a general population decline, and in fact expressed himself very vaguely on such a possibility. Thus, for example, "die Landbevölkerung Deutschlands nahm im ausgehenden Mittelalter, zumindest zwischen der Mitte des 14. und der Mitte des 15. Jahrhunderts ab". Abel, op. cit., quoted here from the third edition, 1976, p. 42. Schreiner wrote of the chronology of the desertion problem that the first cases of deserted farms known from Denmark are from 1340. Yet it was only after the plague year that the situation became more serious; and that in Swedish legislation the problem of the deserted farms appeared for the first time in 1437. The situation in Norway also suggested that the desertion problem only became really serious in the course of the 15th and 16th centuries. But he believed that he could see that the desertion problem had begun long before the Black Death "from the large collection of royal decrees whose purpose was to safeguard agriculture against labour scarcity". But it is important to emphasize that the explicit background given in these royal decrees was the high migration of labour from agriculture to other industries, especially urban ones. But "we lack the means to determine how great a role this in fact played, but can hardly deny the view all value. The tendencies that can be demonstrated elsewhere existed on Norwegian soil too" - according to Kosminsky, in England, cf. note 13 below. At any rate it is clear that Schreiner did not directly associate the beginning of the drop in the volume of labour in agriculture in Norway before 1350 with the beginning of a general population decline. He does however attribute to the famine years in the second decade of the fourteenth century a certain negative influence on population development. Johann Schreiner, *Pest og Prisfall i Senmiddelalderen*, Oslo 1948, pp. 58-64, 89-93.

8. H. J. Habakkuk, "The Economic History of Modern Britain", in *The Journal of Economic History*, Vol. XVIII, 1959, p. 487.

9. M. M. Postan, "Some Economic Evidence of Declining Population in the Later Middle Ages", in *The Economic History Review*, 2nd Series, Vol. II, 1950.

10. W. Beveridge, "Wages in the Winchester Manors", in *The Economic History Review*, Vol. VII, 1936.

11. Postan, "Some Economic...", op. cit., p. 229.

12. Ibid., pp. 226-27.

13. Ibid., pp. 230, 233. Thus Postan rejects Kosminsky's criticism of the Cambridge historians' view of crisis and his claim that they neglected the extent of the new growth outside the agricultural sector - "the cloth industry, the expanding towns, the rising number of free peasantry and proletariat" - and that the rising agricultural wages might be due precisely to this expansion "outside the feudal villages". E. A. Kosminsky, "Problems of an English Agrarian History in the XVth Century", in *Voprosi Istorii* No. 3, 1948 - quoted here from Postan, "Moyen Age", op. cit. On the other hand, Postan's point was supported by E. Carus-Wilson's view of the wool industry's development in the fourteenth century. In the first place the crisis or transformation which she saw in the industry from the beginning of this century was a "shift in the location of the industry"; and in the second place this was "due very largely to a technical revolution in one of its main branches as a result of the introduction of the water fulling mill". Because of better access to powerful enough watercourses, textile manufacturing spread to the rural districts. The crisis in the wool industry was therefore associated with the old wool manufacturing centres. And the technical innovation in wool manufacturing represented by the fulling mills "greatly reduced the labour force required". Against this background the new wool manufactures in the rural districts can hardly have had a stimulating effect on rural wage levels. E. Carus-Wilson, "The Woollen Industry", in *The Cambridge Economic History*, Vol. II, 1952, pp. 409-11.

14. Thus an indexed average wage development for "Daily Wages of Artisans" between 1300 and 1349 exhibits only a rise from 100 to 105, while the rise in the next half-century, after a sharp rise in the years just after the Black Death, is from 138 to 150. Postan, "Some Economic...", op. cit., Table II, p. 233. Similarly, according to Beveridge, the wage in silver pence for agricultural labourers on the Winchester manors in the decades before 1348 did not rise steeply compared with the subsequent period. Ibid., Table I, p. 226.

15. Ibid., pp. 237-38.

16. Ibid., pp. 239-40.

17. Ibid., p. 245.

18. M. M. Postan & J. Titow, "Heriots and Prices on Winchester Manors", in *The Economic History Review*, 2nd Series, Vol. XI, 1959.

19. Death duties in the thirteenth century, imposed on villein holdings, originally a feudal service consisting of weapons, horses or other military equipment, restored to a lord on the death of his tenant.

20. Ibid., pp. 393, 399.

21. Ibid., pp. 397-98.

22. Ibid., pp. 398-99.

23. Ibid., p. 401.

24. Ibid., p. 402.

25. Ibid., pp. 408-09.

26. Ibid., p. 410.

27. See note 22.

28. A. R. Lewis, "The Closing of the Medieval Frontier 1250-1350", in *Speculum*, Vol. 33, No. 4, Cambridge, Mass. 1958.

29. W. G. Hoskins, "The Making of the English Landscape", 1955, p. 92.

30. W. G. Hoskins, "The English Landscape", in A. L. Poole (ed.), *Medieval England*, Vol. I, Oxford, 1958, p. 23.

31. R. H. Hilton. "Mediæval Agrarian History", *V. C. H. Leicestershire*, II, London 1954, p. 166.

32. Robert Trow-Smith, *A History of British Husbandry to 1700*, London 1957, pp. 99-100.

33. W. G. Hoskins, "The Deserted Villages of Leicestershire", in *Transactions of Leicestershire Archaeological Society for 1944-45*, XXII, 1946, pp. 241-64.

34. C. W. Foster & T. Longley, "Lincolnshire Domesday", in *Lincolnshire Record Society*, XIX, 1924.

35. Maurice Beresford, *The Lost Villages of England*, London 1954.

36. Maurice Beresford & John G. Hurst (ed.), *Deserted Medieval Villages*, London 1971, pp. 30-34, 213-26.

37. Beresford, *Lost Villages...*, op. cit., pp. 151-54.

38. Ibid., pp. 137-38.

39. For the Wapentake of Ainsty, a small area west of York, it can be seen from a comparison of Domesday, the so-called *Nomina Villarum* drawn up by the Exchequer in 1316, and fiscal rolls from 1377, that between 1086 and 1377 only two new villages grew up in the area. In the same period five disappeared - i.e. a net loss of three out of a total of 39 villages. More or less the same trend can be traced in Canon Foster's study of Lincolnshire. Ibid., pp. 155-56. See also Table 16, pp. 404-06. We find more detailed information on this question in the studies of Leicestershire (Hoskins, op. cit., 1946), East Riding (Beresford), Norfolk (K. J. Allison, *The Lost Villages of Norfolk*, 1952), and Warwickshire (Beresford). In Leicestershire 61% of later lost villages were still registered in 1334, 18% were listed as one of a pair treated as one unit in the list, 9% which did not feature in 1334 appeared in *Nomina Villarum*. This means that 12% of the abandoned villages were deserted before 1316, and that 9% vanished between 1316 and 1334. In the East Riding 30% of the lost villages had presumably already been abandoned before 1334. In Norfolk 35 villages were not listed in 1316 out of the 726 to be found in Domesday; and 58 villages later abandoned were still listed in 1334. In Warwickshire one can only trace the desertion of six villages before 1400. Ibid., pp. 157-58. Comparing this information with the proportion of lost to still-registered villages in the valuations of 1344 and 1377 respectively - see Table 17, "Proportion of to-be-lost villages in various size-groups by tax quotas of 1334", ibid., p. 407-08 and Table 18, ditto for 1377, ibid., p. 409, - one must agree with Beresford that "it is crucially important to show that so many to-be-lost villages do answer adsum not only in 1334, but also in 1352-53-54 and again in 1377...". Ibid., p. 158.

40. Beresford & Hurst, *Deserted...*, op. cit., p. 7.

41. Beresford, *Lost Villages...*, op. cit., p. 166. Cf. Beresford & Hurst, *Deserted...*, op. cit., pp. 11, 14.

42. Ibid., pp. 8-10; Beresford, *Lost Villages..*, pp. 158-66.

43. Beresford & Hurst, *Deserted...* op. cit., p. 29.

44. Ibid., pp. 8-17.

45. Ibid., p. 11.

46. Postan, "Moyen Age", op. cit., p. 232.

47. Postan, "Trade and Industry...", op. cit., p. 213.

48. Ibid., pp. 195-97.

49. Postan saw the *famulus* - the wage-dependent, landless agricultural labourer - as a consequence of the population growth of the high middle ages. Perhaps *famuli* existed as early as the eleventh century, but it was not until the thirteenth century that they became an important precondition of the manorial system. M. M. Postan, "The Famulus, the Estate Labourer in the XIIth and XIIIth Centuries", in *The Economic History Review*. Supplement 2, London 1954, pp. 1-2, 5, 28-33, 35-37.

50. Johan Schreiner, "Wages and Prices in England in the Later Middle Ages", in *The Scandinavian Economic History Review*, Vol. II, No. 2, 1954, pp. 63-65.

51. Ibid., p. 70.

52. Ibid., pp. 67-68.

53. Ibid., pp. 72-73.

54. Ibid., p. 69.

55. Ibid., p. 70.

56. W. C. Robinson, "Money, Population and Economic Change in Late Medieval Europe", in *The Economic History Review*, 2nd Series, Vol. XII, 1959, p. 73.

57. Ibid., p. 70; "Note" by M. Postan, ibid., p. 77.

58. "Fertility has two kinds of checks which prevent its reaching its biological maximum: first, institutional checks such as age of marriage, proportion single, and such contraceptive habits as may be part of the culture, including infanticide; second, the check imposed by periodic catastrophes or disastrous interruptions to the usual process of reproduction within a society, including wars, disorders, even plague and famine themselves in so far as they result in separation of families and a breaking of the social patterns which lead to the usual levels of fertility". Ibid., p. 72.

59. Ibid., pp. 72-73.

60. Ibid., p. 76.

61. Postan, "Moyen Age", op. cit., p. 231.

62. Schreiner, "Wages...", op. cit., p. 69; Robinson, "Money...", op. cit., pp. 65-66.

63. Robinson used the so-called Fisher Equation, $MV = PT$ (M = stock of money, V = velocity of circulation, P = general price level, T = output of goods and services). Ibid., p. 66.

64. Ibid., p. 74.

65. Postan, "Trade and Industry...", op. cit., p. 211. See also pp. 166-67. and "Note" by Postan, op. cit., pp. 78-79. It should be noted that Postan presented these views in *The Cambridge Economic History...*, Vol. II, in connection with, and partly as an explanation of, decreasing trade relations in the late middle ages.

66. Ibid., p. 81.

67. In a 1958 work Russell repeated his rather vague assumption that the population levelled out in the first half of the fourteenth century. It will be recalled that in his account of medieval English population development he inclined to a "neo-Malthusian" explanation - among other things firmly rejecting the climatic factor. See Chapter V, p.

160. In assessing the general demographic trend in Northern Europe in the first half of the fourteenth century ten years later, he concluded that the "Malthusian check" which came with the great famine from 1315-1317 "probably...was a case when subsistence was lowered by the worsening climate..." J. C. Russell, *Late Ancient and Medieval Population*, Philadelphia 1958, p. 142.

68. May McKisack, "The Fourteenth Century 1307-1399", Vol. 5 of *The Oxford History of England*, Oxford 1959, p. 329. It is interesting to note how in the disposition of precisely this volume, room was found for an account of "Rural Society", Chapter XI, pp. 312-48, and of "Trade, Industry and Towns", Chapter XII, pp. 349-83, since *The Oxford History of England* must otherwise be emphatically classified as political history - even manifestly divided into periods according to the succession of kings. Nor do we find any chapters devoted to the field of economic history in the preceding Volume 4, by Sir Maurice Powicke, "The Thirteenth Century, 1206-1307", Oxford 1953. In the following one, Volume 6, written by E. F. Jacob, "The Fifteenth Century, 1399-1485", Oxford 1961, two issues of economic history are taken up. Chapter VIII(b) "The Trader and the Countryman", pp. 346-84, and Chapter VIII(c) "The Towns", pp. 385-405. This organization excellently illustrates how the economic history of the late middle ages became the object of increasing interest after the Second World War.

69. McKisack was not unfamiliar with certain of the "neo-Marxist" positions, and certainly not with the technological check Postan saw as an integral part of the "neo-Malthusian" theory.

70. R. H. Hilton, "Y Eut-Il une Crise Générale de la Féodalité?", in *Annales - E.S.C.*, No. 1, 1951, pp. 28-29.

71. R. H. Hilton & H. Fagan, *The English Rising of 1381*, London 1950, p. 22. See also E. A. Kosminsky, "The Evolution of Feudal Rent in England from the XIth to the XVth Centuries" in *Past and Present*, No. 7, 1955, pp. 20-22.

72. Hilton "Y Eut-Il une Crise...", op. cit., p. 29.

73. Hilton & Fagan, op. cit., p. 23.

74. Kosminsky, "The Evolution...", op. cit., p. 22.

75. Ibid., p. 24.

76. Ibid., p. 29.

77. Ibid., p. 28.

78. R. H. Hilton, "A Study of the Pre-History of English Enclosure in the Fifteenth Century", in *Studi in Onore di Armando Sapori*, Vol. I, Milan 1955, p, 685.

79. Although in the thirteenth century there already existed a certain social and economic stratification in the individual village - Hilton & Fagan, op. cit., p. 29 - and although it might be aggravated by the slow fluctuations of rents, the social and economic unity of the village against pressure from the lord did not diminish. R. H. Hilton, "Peasant Movements in England before 1381", in *The Economic History Review*, 2nd Series, Vol. II, 1949, pp. 118, 120-22.

80. Ibid., p. 22.

81. Kosminsky, "The Evolution...", op. cit., p. 24.

82. Hilton & Fagan, op. cit., p. 13.

83. Kosminsky, "The Evolution...", op. cit., p. 26.

84. Hilton & Fagan, op. cit.

85. Kosminsky, "The Evolution...", op. cit.

86. Ibid., p. 25. See also R. H. Hilton, "L'Angleterre économique et sociale des XIVe et XVe siècles - Théories et monographies", in *Annales - E.S.C*, Vol. 13, 1958, p. 548.

87. Hilton, "Peasant Movements..", op. cit., p. 131.

88. Ibid., pp. 132-33.

89. F. W. Maitland, "The History of..", op. cit., pp. 418-19.

90. K. G. Feiling, "An Essex Manor...", op. cit., p. 333.

91. Vinogradoff, "Villainage...", op. cit., p. 204.

92. Hilton, "Peasant Movements..", op. cit., pp. 122-30.

93. Ibid., pp. 122-23.

94. Thus Hilton thought that the judgement was often "the concluding stage of a dispute about increased services", ibid., p. 124; Postan, "The Chronology...", op. cit., pp. 187-88.

95. Hilton, "Peasant Movements...", op. cit., pp. 127-30; A. E. Levett, *Studies in Manorial History*, Oxford 1938, pp. 203-04.

96. Kosminsky, "The Evolution...", op. cit., p. 21.

97. Vinogradoff, "Villainage...", op. cit., p. 175.

98. See Chapter V, p. 149. Kosminsky does not depart from his view that "the labour service increased side by side with money rents" from 1935. Kosminsky, "Services...", op. cit., p. 42. In this respect it is worth noting that "usually it is not the basic element of the money rent, the redditus assisae, which is increased, but such things as tallage and entry fines". But in the 1950s he arrived at a position where the increasing exploitation of the primary producers was mainly related to the spread and strengthening of villeinage and the intensification of the labour rent. E. A. Kosminsky, *Studies in the Agrarian History of England in the Thirteenth Century*, Oxford 1956, pp. 327-28, 345-50.

99. Ibid., pp. 191-93. In this major work, Kosminsky elaborated and consolidated his views on the development and geographical distribution of the feudal rent forms.

100. The list shows that the manors took 48% of the harvest to market, while 44% was used as seed corn, and only 8% for immediate consumption. But these corn sales earned less than the money rents levied on the peasants. Kosminsky concluded that, provided not all the money the peasants obtained via the market was swallowed by rent payments, the peasants' market share was greater than that of the manors. Ibid., pp. 324-25.

101. Ibid., pp. 356-57.

102. For wage labour's general position in the feudal agrarian relations of production, see Hilton & Fagan, op. cit., p. 14. It should be stressed that Postan's view of the increasing importance of wage labour in manorial production up until 1300 was based on a supposition that was empirically rather weakly documented. He found no indications on the Winchester estates of increasing use of wage labour, and the changes in this respect he thought he could observe at the Glastonbury and Peterborough Abbey estates were weak. M. Postan, "The Famulus...", op. cit., p. 28. See also Table I, p. 45 and Tables II and III, p. 46. Postan therefore thought that "it is possible, though by no means certain, that with the changing conditions in the countryside the proportion of wage-labourers on the demesne grew, or that the relative weight of the lord's outlay on wages increased" (ibid.). It is further worth noting in this connection that Postan and Hilton placed wage labour differently in the manors' technical consumption of labour. Postan thought that "operations which required specialized skill" and "operations which demanded continuous application throughout the season and through the working week"

were done by *famuli*; while "operations which moreover were seasonal and otherwise discontinuous" were done by villeins subject to labour service. Ibid., p. 2. Against this, Hilton thought: "The labour on the demesne was mainly supplied by the peasants and their families, who owed a variety of unpaid services...as part of their rent. This was supplemented by the work of permanent hired men, and casual labour at harvest and other peak periods". Hilton & Fagan, op. cit., p. 20.

103. Postan, "The Famulus...", op. cit., pp. 20-27.

104. But it should be mentioned here that Postan found an example where, "propter caristiam hoc anno", the payments to agricultural labourers of subsistence goods were reduced on the Winchester estates. He further related the replacement of payments in kind with money wages on some of the Winchester estates - mainly in the fourteenth century - with the high corn prices. Ibid., p. 37.

105. Ibid., pp. 35-36.

106. Ibid., pp. 22-23.

107. Hilton, "L'Angleterre...", op. cit., p. 562.

108. Kosminsky, "The Evolution..", op. cit., p. 30.

109. Hilton, "L'Angleterre...", op. cit., p. 562.

110. Kosminsky, "The Evolution...", op. cit., p. 30. Cf. Chapter I, p. 24-25.

111. R. H. Hilton, *The Economic Development of some Leicestershire Estates in the 14th and 15th Centuries*, Oxford 1947, pp. 94-95.

112. Hilton & Fagan, op. cit., p. 30.

113. For the importance of trade and commodity production for the increasing pressure on the peasants, see Kosminsky, *Studies in the Agrarian...*, op. cit., pp. 326-28.

114. Hilton, *The Economic Development...*, op. cit., p. 122.

115. Ibid., pp. 122, 147.

116. Hilton & Fagan, op. cit., pp. 31-32.

117. Ibid.

118. R. H. Hilton, "The Content and Sources of English Agrarian History before 1500", in *The Agricultural History Review*, Vol. III, Part I, 1955, pp. 6, 15-16.

119. Kosminsky, "The Evolution...", op. cit., p. 26.

120. Ibid., p. 34.

Chapter VII

Revision and Fusion
(1960-   )

VII.1 Economic Expansion after 1350

Along with Edward Miller one can say that in a work from 1962 Anthony Bridbury "has turned some of the current learning upon its head"[1] - not because Bridbury described the thirteenth century as a period of acute economic crisis, but because he thought he could demonstrate that the fifteenth century was a period of general economic expansion. He believed, along the lines we saw suggested by Kosminsky, that agricultural productivity at the end of the fourteenth century increased greatly, not least because the marginal lands were no longer cultivated and this rise in productivity compensated for the population losses.[2] By contrast, the rising population in the thirteenth century, the increasing scarcity of land and the intensified exploitation of the peasants led to a drop in agricultural productivity.[3] Not only did the conditions of the peasants thus improve between the thirteenth and fifteenth centuries: it is claimed that socioeconomic crisis turned into general economic expansion because the threat of overpopulation was warded off by famine and the ravages of epidemic diseases.[4]

Apart from the familiar theory of the withdrawal from marginal lands, Bridbury produced no direct documentation for this revisionist interpretation. His claim for the increase in agricultural productivity in the fifteenth century was mainly based on indications of manufacturing, commercial and urban developments and on conjectures about the necessary relationship between these non-agrarian activities and the results of agricultural production.

The production of tin in Devon and Cornwall does not seem to have been affected by the population decline,[5] and although wool exports fell at the end of the fourteenth century,[6] the question for Bridbury is whether increasing textile production did not compensate for the drop. At any rate he concludes that both exports of textiles and domestic consumption of cloth rose substantially despite the drop in population. He reaches the inevitable conclusion that overall industrial production per

217

capita rose dramatically after the Black Death.[7] A comparison of taxation in the urban and rural districts in 1334 and 1524 respectively,[8] the observation of constitutional reforms in the cities,[9] and a rise in the percentage of citizens in the cities[10] all suggest to Bridbury that urban developments did not come to a stop in the fifteenth century either.

These factors, taken together with the declining population figures in the countryside, indicate for him that agricultural productivity must have been rising. Thus for Bridbury the fifteenth century was not only "the golden age of the middle classes and the peasantry",[11] nor was it only a prosperous period for English agriculture. It was a century of general economic growth throughout English society - a growth not necessarily of the gross national product, but certainly of the per capita product.

Bridbury's concept of crisis is tied up with reflections on the development of national productivity, especially agricultural productivity. Even though he acknowledges that there had been a population crisis in the first half of the fourteenth century, he firmly dissociates himself from the general interpretation of the late medieval crisis that had been offered by Postan in the 1930s. He does not connect the crisis with an upheaval in the agrarian relations of production or with the development of the national product. At the core of Bridbury's crisis definition lies the per capita product.

Largely the same interpretation can be found from J. R. Lander, who however, unlike Bridbury, also describes the fall of the manorial system as a sign of crisis. Yet this does not mean that the breakdown of the manorial order in itself necessarily meant a general drop in income for the landlords. "The golden age of the peasantry" was not inevitably, as so many historians had imagined, "a nightmare for the landlord".[12]

Lander viewed the fifteenth century, compared with the century before, as a prosperous and happy time, first and foremost because the people-to-land ratio after the Black Death worked to the advantage of the survivors, whether we see the problem from the estate-holders' or the peasants' angle.[13] The deciding factor, according to Lander, was the individual's luck and ability to adapt to the changed economic circumstances that accompanied the reduction in population.[14]

True, for the landlords, the plague and the population decline brought crisis to their holdings in the first instance - a crisis which was also manifested by falling incomes, and which was especially prevalent in the first decades of the fifteenth century on the central and eastern English manors. But for others, the reorientation meant rising incomes. "Like any other generation, the generations of the fifteenth century boasted both their competent and incompetent members...[and]...as in any other age, the incompetent landlord suffered".[15]

VII.2 The incomes of the landlords after 1350

It is true that the incomes of certain landlords do not seem to have been affected by the reorientation process that followed the Black Death. F. R. H. Du Boulay's study of the holdings of the Archbishop of Canterbury, for example, suggests that his income did not fall in the century that came after the Black Death and preceded the rise he thought was observable after 1475.[16] Levett had already found relatively high incomes at Winchester in the period after 1350,[17] and we can mention that similar examples can be found in the literature from the period immediately before 1960.[18]

John Hatcher's study of the rent incomes on six of the manors of the Duchy of Cornwall also shows that a landlord's income could be maintained, even increased in the fifteenth century.[19] But neither he nor Du Boulay thought that these many examples could shake the established view that landlords' incomes in this century largely fell.[20] A. J. Pollard, in his investigation of "estate management in the Later Middle Ages" on the manor of Whitchurch in northern Shropshire, agreed with this, and further thought that in this area external economic circumstances - not the luck and ability of the landlords - determined their incomes.[21]

As pointed out by John Hatcher, Postan was not unaware either that the fifteenth century did not mean falling incomes for certain of the bigger landlords.[22] In the first place, he had proposed the not unfamiliar view that the concentration of the holdings of the barons that began in the 1340s and was strengthened as a result of the Wars of the Roses must have modified the general decline in income for the individual estate-holder. Secondly, he thought that "a landowner well provided with investable resources and able to concentrate on sheep farming could do well even in the midst of a profound agricultural depression". Thirdly, he had stressed an example of how a successful magnate, Sir John Fastolf, had in the fifteenth century not only been able to invest in agricultural activities, but also "to exploit the prosperity of clothworking villagers on his land".[23] But in general Postan still maintained that the fifteenth cen-

219

tury was a period of depression for the big landlords, one which the majority were unable to struggle against.

The real losers in the agricultural crisis of the fourteenth and fifteenth centuries were the big landlords. But Postan found it "much easier to diagnose the good health of the late medieval gentry than to demonstrate the ailings of the magnates", and very easy to show how certain big landlords managed excellently during the crisis, although "in general most of the greater estates will be found among the losers".[24] Postan's statement of this rule, then, was not based on empirical proofs of actual income drops on specific manorial complexes, but on the general and theoretical reflections of past times.

He stated that the fourteenth and fifteenth centuries were "a time of falling land values, declining rents, vacant holdings and dwindling profits of demesne cultivation". So "we may...presume that the class whose income from land took the form of rents or farms must have suffered from the new dispensation: indeed must have been its main casualty"; while "the estates of the Benedictine abbeys which happened to retain grain-growing demesnes to the last possible moment suffered from high wages and sagging prices more than most other landowners". In other words, whether the big landlords carried on with demesne cultivation or let out the land, they were the losers. They were the victims of the market fluctuations that had their roots in the population decline. It is therefore not so surprising that Postan supposed that the smaller estate-holders who "consumed a large proportion of their produce and presumably farmed away little...may have suffered less".[25]

As we have seen from Pollard, Lander, Hatcher and Du Boulay, there is plenty of documentation to contradict this classic, generalizing interpretation. However, other empirically specific research seems to confirm it - for example, Christopher Dyer's study of the estates of the Bishop of Worcester, which however directly documents not crisis conditions, but a declining trend in the real wages on this West Midlands complex.[26]

Unfortunately, in the detailed studies of large manorial complexes, like Barbara Harvey's[27] and Eleanor Searle's,[28] there are only rather incomplete attempts to assess the income conditions of the landlords between the thirteenth and fifteenth centuries. So far, then, we must resign ourselves to the fact that a clear general interpretation of the pecuniary situation of the big landlords in the late middle ages is not possible. We still have too few studies that clarify the incomes of the big manorial complexes in the period when manorial cultivation was abandoned in earnest.

VII.3 The manorial economy after 1350

Hatcher and Du Boulay thought that the situation of the landlords in the fifteenth century should be understood in the light of local economic conditions. Hatcher found an explanation of the relatively favourable development of incomes on the Cornwall manors he studied in the fact that agriculture in this area was part of a many-faceted occupational structure where mining, fisheries and shipping played a prominent role.[29] Similarly, Du Boulay took the view that an evaluation of England's economic development in the late middle ages - and thus also that of the landlords' incomes - should be related to the geographical distribution of wealth that R. S. Schofield had drawn up on the basis of tax rolls from 1334 and 1515. From these it emerges that the taxation basis rose most in the southwestern and southeastern parts of the country, and that the area around London continued to become wealthier, relatively speaking, than the rest of England.[30]

Another familiar position we have seen was that the smaller landlords were less affected by the crisis - especially those who consumed the bulk of their own product, and who had not leased out most of their holdings. In fact, it was this social stratum which, according to Postan, profited from the falling land rent, the deserted farms and the falling incomes from demesne production, since precisely this group, which was swelled by the influx of the class of upward-moving peasants, leased the abandoned land on ever more favourable terms.[31]

There was general agreement that the smaller landlords - the gentry - were in a favourable economic position in the fifteenth century. It was similarly generally accepted that this class, which cultivated the land in a system of production that we can call "petty bourgeois",[32] expanded as it absorbed prosperous peasants. But there were also authors who thought that the leasing-out of manorial land did not necessarily entail that the interests of the big landlords were neglected.

Du Boulay claimed, for example, that the Archbishop of Canterbury leased out land in a balanced community of interests between himself and his lessees. Du Boulay tried to demonstrate how the lessees were obliged to maintain and sometimes even improve their holdings, and how they still had to ensure the Archbishop the goods and services he had a claim to.[33] On the whole, the Archbishop did not lose by leasing out the demesne. He managed to keep it continuously tenanted, to have the buildings and land maintained and improved without untoward fragmentation of the holdings, and - especially important - he was now purportedly able to collect his rents punctually without having to write off arrears. Du Boulay further concluded that "lease prices were steady".[34]

Barbara Harvey's study of the Westminster abbot's leasing-out of the demesne confirms this impression. "The rents for most of the abbot's demesnes...were stable over long periods of time"; and she considers it proven that there was a requirement that the lessee "should keep the land in good condition and repair demesne buildings".[35]

In both these contributions there is a recurrent point that clashed with the prevailing view that the development of the lease rent should be seen in relation to the increasing supply of land (Postan) and that the leasing system constituted a financial relationship between lord and lessee that broke with the non-financial, personal nature of the feudal relations of production, and which situated rents in a context of market competition (Hilton).[36] Apart from the fact that Harvey explained the wide fluctuations in lease prices at the end of the fourteenth century, and the tendency of some of them to fall at the beginning of the fifteenth century, in terms of the gradual decline of the stock and land lease,[37] both authors link the stable, fairly low rent level of the fifteenth century precisely with a non-financial, non-competitive relationship between lord and lessee.

Thus Du Boulay takes the view that the Archbishop's choice of lessees was generally determined by two not always compatible motives: "The wish to bestow patronage upon those who had some claim upon him by reason of kinship or familiar service, and the need to place a demesne in the hands of someone who was suitable from the economic and financial point of view".[38] By contrast, Harvey thought that the Winchester abbot's choice of lessees was determined by "the need to favour the families on whom his local administration had come so largely to depend..." and that this "was a foremost consideration when it came to arranging terms".[39]

The revision of the view of the fifteenth century rested on two crucial premises. First, any talk of the crisis of the big landlords was rejected. This could be defended inasmuch as there really was documentation that the reorganization of manorial holding as leaseholding did not necessarily lead to falling incomes. Secondly, it was asserted that the agricultural per capita product was rising, and that this rise could be observed by linking

it with purportedly progressive features of the development of manufacturing, trade and the cities.

The revisionists from Bridbury on saw these premisses as having developed from a neo-Malthusian interpretation of the precapitalist relationship between society and nature, since "the widely diffused prosperity of
the fifteenth century was indeed the result of the fluctuations of nature
rather than of economic progress".[40] The period was thus regarded in the
same way as Fussell later described the fourteenth century, as a period
of "social change but static technology". In the fifteenth century "social
change was not caused or accompanied by any advance in methods".[41]

The closest thing to an explanation of how agricultural productivity
could rise in these circumstances was, as mentioned before, Bridbury's
argument from the withdrawal of the marginal land, and vague remarks
in Du Boulay that the leasing of demesnes made for a more flexible system of cultivation.[42]

VII.4 Agricultural productivity after 1350

It has been pointed out several times that there are few studies of the
development of agricultural techniques. We come all the way up to 1977
before we find a work that sheds light on the above hypotheses. On the
basis of a number of accounts from the fourteenth and the beginning of
the fifteenth century from the manor of Tolleshunt in northeastern Essex,
R. H. Britnell dealt here with the relations between agricultural techniques
and the margins of cultivation.

The intensive tillage of the first half of the fourteenth century was said
to have so deteriorated the quality of the soil on this manor that "an
unusually large portion of the soils of the manor were marginal for the
cultivation of cereals".[43] This, along with the immediate consequences of
the population decline, did not however mean that, faced with the
problems in the 1350s, the cultivators simply abandoned the marginal soils.
"Instead of concentrating on grain production on the nucleus of the
demesne and abandoning marginal, they took quite the opposite course
in order to give the better land time to recuperate".[44] Thus we see that
Britnell, in his study of Tolleshunt, argued against Postan's and Bridbury's
general assumption that it was the marginal lands that were deserted
when the shrinkage of the arable began.

This reduction of the cultivated area both enabled and necessitated changes in agricultural technology. The accounts of Tolleshunt are thus in Britnell's opinion an extreme example of how "a change in the techniques of cultivation accompanied the decline of the arable after 1349". the study shows that the attenuation of the arable area on this manor from the latter half of the fourteenth century apparently ran parallel with a slackening of the rotation system.[45] This was followed by a tendency to extend the fallow period and perhaps to fertilize better[46] - "perhaps", because livestock farming did not increase much after 1350, although the ratio of arable to pasture at the beginning of the fifteenth century had changed substantially since the period before 1350.[47]

Britnell's example does not confirm Du Boulay's suggestion that the leasing system gave rise to a more reasonable, flexible rotation practice; only that it was made possible and necessary by the reductions in the cultivated area that took place. This contradicts, as pointed out above, Postan's and Bridbury's blanket assumption that the marginal land was abandoned, but not that agricultural productivity may have risen in the fifteenth century as the best soil was regenerated and cultivated again at Tolleshunt. The study contains no clearer statements on agricultural productivity after 1350.[48]

Studies that say anything directly about this issue are thin on the ground; but we can draw attention to two. For good reasons these, like Britnell's, only deal with manors where demesne cultivation was maintained, but this makes them all the better for clarifying aspects of the fate of the manorial system in the late middle ages. In 1977 D. L. Farmer took up the thread after Beveridge and Titow,[49] who had dealt with the development of grain yield on selected Winchester manors in the period before 1350, when he published his interpretation of the same development in the following century. P. F. Brandon took up the same issue in a 1972 article on the holdings of Battle Abbey in Sussex.

Given the fact that Postan and Bridbury assumed that the marginal lands were abandoned after the Black Death, and in particular the latter's view that precisely this led to a general increase in agricultural productivity, it is surprising that Farmer observes that the rising grain yield of the Winchester estates in the century after the Black Death was not most striking on the estates that had reduced their tillage most. This makes him doubt "whether improved yields owed much to the abandonment of inferior land"; all the more so as he could conclude from a comparison of fields itemized as "ploughland" before and after 1350 that "the reeves were sowing grain very largely where their predecessors had sowed it..."[50]

What he does find is a striking relationship between livestock farming and grain yield. The ratio of animal farming to the area sown changed considerably in the period,[51] and what is more, Farmer can prove that livestock farming rose substantially more on the most productive estates than on those where the grain yield developed less favourably.[52] These discoveries lead Farmer to conclude that the contraction in manorial production was very slow, and that grain production rose most on the estates where there was most livestock farming and the amount of seed sown per acre was least. He further noted that the use of legumes was relatively negligible.[53]

On the other hand, the relatively high grain yields documented by Brandon are partly explained by the fact that the Benedictines of Battle Abbey to a great extent used legume cultivation more or less as an alternative to the practice of rotation associated with their sheep farming and grain cultivation.[54] Although Brandon will not exclude the possibility that a tendency towards a slight rise in grain yield shown by his table towards the end of the fifteenth century can be explained by a concentration of tillage on the very best soil, he concludes that the generally big harvests of the period must be attributed to "the high technical efficiency of the monastic farming".

He finds this example inspiring because it shows, in direct contradiction of the conventional view of medieval agriculture, that it did not always teeter on the brink of a crisis due to the progressive exhaustion of the inadequately manured and overexploited soil. The constant high yield from these continuously cultivated soils over a long period which he believed to have included disproportionately many years with bad weather, are for him the proof that medieval agricultural technology did not make people quite as dependent on the vagaries of nature as certain scholars - for example, Lander - would have it.[55] But, according to Brandon, it is also proof that productivity on these monks' estates was higher than what he found on smaller lay estates in Sussex.[56]

The material from Battle Abbey's estates unfortunately provides no basis for comparisons between the periods before and after the Black Death. Yet, with Brandon, one can observe that trends in average yield on the Battle Abbey estates between 1353 and 1494 are on the whole comparable with what we know of grain yields from medieval agriculture, including the data Beveridge supplied in the 1920s for five Winchester estates in the period before 1350.[57] Furthermore, Brandon's calculations show that there can be no basis for speaking of a progressive development of grain yield in the period he studied - although the yield perhaps rose slightly at the end of the period.

We are rather better off as regards the development of productivity on the Winchester estates. Farmer has constructed a table for the development of grain yield on a number of estates between 1325 and 1453, where he includes material from the above-mentioned survey of Winchester yields in 1209-1349 published by Titow in 1972.[58] That this table shows no dramatic development in grain yield, but perhaps a tendency towards a rise, is because the drop in yield that Titow thought he could document can be traced in the period just before 1325 - that is, the last decades of the thirteenth and the first decades of the fourteenth century.[59]

Nevertheless, Farmer's Table 3 - "Mean Gross Yield Ratios 1209 - 1454", which also includes data from Titow's study up to 1349 - suggests a fairly stable yield throughout the period, especially for wheat.[60] We will refrain from considering Titow's study in detail until we come to its proper context, but at present we can conclude that the present efforts to calculate the fluctuations of grain yield in the late middle ages cannot be regarded as definitive documentation that agricultural production rose in the fifteenth century.

Nor have we found documentation for this in Britnell's exemplary demonstration of how the reduction of tillage at Tolleshunt led to the introduction of a more flexible rotation system. The scanty documentation for the development of agricultural production that exists rather suggests that any changes in agricultural techniques were manifested as a stabilization of grain yield and perhaps a slight rise in the final years of the fifteenth century.

VII.5 Summing-up

With Bridbury's revision of the history of the fifteenth century a frontal attack was launched on fixed ideas about socioeconomic development in that century. The clash should not be seen as much in the fact that he spoke of increased productivity as against Postan's older view that the agrarian crisis was expressed by a waning national - especially agricultural - product, since these two views did not necessarily exclude each other in a situation where the population was declining. What was really dramatic in Bridbury's revision was that he flew in the face of the dominant opinion by claiming that the fifteenth century was a period of general economic expansion.

With this he denied Rogers' classic view that "the golden age of the peasantry" was a period of "distress" for the landlords, although contemporary and later studies of the landlords' incomes and the reproduction of the manorial economy after 1350 largely seemed unable to confirm his claim. Nor do the same studies, however, generally provide grounds for seeing the century after the Black Death as a crisis of the landlords.

Notably, Postan's explanation of the development of the agrarian product and Bridbury's documentation of the development of agrarian productivity have the same main premiss - that from the end of the fourteenth century on there was "a net contraction of the area under cultivation". Bridbury combined this premiss with the marginal land theory later defended by Postan when he claimed that it was precisely the marginal land that fell into disuse after the inception of the population decline, and that this withdrawal of less fertile land increased agricultural productivity.

However, we have seen indications that the contraction of the arable area after 1350 did not necessarily lead to an abandonment of the marginal lands. In fact, we have proof from Tolleshunt that precisely the opposite was practiced when the arable was reduced. There is still little documentation that the marginal land actually was abandoned, at least in the tillage areas proper, and it should be mentioned that there are signs that agricultural productivity on the Winchester estates after 1350 did not rise markedly on the estates where tillage had been reduced most. Furthermore, the small amount of existing direct documentation of the development of agricultural productivity in the fifteenth century, if anything, indicates a stabilization of the grain yield.

So these facts seem neither to confirm Bridbury's assumption that the marginal land was deserted after the population decline had made its mark, and that this led to an increase in productivity, nor his general claim that agricultural productivity rose from the end of the fourteenth century on. On the other hand, Postan's assertion that an agrarian crisis held sway in the same period seems meaningless against this background: with static productivity and a declining population it seems pointless to talk of an agrarian crisis even if the agrarian product may have been falling.

VII.6 The "Malthusian check"

Two of the minimum requirements we posited above (see Chapter VI, p. 196) as partial preconditions for accepting the "neo-Malthusian" overpopulation theory, were convincing proof that agricultural productivity was in serious decline, and that the natural resources of agriculture tended towards exhaustion around 1300. We did not find these preconditions met before 1960; but were they met later?

In the *Cambridge Economic History* Postan argued that on the one hand there was "land hunger", and on the other a "deterioration of the land" towards 1300. He found proof of the former in the fact that "land was dear and getting dearer". He pointed out that entry fines rose and extraordinary levies - tallage - became more frequent. He also maintained that livestock farming was disproportionately low compared with tillage, which was expressed as higher land values for meadow and other pasture areas than for arable.[61]

We have before criticized the assumption that a rise in entry fines and the like can be taken as proof of a rising rent burden. Similarly, one may doubt whether these purportedly rising "land values" reflect a real "land hunger" in a period of rising prices for agricultural products. Again we can object that the rising rents may simply express the attempts of the landlords to compensate for the fixing of the customary rent in a period of progressive price development.

In his study of "Demesne Farming in the Chiltern Hills" Davis Roden, like Postan, has noted that "scarcity of grassland was reflected in the high values it commanded: meadow was consistently assessed in manorial extents as worth four or five times as much as arable, while the better pastures were twice as valuable as arable on the farm".[62] But this did not mean, although "arable land was...the major component of all demesne farms" in the Chilterns, that arable was not manured. "Marling was widely practiced, legumes were grown on most farms", and moreover the soil was fertilized with waste, leaves, animal manure and through the pasturing on fallow fields that was made possible by the rotation systems and the big sheep flocks on the manors.[63]

Against this Postan thought that the disproportion between tillage and animal farming he had also sought to document elsewhere[64] reflected not only the scarcity of land in the thirteenth century, but just as much the scarcity of manure and the increasing exhaustion of the arable. This agrotechnical problem, which Hilton and Trow-Smith thought was peculiar to

228

the peasants, was thus turned into a general problem "in many parts of England".[65]

The relatively few studies we have of the medieval colonization process in England all indicate that colonization stopped around 1300. B. K. Roberts' study of medieval colonization in the Forest of Arden from 1968 also confirms this dating.[66] On the other hand, there is some confusion as to when the halt to colonization turned into the contraction in the cultivated areas that is typical of the late middle ages[67] - a contraction that Postan and for that matter also Titow associate with the gradual exhaustion of the soil.[68]

Postan thought in *The Cambridge Economic History* that colonization took place not only because of the population increase but also because parts of the formerly cultivated areas were exhausted; similarly, he thought that colonization not only stopped because the pressure of population decreased, but also because "all the colonizable reserves had been exhausted".[69] Here we see that between 1950 and 1960 there had been a crucial shift in Postan's view of the causes of the late medieval agrarian crisis.

While he argued in the years just before and after 1950 that the "Malthusian check" set in because, among other reasons, the cultivation of the marginal lands failed very quickly after their incorporation, he elaborated on this argument in his article in *The Cambridge Economic History*. He now also maintained the opposite: that the desertion of arable land can be documented by, among other things, the fact that "assarts were valued more than the fields anciently occupied and cultivated."

Apart from Saltmarsh's observation that marginal lands in Norfolk went out of cultivation throughout the late middle ages, which Postan's marginal land theory seemed to rely on heavily around 1950, the investigation of the colonization problem we have seen in the preceding chapter seems to contradict the marginal land theory in its original form. Whether this was why Postan supplemented his theory with an opposite variant we cannot say; but we must remark that colonization research neither confirms nor refutes the new variant of the theory.

Thus Postan was saying on the one hand that the population expansion drove tillage out to the marginal lands, which were exhausted relatively quickly; on the other that there was a progressive exhaustion of the old land and new and more valuable tillage areas were taken under cultivation. The former variant is directly related to the "neo-Malthusian" interpretation of the reasons for the late medieval crisis; the latter can in principle be defended independently of it. There are good reasons to stress this difference, as it illustrates how Postan during the 1960s attempted to look behind and explain the "Malthusian check" in the light of the agrotechnical problem he believed had arisen because of the increasing disproportion between tillage and cattle farming. But these speculations and attempts to verify the old theory of a disparity between tillage and livestock farming were not in themselves proof that agricultural productivity was on the wane around 1300.

VII.7 The soil exhaustion theory

It will be recalled how the exhaustion theory was launched early by Denton; how it was subjected to intensive research involving fluctuations in the medieval and late medieval grain yield by Bradley and Ernle, and later by Lennard, Beveridge and Bennett; and how this research resulted in a refutation of the theory. Given this, Postan's persistence in proffering the views above was probably due to two things. In the first place he had certain general critical objections to Beveridge and Bennett; and secondly, he was familiar with the study of the Winchester yields being carried out by Titow.[70]

Besides claiming, without going into detail, that Beveridge's and Bennett's statistics were not of course technically perfect, he thought that they demonstrated the dangers of extrapolating data from medieval sources regardless of their backgrounds. First, Postan wanted the stagnating grain yield demonstrated by Beveridge and Bennett to be seen in the light of the fact that the data they used for their calculations all came from demesne cultivation at Winchester, and that this was organized by progressive lords who understood how to occupy the best soils, and who had far better means of fertilization than the peasants. Secondly, he thought that Beveridge and Bennett had forgotten to take into account the way the bishops, in the period covered by the latter part of their studies, were busily concentrating demesne production on the very best soils and leasing out the worst. "So if even in these circumstances their yields did not rise but remained stationary, this would denote not a stable but a declining fertility of the soils in the countryside at large".[71]

It should be noted that Beveridge was not in fact unaware of this. But it should also be mentioned that he did not venture to speculate about which parts of the demesne were withdrawn from cultivation, and whether the poorest land was necessarily leased out. We have been unable to find any direct documentation for this, but should recall that Britnell's example from Tolleshunt in Essex - from a later period, it is true - provides no grounds for thinking that reductions in tillage on this manor led to the abandonment of marginal and/or exhausted soil.

It is repeatedly assumed by several authors that manorial cultivation was concentrated on the best soils, and that, compared with the peasants' holdings, it was favoured by better fertilization possibilities. Thus we have seen how Denton, Bennett, Hilton and Trow-Smith regarded the disproportion between tillage and animal farming as a particular problem for the peasants. Yet this cannot be established with certainty and as a general rule. For example, one cannot necessarily interpret the *jus faldae* rights of the lords as a direct indication of this. These rights could easily be imagined to reflect quite the opposite: that the peasants kept larger flocks of animals than were kept in strictly tillage-oriented demesne production; and that the *jus faldae* may have been the lords' attempt to solve their own fertilization problem by exploiting the peasants' bigger animal holdings. We do not know. But we can object that in certain areas livestock farming may have been more widespread among the peasants than in demesne farming - for example, perhaps, in the Chilterns.[72]

At any rate, according to R. H. Alan Baker, there is little to suggest that the contraction of the peasants' arable land between 1291 and 1341 in Cambridgeshire, Buckinghamshire and Bedfordshire necessarily had anything to do with the exhaustion of the soil. Baker thought the main reasons for these desertions should be sought in the poverty of the peasants, scarcity of seed and a declining population.[73]

Postan also rejected the possible illustrative value of the Rothamsted experiments for the development of grain yield using medieval methods of cultivation. His main objection was that the soil on which these experiments were done must in the first place be considered very rich in mineral nutrients, and in the second light enough to ensure reasonable drainage conditions. These well-nigh optimal conditions can hardly have existed everywhere - and certainly not on the marginal lands brought under cultivation in the thirteenth century.[74]

VII.8 The grain yield, 1200-1350

Nevertheless, as Titow points out, there is quite broad agreement between a few of the Rothamsted experiments' principal results and a similarly broad tendency in Titow's survey of the grain yields on the Winchester estates. Just as it emerged at Rothamsted that wheat seemed better able than barley to maintain yield levels under continuous cultivation without manuring, Titow's third main conclusion was that there was "a striking difference between the behaviour of wheat and that of other crops....Thus it is quite clear that on the whole, wheat was doing much better than other crops..."[75]

The average grain yield for wheat - the ratio of grain harvested to seed sown - on the 39 Winchester estates used in Titow's study turned out to have remained constant at 3.8 throughout the thirteenth century, and then to have risen insignificantly in the first half of the fourteenth century. It is therefore the development of yields for the other grain types - barley, oats and "mancorn" - that leads Titow to his other main conclusion, that "when changes in productivity are considered, deterioration in yields is found to be far more common than improvements"; and to remark that "the most usual pattern was for the yields to reach their lowest level in the last quarter of the thirteenth century and then to improve gradually..."[76] The grain yield for barley fell between the 1200-1249 period and the 1250-1299 period from 4.4 to 3.5, while the yield for oats fell from 2.6 to 2.4 and for "mancorn" (a mixture of rye and wheat, "maslin") from 3.9 to 3.1.

These changes in grain yield can be regarded as serious because, as we have already mentioned in our discussions of the fluctuations of medieval grain yield, they should be seen in the context of what Titow considers his first main conclusion: that "the general level of productivity of all crops appears to have been very low by any standard..."[77] Against this background the relative effect of the changes mentioned may well have been very serious.

We will not enter into a discussion here of whether the general interpretation of the Winchester yields arrived at by Titow is sound or not: that is, to what extent it can be explained by and related to changes in the amount of seed sown throughout the years. We will accept his claim that its basis is sound, and that the crop yield per acre shows more or less the same tendency as the grain yield. Instead we will turn to his explanations of why the grain yield fell in the latter half of the thirteenth century, and developed positively again in the first half of the fourteenth.

Although the Winchester estates provide no basis for general assessments of the influence of the climate on the development of arable productivity, Titow concludes that he must reject the idea that this external factor had any great importance for the trend discovered. It is reasonable enough that Titow does not want to discuss the climatic changes that Scandinavian researchers in particular have found indications of around 1300 - among other reasons because these were long-terms effects between 1300 and 1600, and because we know very little about how these changes affected the English weather. But it is surprising that he does not reckon with "the worst climatic shocks [which] seem to have been those which came in the years 1314 to 1325..."[78] - particularly surprising since this knowledge supports Titow's main point, to which we will return shortly.

Nor does Titow think that agrotechnical measures such as marling, the cultivation of legumes and changes in the rotation systems used can have had any great influence on the grain yield fluctuations shown by his study.[79] He discovers the determining factors in the incorporation of marginal lands, and to a lesser degree in the exhaustion of original arable because of the increasing disproportion between tillage and animal farming.

He finds confirmation of a striking relationship between the colonization of marginal lands and falling grain yields in three things: first, that the general drop in grain yield took place in the period between 1250 and 1300 just after the biggest expansion of Winchester's demesne cultivation; secondly, that the wheat yield was generally far more stable than the barley and oats yields, and that these proportions corresponded with those of the Rothamsted experiments; and thirdly, that the drop in grain yield was most marked on the biggest, most expansive of the Winchester estates.[80]

He seeks to prove in two ways that the falling yield was also connected with the exhaustion of the arable, and that this was due to the disproportion between tillage and animal farming. On the one hand he compares the development of grain yield on the individual manors with the overall contraction of demesne that was a marked tendency after 1270. And on the other he compares the number of animals per acre with grain yield fluctuations and arable area on a number of selected manors throughout the period.

For the first of these constructions Titow builds on the premiss that any reduction in the manorial land meant a concentration of cultivation on the most fertile land. It follows from this that any observed contraction in demesnes must have resulted in rising yield, and any claim for falling yield on manors with decreased arable areas plays down the actual degree of exhaustion - all other things being equal, since it of course affects the plausibility of this method whether the individual manor had a soil which was homogeneous or very heterogeneous in terms of fertility. This is something Titow is aware of, but which he cannot, for good reasons, take into account within the comparative schemes he sets up.

Partly for this reason, but primarily because the above main premiss for the method can be called into question (cf. our scepticism above as regards Postan's view that it was precisely the marginal and/or exhausted soils that were abandoned during the contraction of demesnes in the late middle ages), we should be sceptical about what can be read out of the comparisons Titow makes.[81] This, despite the fact that they show convincingly that his calculations for the drop in grain yield in the latter half of the thirteenth century are only conservative estimates of the real degree of exhaustion, and that the improvements he demonstrates as having taken place in the course of the first half of the fourteenth century largely do not express the real rise in productivity - only that tillage was concentrated on the best soils.

Yet there are indications that the same practice may have been used on the Winchester estates as Britnell documented at Tolleshunt a century later. It is conceivable that the choice was made on the Winchester estates to give the best land a chance to regenerate its full fertility, precisely by fallowing that rather than the poorest, most exhausted soil, since Titow demonstrates that the lowest grain yields appear precisely in the period just after the contraction of the demesne began, and since the rising grain yield only developed gradually towards the second quarter of the fourteenth century.[82]

If this reading is correct, it means that the drop in yield Titow claims to have taken place in the latter half of the thirteenth century, and on certain manors in the first decades of the fourteenth century, expresses an exaggerated degree of exhaustion. But this does not necessarily mean that the rising grain yield after this was the same thing as a genuine rise in productivity.

We can now turn to Titow's calculations of livestock holdings and his assessments of the existing fertilization possibilities. His general result here is that total animal stocks appear to have peaked around 1260. But this does not mean that the number of animals per acre of arable was falling everywhere, as the contractions in demesnes after 1270 followed in the wake of stagnating or falling livestock holdings on many manors.

On the whole Titow stresses that "one must...not exaggerate the effect on yields of the change in the animal ratio since in no case did the change amount to more than half an animal per acre". All the same, he assigns these fluctuations some importance on certain manors: those where more significant fluctuations in animal stocks correspond to fluctuations in yields - that is, where falling livestock holdings correspond to falling grain yield.

Although the material provides no basis for assessing the supply of manure that may have come to the demesne by virtue of the lord's *jus faldae* rights. Titow comes to the major conclusion that his calculations show how livestock holdings throughout the period were extremely unsatisfactory in terms of the need for reasonable manuring of the arable areas so used; but that this animal-to-arable ratio shows a progressive tendency with the contraction of the arable areas of the estates. Thus, although we should be cautious of explaining fluctuations in grain yield in terms of the development of cattle farming, according to Titow we can go a long way towards an explanation of the generally low yields on the Winchester estates throughout the period by considering that this arable land was chronically under-supplied with animal manure.[83]

For supporters of the "neo-Malthusian" body of theory Titow's interpretation of the "Winchester Yields" was documentation of certain ecological relationships which they thought lay behind or reciprocally caused the reduction in population that is seen in the last analysis as the reason for the crisis of the manorial system and society. The exhaustion of both marginal and normally fertile soil accelerated the increasing disproportion between cultivable land and population, but did not immediately lead to falling incomes for the landlords.

Their period of prosperity derived from the rising pressure of population and the accompanying rises in prices, rents, labour services and revenues from mills and courts, and from the falling wages. Only when the increasing disparity between land and population reached the point where it provoked a levelling or a direct decline in the population, and when this decline made its impact felt economically as rising wages and falling prices, did the fall in the landlords' incomes occur which, along with the subsequent restructuring of manorial production, is called the crisis of the manorial system, from which the subsequent social crisis is seen as deriving.

VII.9 The landlords' incomes in the thirteenth century

However, the question is, what is specifically thought to be known about landlords' incomes in the thirteenth century; and further, what are thought to be the reasons why the above-mentioned ecological problems were allowed to take place? Apart from the classic assumption, that the colonization of the thirteenth century, the rising corn prices and rents etc. and falling wages in themselves led to general income increases for the landlords - which is repeated again and again -[84] we must admit that we know very little about the landlords' financial situation from specific studies.

Of course, not all manorial complexes managed equally well;[85] and it is, as we have noted, a common view that "the high farming era" to a certain extent brought with it a number of difficulties for the smaller lay landlords.[86] Moreover, J. R. H. Moorman, who has studied a number of monasteries' financial positions, thinks that "all the evidence which we possess - and which is considerable - goes to prove that almost every English religious house in the thirteenth century was deeply in debt".[87] This general view appears to fit excellently with the way the Archbishops of Canterbury were burdened throughout most of the thirteenth century with substantial debt.

It is characteristic of this debt that it was not due, as R. A. L. Smith supposed, to productivity-promoting investments in the Canterbury estates,[88] but primarily to rising expenditure connected with the investment of new Archbishops, court cases, travel and payment of royal taxes.[89] It is of course only a formal problem whether the loans taken were used for one purpose or the other, as long as we know that the Archbishops also made productive investments - which they in fact did "out of their normal manorial revenues".[90] But it is a real problem to interpret what the Archbishops' debt was an expression of.

On the face of it one might think like Moorman that the debt in itself was an indication of financial problems and an economy where ends did not meet. Moorman is partly justified in such an interpretation. At any rate it seems - witness Walter of Henley - to have been well in keeping with medieval economic thinking.[91] But the question is whether the constant borrowing of the Archbishops of Canterbury suggests an expanding economy - production in such a rapid process of expansion that its own accumulation of funds was insufficient.

We do not know. But we can note that Mate's study does not exclude this possibility; in fact it rather suggests it. In the first place the very fact that they were constantly able to borrow is presumably a sign of the Archbishops' credit-worthiness. A second and related factor is that the loans seem to have been paid back, although sometimes in arrears and by taking out new loans to redeem the old debts.[92] Thirdly and finally, the loans, which were borrowed not from Italian merchants, but from secular and religious neighbours, were perhaps not loans at all but deposits in an expanding concern. "Thus the priory was in effect acting as a deposit banker..."[93]

If one concurs in this interpretation and the prevailing view that the thirteenth century was, at least for the big landlords, a general boom period, one must ask how the previously-mentioned ecological contradictions could make any impact at all - insofar, of course, as one accepts that they did in fact make an impact, and no attempts were made to counteract them, a probability we saw in Titow's reading of the "Winchester Yields".

VII.10 Investment levels in the thirteenth century

For Titow the answer is very simple. He thinks that medieval agriculture was hampered by certain technological limitations which could not be overcome. "Unlike the position nowadays, the technical limitations of medieval husbandry seem to me to have imposed their own ceiling on what could be usefully spent on an estate".[94] This means that the landlords' incomes, all other things being equal, were within certain limits unimportant for the development of agricultural productivity; and that the manorial economy was hemmed in by technical shortcomings that sooner or later must lead to an ecological check, independently of the lords' ability and will to make productive investments.

As a logical consequence of this "technological position", which he too argues for in his contribution to *The Cambridge Economic History*,[95] Postan tries in a 1967 article to say why this problem arose. Here he claims that gross investments on the Winchester estates rarely exceeded five per cent of total income, which was more or less the same investment level as on the Duke of Lancaster's southern estates in the thirteenth century; and furthermore that some of the neighbouring estates to the Winchester ones - those of St. Swithin's Priory - invested an even smaller percentage of their revenues - about three per cent - and that the level on Glastonbury's southwestern estates varied between two and four per cent.

But why was the investment level so low? According to Postan, at any rate, it was not because the overall revenues of the manorial complexes in question were poor - on the contrary.[96] Although Postan persists with the rather tautological observation that the poor level of investments was partly due to the fact that the technology and equipment necessary to develop production only existed to a very limited extent,[97] he thinks that the main reason was that "very little was actually saved: that the bulk of the profits was squandered, or that such savings as were made were not devoted to productive investment".[98] It is notable that Postan in this work arrives at a confirmation of Dobb's general view of the principal reason for the fall of the manorial system - see Chapter V, pp. 152-153.

Through Postan's life work one can trace the following development of his view of the reasons for the late medieval crisis. In his works from the 1930s and 1940s there are no direct attempts at explanation, only suggestions that the population development must have been a decisive factor. This suggestion is developed into a "neo-Malthusian" explanatory model in the 1950s. His understanding of the reasons for the crisis now focuses directly on the demographic reduction and its triggering-off of a "Malthusian check", which through time was rooted in the thirteenth century's exclusively extensive development of agriculture and the ecological limits this involved. Finally, in his works from the 1960s, Postan tries to penetrate behind these agrotechnical problems. He arrives at the above-mentioned "technological position" and ultimately thinks that the main reason for the lack of technological development must be sought in the landlords' disproportionately large unproductive consumption.

Against this background it is further interesting to see how Postan's study of investments in medieval agriculture fits in with the analysis of the same problem that Hilton had offered earlier at the Second International Conference of Economic History. On the face of it, the results of this study seem to support Bridbury's and Kosminsky's assumptions of a rising agricultural productivity between the thirteenth and fifteenth

centuries. At least Hilton thought he could demonstrate how the investment rate on a number of manors in west and south west England rose
very strikingly, from around the level Postan had found in the thirteenth
century.[99]

For Hilton, the question of the investment level of the manors was an
obvious area for empirical research: since his earliest writings he had
agreed with Dobb that one main reason for the fall of the manorial
system was precisely the landlords' extensive unproductive but necessary
consumption,[100] and their lack of the ability and will to invest in
productivity development.

Seen in retrospect, it is not remarkable that both in Hilton's and Postan's studies of the medieval and late medieval investment levels we find
matching views of the peasants' investments. Even if these cannot be judged on the basis of direct information about them, the authors agree that
the peasants' opportunities for investment were in general extremely slight
- first and foremost because right up to the beginning of the fourteenth
century they were obliged to bear very heavy rent burdens.[101] But
according to both Postan and Hilton, not all the peasants were in this
desperate situation. As early as the thirteenth century, but especially in
the fifteenth, the bigger peasants were able to accumulate and invest - a
fact that is most clearly reflected in the peasants' rising stocks of
animals.[102]

With Hilton, one can sum up as follows. Investments in thirteenth-century manorial production were limited to an absolute minimum, in reality
almost to reinvestments. The peasants' investments, including reinvestments, were made impossible primarily by the high rent level and
secondarily because of the scarcity of pasturage, although for the big
peasants opportunities for investments did exist. The changed economic
circumstances after the mid-fourteenth century "operated in favour of
tenants as against landlords". Bigger plots "made possible a build up of
peasants' flocks and herds". Smaller rents "left greater surpluses in tenants'
hands". And the peasants' pressure on the landlords enabled by the
changed land-manpower ratio "caused landlords to invest a greater proportion of their incomes. The physical volume of investment may not have
been greater than in the thirteenth century, but it may well have been
greater per head or per acre in use".[103]

VII.11 Farm sizes and social stratification

We have already dealt with the way several authors sought documentation of the rising rent burden in the thirteenth century. We have also seen how the size of peasant plots had earlier been the object of debate, partly in relation to the question of population development in the thirteenth century, partly as an indication of the peasants reproductive conditions during that century. Both Titow and Hilton[104] took up the latter issue in articles from the 1960s.

Titow analysed payments of so-called "tithings-penny" from the Winchester account rolls with a view to clarifying the population growth in the thirteenth century and relating it to the peasants' reproductive conditions, by comparing population figures with available arable between 1248 and 1311.[105] Hilton occupied himself with the reduced arable areas on which an increasing number of the peasants were said to have had to produce the means of subsistence between 1250 and 1300 - but without claiming with any certainty that this necessarily reflected "a process of peasant impoverishment in the second half of the thirteenth century" on the lands of the Worcester and Berkeley Barons in the West Midlands.[106]

Hilton, on the other hand, is convinced that the institution of villeinage, which was introduced in earnest during the reign of King John (1199-1216) had its natural result, because of the rising consumption of the ruling class, in increased exploitation of the primary producers. He took up the thread after Vinogradoff, although unlike the latter he insisted on this fairly late beginning for villeinage,[107] observing with Vinogradoff that "by the side of the freeholder recognised by later law there stands the villain as a customary freeholder who has lost legal protection".[108]

When Hilton connects increased exploitation with the introduction of villeinage, it is not because the villein in the thirteenth century was always worse off than the free yeoman, but because, like Vinogradoff, he does not simply associate villeinage with labour rent obligations.[109] The villein was also subject to arbitrary dues and deprived of protection against the exploitation of the appropriators, while the free yeoman was protected by law against increased exploitation, was free to buy and sell land, and in general his free mobility was safeguarded.[110]

It will be recalled that Hilton did not only think that the thirteenth century was characterized by increasing exploitation and impoverishment of the primary producers. Although he took the view that an increase in social stratification within this class is a significant factor in the late medieval centuries, he had since his earliest writings been no stranger to the idea that this stratification had also existed in the thirteenth century - and that the social divisions did not necessarily match the boundaries between villein and freeman.

Titow dealt with social stratification on some of the Winchester estates. Here he arrived at the conclusion that an assessment of the conditions of the peasants in the thirteenth century must be seen in relation to the specific type of manor where they were villeins. He found several significant factors in this respect: whether the manors practised more or less monocultural production, the location of the estates and corresponding opportunities for emigration, but most importantly the extent to which the manors were able to colonize.

Titow thus compared the population-to-land ratio, conditions of inheritance and the development of entry fines in the thirteenth century on the still-expanding manors of Wargrave (Berkshire) and Witney (Oxfordshire) with the above-mentioned Taunton (see note 105). As regards the entry fines, "there was a very marked upward trend in the level of entry fines at Taunton from the beginning of the thirteenth century onwards, while the level of entry fines at Wargrave and Witney had hardly risen at all".[111]

More direct evidence of the stratification of the peasant population could be found as early as Kosminsky, who saw even in the rising money rent he had documented in the thirteenth century "a considerable degree of stratification among the English peasantry", but who also took examples of land transactions among the peasants to indicate the same thing, and stated explicitly that he could observe extensive social stratification in the hundred rolls.[112]

The development of the land market among the peasants in the thirteenth century has later been discussed by Postan and Paul R. Hyams. The disagreements between them were on the one hand about the dating and the geographical spread of this land market,[113] and on the other about why it arose. Postan mentioned two possible answers to the latter question: first, "as good land was getting short, and rent and land values were rising, many a wealthy villager may have been tempted to cash in on the rising market and to offer his lands for sale in parcels"; and secondly, "we must assume that in societies in which the family is the unit of ownership and exploitation, the needs and resources of individual families are too unequal and too unstable to allow the family holding to remain uniform and unaltered in use and size..."[114] Hyams limits himself to criticism

241

of Postan and Kosminsky, but tends to attempt to combine these not quite compatible views.[115]

Both contributions confirm the impression that the land transactions among the peasants increased in the course of the thirteenth century. Whether one inclines to one or the other author's view of the reasons why there were apparently increasing purchases and sales - and also subletting - of land among the peasants, one is left with the impression that the class of primary producers in this century was far from homogeneous in social and economic terms.

The very existence of the land market seems to suggest that not everyone at least can have been living at subsistence level. As Postan sees the problem, some people were so prosperous that they saw a chance to profit from the rising prices by selling land. Others must correspondingly have possessed the means to buy land. But this reasoning does not of course exclude the possibility that some people must have had to give up land out of harsh necessity, while others must equally have been forced to buy a smaller plot where they could eke out their existence at a very frugal level.

The existence of the peasant land market also suggests, given Postan's reasoning, an adaptation of farm sizes to the individual family. It must have been crucial for the single family to have a farm which in size and quality was optimally suited to the size of the household and its technological resources. In this connection it is interesting to see that Hilton, in another context, explained the decreasing sizes of farms in Europe between the ninth and thirteenth centuries by a possible drop in the size of families, at least in the thirteenth century, when they no longer included a few slaves.[116]

VII.12 Fines and social unrest in the thirteenth century

In support of the view that there was an increasing impoverishment of the peasants in the thirteenth century, Alfred N. May has tried to draw up "An Index of Thirteenth-Century Impoverishment". Reviewing Winchester pipe rolls covering the period between 1208 and 1300, supplemented with manorial legal sources from the monastery of Bec from the period 1246-1321, May discovered that fines for criminal offences rose in volume,[117] while the actual sizes of the fines fell - at Winchester from about the mid-century, at Bec continuously.[118]

May thought that both these facts reflected what he believed was generally agreed: "that the thirteenth century witnessed an economic crisis that led to the impoverishment of the population....This would fit...in with the growth of court business. The propensity to "crime" would be increased by general impoverishment and pressure on resources etc."[119]

We observe that May has no qualms about talking about an "economic crisis" in the thirteenth century; but also that he equally freely establishes direct links between the alleged crisis and the increasing number of court cases, without reflecting on possible administrative reasons for this rise.

For he does not think either that the rising number of trials and fines served to stimulate the incomes of the manors - only to prevent increasing crime. In this light, it seems surprising that the fines fell in size; but it does not surprise May, who concludes that "the punishment must fit the crime, but it must also fit the man". He therefore sees the falling fine sizes as a direct reflection of "a decline in peasant prosperity".[120] "It seems clear that the rate of fine levied in manor courts provides a reasonable index of peasant prosperity in thirteenth-century England".[121]

This seems a rather bold conclusion. Criminological common sense says that lawbreakers are mainly found in the very lowest social strata of a society. Surely this was also the case in the thirteenth century? May's study provides no basis for answering these questions. Nor, on the other hand, does it give a basis for a general assessment of the peasants' overall reproductive conditions. It is rather an indication that the conditions of a more or less marginal group, the most disadvantaged, may have deteriorated in the period up to 1300. The falling fine rates might simply indicate more clemency and a humaner attitude to these disadvantaged lawbreakers - a possibility May does not exclude, but which he absolutely refuses to concede any explanatory value.[122]

We have already seen that Hilton, in one of his first works, considered the revolt of the peasants against growing exploitation as both a consequence and a cause of the incipient decline of the feudal relations of production. The scattered uprisings of the thirteenth century thus reflect the way the peasants tried to protect themselves against the landlords' successful attempts to load extra burdens on their shoulders.[123] In other words, the insurgencies were unsuccessful, and for that reason among others are seen by Hilton as a reaction against the increased exploitation that was rooted in the lords' growing necessary but unproductive consumption and their failure to invest in production.

The situation is different with the revolts of the late middle ages. Hilton explains the successful outcome of the 1381 Revolt by several factors. The Revolt was far more widespread geographically than the earlier local risings because of the development of state power and state taxes, the breakdown of local isolation by trade, and the links the Revolt had with the religious mass movements.[124] But most importantly, the Revolt took place against the paradoxical background that "the distribution of incomes between landowner and peasant was [now] in favour of peasants, whether in their capacity as tenants or in their capacity as labourers".[125] The background of the Revolt was that the balance of strength between lord and peasant after 1350 had swung in favour of the peasant because the population decline had taken effect and changed the population-to-land ratio.

As we have seen, Hilton is far from alone in thinking that this was a determining factor for the socioeconomic transformation of the late middle ages.[126] Hilton's assessment of the historical significance of the 1381 Revolt, therefore, on closer inspection, does not differ substantially from the classic denial that it had any crucial importance for the late medieval transformation process.[127]

VII.13 Summing-up

Hilton is fond of describing the battle over the feudal rent as feudalism's "prime mover". This motive force was grounded in the class conflict between lord and peasant and in the internal rivalries of the ruling classes; but the development of population had a crucial influence on the outcome of this struggle. Thus the battle seems in Hilton's interpretation to have benefited the lord as long as it took place under the influence of a rising population and as long as the consequent rise in the rent burden and impoverishment of the peasants did not result in a demographic check - that is, a population drop.

The rents could apparently rise until exploitation had so impoverished the peasant population that, vulnerable to the influence of climatic and epidemic factors, it was seriously reduced. After this the battle of the rents seems to have gone in favour of the peasants. Even though Hilton does not directly dispute that the population growth of the high middle ages levelled out or perhaps turned around at the beginning of the fourteenth century, the Black Death and the resulting drastic demographic decline therefore assume very central importance for him.

It is characteristic of Hilton and Kosminsky that they regard the primary producers as a progressive class whose demands for legal and economic freedom were to some extent met, as a result of the population decline, by the end of villeinage and the spread of the leasing system. It is therefore not so surprising that they and those who incline towards their views - Bridbury and others - have a clear if not always well-defined tendency to regard the end of the thirteenth century and the beginning of the fourteenth as a period of crisis. It was in this age that the basic conflicts of the feudal relations of production, by virtue of the population growth, opened up the dark road towards the fulfilment of the progressive peasant class's demands for legal and economic freedom, through increasing exploitation, impoverishment and ultimately the annihilation of large numbers of the members of this class.

By contrast, the period after 1350 is seen, if not directly as a boom period, at least as a flourishing period when the retardation of social productive forces by the medieval relations of production was broken down; a period when the power of the reactionary ruling class was broken in favour of the growth of a progressive middle class - yet with due consideration of the fact that this progress took place in the shadow of one of the most serious demographic crises in world history.

Against this, we have seen how the "neo-Malthusian" school, with Postan at its head, described the period from the second decade of the fourteenth century - and even more so the fifteenth century - as the crisis period *par excellence*. This dating is in the first place rooted in an emphasis on the demographic crisis as an indisputable sign of society's inability to reproduce its members; and in the second place in an attribution of extraordinary socioeconomic importance to the manorial system and the big landlords.

These two different views of when one can speak of a late medieval crisis are not due to critical differences of opinion on the causes of the crisis, nor to any real dispute over its socioeconomic consequences. They are primarily due to different emphases on who was affected by the crisis, derived from normative and ideological disagreements.

Thus the "neo-Marxists" typically assess the agrarian relations of production in the thirteenth century in terms of diachronic norms - a view of history that recalls the prevailing one in the latter half of the nineteenth

245

century. Although Rogers, for example, stressed the crisis of the manorial system, he and many of his contemporaries viewed this socioeconomic system in the same way as the "neo-Marxists" - as an unfree, retarded system of production that had to be abolished to pave the way for the legal and economic freedom they rightly thought was the first step on the road towards the establishment of modern capitalist relations of production.

Postan, on the other hand, takes a more synchronic approach where, although he is far from blind to the irrationalities of the manorial system in terms of economic progress, he develops his view of the crisis in direct relation to the inability to survive of this production system and the socioeconomic consequences of that inability. Postan sees our problem area in terms of the universe of the feudal mode of production and the big feudal landlords, while the "neo-Marxists" stand by the result of the historical process and retrospectively describe the late medieval transformation process, applying their own diachronic norms to the alleged class of primary producers. They see our problem area in terms of what they, with their own attitudes to the feudal relations of production, think must have been the point of view of the peasants.

In views of the causes of the late medieval crisis, there is a tendency for the "neo-Malthusian" and "neo-Marxist" positions to merge to some extent from the 1960s on. Yet there are still clear dividing lines between the two bodies of theory.

They agree that the degree of exploitation increased throughout the thirteenth century; that the landlords' volume of investments was inadequate; and that the pressure of population on available land increased in the same period. They also agree that agricultural productivity tended to drop; and that climatic and epidemic factors had an important socioeconomic effect because they had easy play in an already sensitive socioeconomic situation. Where their paths diverge, and what justifies us in continuing to speak after 1960 of "neo-Malthusian" and "neo-Marxist" theories, is in their different emphases on which of the factors mentioned should be seen as determinants of the transformation process.

From the 1960s on the "neo-Marxists", represented by Hilton, linked the conflicts in the feudal relations of production with the development of population; this was not the case in the 1950s, when Kosminsky directly claimed that one could disregard the demographic factor - although it was given some consideration by Dobb in 1946. Hilton's gradual rapprochement with the "neo-Malthusians" consists of a greater emphasis on the importance of the demographic factor.

The latter school, on the other hand, represented by Postan, had ever since the 1930s and 1940s been fuelling the fires that the "neo-Marxists" saw, and still see, in the growing exploitation of the thirteenth-century peasants. But in reality they never used this knowledge to attempt to unravel the reasons for the late medieval crisis. The crisis was primarily due to the "demographic check" that the mainly quantitative expansion of the high middle ages resulted in, in their opinion. But after 1960 we saw the "neo-Malthusians" producing ecological, technological and finally financial reasons for this "check". Postan's gradual rapprochement to the "neo-Marxists" involves a stronger emphasis on the unproductive consumption and lack of investment of the landlords as underlying causes of the "Malthusian crisis".

Yet it is still clear that the "neo-Malthusians" continued to seek the reasons for the late medieval crisis in an insoluble conflict between nature and society, whereas the "neo-Marxist" investigation of the causes of the crisis is rooted in a social conflict between lord and peasant.

Notes

1. Edward Miller, "The English Economy in the Thirteenth Century: Implications of Recent Research", in *Past & Present*, No. 28, 1964, p. 30.

2. A. R. Bridbury, *Economic Growth, England in the Late Middle Ages*, London 1962, pp. 52-54.

3. Ibid., pp. 71, 83.

4. Ibid., p. 89.

5. Ibid., p. 25.

6. Ibid., Table, p. 32.

7. Ibid., p. 36.

8. Ibid., Appendix II, p. 111.

9. Ibid., pp. 60-61.

10. Ibid., pp. 65-69.

11. Ibid., p. 103.

12. J. R. Landers, *Conflict and Stability in Fifteenth-Century England*, London 1969, p. 28.

13. Ibid., pp. 20-22.

14. "Poorer villeins fled the manor or sold out to become landless wage-earners. The more fortunate and the more able, by adding strip to strip and lease to lease, became prosperous substantial farmers..." Ibid., p. 25.

15. Ibid., pp. 30-32.

16. F. R. H. Du Boulay, *The Lordship of Canterbury: An Essay on Medieval Society*, London 1966; "A Rentier Economy in the Later Middle Ages: the Archbishopric of Canterbury", in *The Economic History Review*, 2nd Series XVI, 1963-64; "Who were farming the English demesnes at the end of the Middle Ages?", in *The Economic History Review*, 2nd Series XVII, 1964-65.

17. A. E. Levett, "The Black Death on the Estates of the See of Winchester", op. cit., pp. 121-23.

18. G. A. Holmes, *The Estates of the High Nobility in Fourteenth-Century England*, Cambridge 1957, pp. 114-15; J. A. Raftis, *The Estates of Ramsay Abbey - A Study in Economic Growth and Organisation*, Toronto 1957, p. 256; H. P. R. Finberg, *Tavistock Abbey*, 1951.

19. John Hatcher, "A Diversified Economy: Late Medieval Cornwall", in *The Economic History Review*, 2nd Series XXII, 1969.

20. F. R. H. Du Boulay, *An Age of Ambition - English Society in the later Middle Ages*, London 1970, p. 40.

21. A. J. Pollard, "Estate Management in the Later Middle Ages: The Talbots and Whitchurch, 1383-1525", in *The Economic History Review*, 2nd Series XXV, 1972, pp. 553-54, 566.

22. Hatcher, "A Diversified...", op. cit., p. 209.

23. M. M. Postan, "Medieval Agrarian Society in its Prime", § 7 "England", in *The Cambridge Economic History of Europe*, Vol. I, Cambridge 1966, pp. 596-97.

24. Ibid.

25. Ibid.

26. Dyer's summing-up of the development of the Bishop of Worcester's revenues is that "by the early fourteenth century they had reached a turning-point, and the trend in the later middle ages was towards a slow erosion of income. The "Indian summer" of the late fourteenth century delayed decline, and administrative measures in the mid-fifteenth century paved the way for a small recovery in the later years of the fifteenth century..." Christopher Dyer, *Lords and Peasants in a Changing Society - The Estates of the Bishopric of Worcester, 680-1540*, Cambridge 1980, p. 377. In an earlier work Dyer dealt with the development of rents on the same manorial complex. Here, the falling rent revenues are seen in the light of the peasants rent refusals in the fifteenth century. By contrast with the thirteenth and fourteenth centuries, the peasants succeeded in this century in winning the "battle of rent sizes" this way, as a result of the population decline after the Black Death. C. Dyer, "A Redistribution of Incomes in Fifteenth-Century England?", in *Past & Present*, No. 39, 1968.

27. But we find the following comment from Harvey: "Rents did not fall equally with the demand on this estate after 1348, if indeed they fell at all....Towards the end of the century, however, even tenants holding on the Abbey in villeinage began to enjoy some of the advantages that normally belong to tenants - to many of them, at least - when land is in plentiful supply..." A drop in rents is perhaps also suggested by the granting of ad hoc rent rebates, which developed administratively in the course of the fifteenth century to a consistent distinction in the Abbey surveys from about 1450 between rent *levabilis* and *non levabilis*. This may only mean that "the monks preferred to settle with their tenant on a rent that was, if anything, too high, and to adjust it subsequently" in a period when rents fluctuated greatly. This may only be an indication of administrative conservatism - the same conservatism that prompted a monk in the reign of Henry VI to draw up a nostalgic review of rent developments on the Laleham estate. Unfortunately, Harvey did not reproduce this review, and she does not comment further on it. Barbara Harvey, *Westminster Abbey and its Estates in the Middle Ages*, Oxford 1977, pp. 268, 292-93.

28. The closest Searle comes to an overall clarification of the incomes of the landlords in the late middle ages is an overview of "Battle rentals". From this it emerges that the highest rent revenues were obtained in 1367, that there was a negligible drop between 1305 and 1433, and a rise between the 1240-1300 period and 1433. Eleanor Searle, *Lordship and Community - Battle Abbey and its Banlieu, 1066-1538*, Toronto 1974, Appendix 13, p. 467.

29. Hatcher, "A Diversified...", op. cit., p. 209.

30. R. S. Schofield, "The Geographical Distribution of Wealth in England, 1334-1649", in *The Economic History Review*, 2nd Series XVIII, 1965, p. 509.

31. Postan, "Medieval Agrarian...", op. cit.

32. The petit-bourgeois organization of production is characterized by the necessary participation of the owners of the land and/or means of production in the work process that otherwise takes place on the basis of wage labour. It is of course disingenuous to use this concept here, as it applies to divergent systems of organizing production under the capitalist mode of production.

33. Du Boulay, "Who were farming...", op. cit., pp. 447-48.

34. Ibid., p. 449.

35. Barbara Harvey, "The Leasing of the Abbot of Westminster's Demesnes in the Later Middle Ages", in *The Economic History Review*, 2nd Series XXII, 1969, pp. 23-24.

36. The leasing system, according to Hilton, led to a loosening of the non-financial ties between lord and peasant. The leasing terms were financial terms that could be influenced by the market. Just as G. Duby thought that economic conditions - no longer inheritance and legal factors - determined people's social life from about 1300 on in France, Hilton thinks that the same was the case in England about a hundred years later. R. H. Hilton, *The Decline of Serfdom in Medieval England*, London 1969, pp. 44, 57.

37. Harvey, "The Leasing of...", op. cit., p. 23.

38. Du Boulay, "Who were farming...", op. cit., p. 451.

39. Harvey, "The Leasing of...", op. cit., p. 24.

40. Lander, "Conflict and Stability...", op. cit., p. 47.

41. G. E. Fussell, "Social Change but Static Technology - Rural England in the Fourteenth Century", in *History Studies*, Vol. 2, 1968, p. 32.

42. Du Boulay, *An Age of Ambition...*, op. cit., p. 55.

43. R. H. Britnell, "Agricultural Technology and the Margin of Cultivation in the Fourteenth Century", in *The Economic History Review*, 2nd Series XXX, 1977, pp. 65-66.

44. Ibid., p. 62.

45. Britnell's interpretation of the development of the rotation system in northeastern Essex partly corresponds to certain modern scholars' general view of the reasons for the progress of the English field and rotation system in the middle and late middle ages. J. Thirsk thought that the open-field system developed slowly because of population growth and the splitting-up of the land by inheritance. J. Thirsk, "The Common Fields", in *Past & Present*, No. 29, 1964, pp. 7-9, 12-18, 24. Against this, Pitkin thought that "in England impartible inheritance was a necessary concomitant of the open fields". Donald S. Pitkin, "Partible Inheritance and the Open Fields", in *Agricultural History*, Vol. 35, 1961, p. 65. In a study of a number of religious estates in the south east, R. S. Faith demonstrated how the formerly dominant practice in inheritance - primogeniture - broke down in the fourteenth and fifteenth centuries, purportedly as a result of the population decline. This observation prompts Faith to link population growth with impartible inheritance, and population reduction with partible inheritance of tenements - that is, exactly the opposite view to Thirsk. Rosamond Jane Faith, "Peasant Families and Inheritance Customs in England", in *The Agricultural History Review*, Vol. XIV, 1966, p. 92. We find the same view in a work by Howell, who thinks that "the areas which [in the thirteenth century] adopted the option of uniginiture coincide remarkably faithfully with areas of open-field cultivation, which happen to be also the area of widespread manorialization in England..."; and further that "it could be argued that uniginiture developed in areas of land shortage...[while]...partible inheritance is found where there was enough land for all sons..." Cicely Howell, "Peasant inheritance customs in the Midlands, 1280-1700", in Jack Goody et al. (eds.), *Family and Inheritance - Rural Society in Western Europe, 1200-1800*, Cambridge 1976, p. 117. As regards the above general linking of the development of population with that of the field and rotation systems, see also Alan H. R. Baker & Robin A. Butlin (eds.), *Studies of Field Systems in the British Isles*, Cambridge 1973,

Chapter 14, "Conclusions, Problems and Perspectives" by Baker & Butlin, pp. 630-32, 653-56. Against this, Homans continued to defend the view introduced by Gray that the "the open field systems of England were old and were brought as institutions from Continental Germania". George C. Homans, "The Explanation of the English Regional Differences", in *Past and Present*, No. 42, 1969, p. 32. Less radically, Titow also said that "there is overwhelming evidence from all over what is usually considered the open-field area that the bulk of ancient land was already in the form of standard customary holdings at least as early as 1086", and further that there is "clear evidence that regulated crop rotation on the basis of two-and three-field systems was known on tenant land in England in the twelfth century". J. Z. Titow, "Medieval England and the Open-Field System", in *Past and Present*, No. 32, 1965, p. 98. We may note that the combatants defend, respectively, Gray's "cultural explanation" (see Chapter III, p. 91-94) and Orwin's "economic explanation" (see Chapter V, p. 167) and further that this discussion, along with Britnell's view of the development of the rotation system, urges us to caution with respect to the monocausal linking of population development with that of the field and rotation systems.

46. Britnell, op. cit., p. 56.

47. Ibid., pp. 62-64.

48. Britnell, on the other hand, has constructed a table showing grain yields for wheat and oats between 1338 and 1351 on a single manor. The table supports his view that the soil was progressively exhausted up until the 1350s. Ibid., Table 3, p. 60.

49. We will later return to Titow's study of the "Winchester Yields" before 1350.

50. D. L. Farmer, "Grain Yields on the Winchester Manors in the Later Middle Ages", in *The Economic History Review*, 2nd Series XXX, 1977, pp. 562-63.

51. Ibid., Table 5, p. 563.

52. Ibid.

53. Ibid., p. 566.

54. P. F. Brandon, "Cereal Yields on the Sussex Estates of Battle Abbey during the later Middle Ages", in *The Economic History Review*, 2nd Series XXV, 1972, p. 405.

55. Ibid., p. 418.

56. Ibid., pp. 419-20.

57. Ibid., pp. 412-13, see Table 3.

58. Farmer, op. cit., Table 2, p. 559.

59. J. Z. Titow, *English Rural Society, 1200-1350*, London 1969, p. 53.

60. Farmer, op. cit., p. 560.

61. Postan in *The Cambridge Economic History*, op. cit., pp. 552-59.

62. David Roden, "Demesne Farming in the Chiltern Hills" in *The Agricultural History Review*, Vol. 17, 1969, p. 12.

63. Ibid., pp. 14-16.

64. M. M. Postan, "Village Livestock in the Thirteenth Century", in *The Economic History Review*, 2nd Series, Vol. XV, 1962.

65. Postan in *The Cambridge Economic History*, op. cit., pp. 556-57.

66. B. K. Roberts, "A Study of Medieval Colonization in the Forest of Arden, Warwickshire", in *The Agricultural History Review*, Vol. 16, 1968. Roberts speaks of three phases of colonization in this area, the third and final "lasting from the mid-thirteenth century until the early decades of the fourteenth..." (p. 108).

67. For example, "the "high farming" period par excellence" on the estates of Canterbury Cathedral Priory comes almost in the same period (1306-1324) as the greatest contraction in the Winchester estates. R. A. L. Smith, *Canterbury Cathedral Priory*, Cambridge 1943; Titow, *English Rural Society*, op. cit., pp. 52-53.

68. Ibid.; Postan, in *The Cambridge Economic History*, op. cit., p. 558.

69. Ibid., p. 560.

70. Ibid., p. 557.

71. Ibid.

72. David Roden, op. cit., p. 23.

73. Baker has studied the so-called *Nonarum Inquisitiones*, carried out with a view to levying taxes in 1342. Parishioners were to assess, under oath, the value of corn, wool and lambs in the kingdom, and were obliged to explain disparities between this list and a previous one drawn up for tithe payments in 1291. The source only accounts for 25 counties satisfactorily, but this provides indications that four areas were hit by widespread desertions between 1291 and 1342 - the North Riding, Shropshire, Essex and a group of counties north and west of London. In the last of the groups, Baker concentrated on the counties mentioned in the text. One can of course object that Baker simply takes the assessors' explanations at face value, as these, as mentioned above, were picked out from parishioners in each parish. On the other hand, one must in this context suppose that any good reason for a drop in the assessment, such as sterile or exhausted soil, would hardly be passed over in many cases. Alan R. H. Baker, "Evidence in the "Nonarum Inquisitiones" of Contracting Arable Land in England during the Early Fourteenth Century", in *The Economic History Review*, 2nd Series, Vol. XIX, 1966.

74. Postan, in *The Cambridge Economic History*, op. cit., pp. 558-59.

75. J. Z. Titow, *Winchester Yields. A Study in Medieval Agricultural Productivity*, Cambridge 1972, pp. 12, 15.

76. See Table I, "Yields per measure of seed", ibid., pp. 4, 12, 14.

77. Ibid., p. 12.

78. H. H. Lamb, "Britain's Changing Climate", in *The Geographical Journal*, Vol. 133, Part 4, 1967, pp. 446, 453-60; Gordon Manley, "Climate in Britain over 10,000 years", in Alan R. H. Baker & J. B. Harley (eds.), *Man Made the Land*, 1973, pp. 18-20.

79. Titow, *Winchester Yields*, op. cit., pp. 30-32.

80. Ibid., pp. 32-33.

81. Ibid., Tables 8-10, pp. 27-29.

82. "The most usual pattern was for the yields to reach their lowest level in the last quarter of the thirteenth century and then to improve gradually, sometimes failing to reach, and sometimes reaching and surpassing, the level of yields in our first period. Yet another group of manors shows a progressive decline in productivity right up to the second quarter of the fourteenth century followed by an improvement in that quarter". Ibid., pp. 14-15; "Practically all changes after 1270 are instances of contraction of the demesne". Ibid., p. 25.

83. Ibid., pp. 30-31. See also Appendix L: "Some data relevant to the assessment of changes in productivity", pp. 136-39, and Appendix M: "Comparison of changes in productivity with changes in the area under seed and the ratio of animals to that area", pp. 140-44.

84. See for example Titow, *English Rural Society*, op. cit., pp. 44-45, and Edward Miller & John Hatcher, *Medieval England - Rural society and economic change 1086-1348*, London 1978.

85. See Postan's discussion of this in *The Cambridge Economic History*, op. cit., pp. 577-81.

86. Ibid., pp. 593-95; R. H. Hilton, *A Medieval Society - The West Midlands at the End of the Thirteenth Century*, London 1966, pp. 51-52.

87. J. R. H. Moorman, *Church Life in England in the Thirteenth Century*, Cambridge 1955, p. 294.

88. R. A. L. Smith, op. cit., p. 18.

89. Mavis Mate, "The Indebtedness of Canterbury Cathedral Priory 1215-95", in *The Economic History Review*, 2nd Series, Vol. XXVI, 1973, p. 193.

90. Ibid.

91. "I pray you, order your life according as your lands are valued yearly by the extent, and nothing beyond that". Walter of Henley, *Husbandry*, p. 3.

92. Mate, op. cit., pp. 187, 192.

93. Ibid., p. 191.

94. Titow, *English Rural Society*, op. cit., p. 50.

95. This point of view was also common in the Continental context at the time. See for example B. H. Slicher van Bath, *The Agrarian History of Western Europe*, 1963.

96. M. M. Postan, "Investment in Medieval Agriculture", in *The Journal of Economic History*, Vol. 27, 1967, pp. 578-79.

97. Postan thinks that on the whole there existed only the following possibilities for productive investments: "additional ploughings of fields and fallow (rebinatio)", "change...of seeds and seed ratios", "changeover from two- to three-field system", and "reclamation and colonization". Ibid., p. 581.

98. Ibid., pp. 579-80.

99. R. H. Hilton, "Rent and Capital Formation in Feudal Society", in *Second International Conference of Economic History*, Vol. II, Paris 1962, pp. 51-53, 67.

100. Ibid., pp. 35-36.

101. Ibid., pp. 55-56; Postan, "Investment...", op. cit., pp. 585-87.

102. Hilton, "Rent and Capital...", op. cit., pp. 58-60.

103. Ibid., p. 67.

104. It is characteristic of Hilton's work that it focuses on the most difficult field - on empirical grounds - in English social and economic history: the peasants' conditions in the high and late middle ages. As early as 1955 Hilton had pointed to this area and the fifteenth century in general as the two most neglected ones. R. H. Hilton, "The Contents and Sources...", op. cit., pp. 5-6.

105. "On all the manors of the bishopric a twice-yearly fine, ad Hundredum Sancti Martini and ad Hundredum de Hockeday, usually called tithingspenny, was collected from a very early date". This duty consisted of a payment of one penny from every male above the age of twelve to whoever presided over the Hundred Court. According to this - said to be the only direct indication of the development of the male population - the annual growth rate on the manor of Taunton had been 0.85% between 1209 and 1311. It should be mentioned in connection with Titow's survey of the population-land ratio between 1248 and 1311 that Taunton was "the most extreme case of an anciently settled

manor" - that is, a manor without significant colonization in the period. J. Z. Titow, "Some Evidence of the Thirteenth Century Population Increase", in *The Economic History Review*, 2nd Series XIV, 1961-62.

106. R. H. Hilton, *A Medieval Society...*, op. cit., p. 114.

107. Vinogradoff, *Villainage...*, op. cit., p. 132. Hilton defines villeinage, in accordance with the thirteenth-century legal sources, as a liability to render labour services and pay merchet, heriot and similar customary dues. R. H. Hilton, "Freedom and Villeinage in England", in *Past and Present*, No. 31, 1965, p. 6. See also Hilton's more detailed account of the villein's duties and dues in R. H. Hilton, *The Decline of Serfdom....* op. cit., p. 24.

108. Vinogradoff, *Villainage*, op. cit., p. 220.

109. Villeinage is mainly a matter of the subjection of the peasants to a manorial form of organization. It is a relationship of personal subjection which, apart from certain aspects related to criminal law (ibid., p. 47), totally subjects the villein to the jurisdiction of his lord as regards rent, mobility etc. Ibid., p. 57.

110. Hilton, *The Decline of Serfdom...*, op. cit., p. 30.

111. J. Z. Titow, "Some Differences between Manors and their Effects on the Condition of the Peasant in the Thirteenth Century", in *The Agricultural History Review*, Vol. X, 1962, p. 6.

112. Kosminsky, *Studies in the Agrarian...*, op. cit., pp. 206, 212-13, 214ff.

113. Hyams thinks that (a) "there is no good evidence of a general peasant land market in the early thirteenth century"; (b) "it is likely that the peasant land market developed earlier in East Anglia (or perhaps more generally over eastern England) than elsewhere". R. Hyams, "The Origins of a Peasant Land Market in England", in *The Economic History review*, 2nd Series XXIII, 1970, p. 19. Postan thinks that "there are strong grounds for believing that the "carte nativorum" were merely a local instance of the new method which some landlords of the thirteenth and early fourteenth centuries began to use in regulating land transactions of the villein tenants". He supposes that the peasants' land market had existed a few decades before Kosminsky could prove its existence in the hundred rolls from 1270 (Kosminsky, ibid.). But this was not the only reason. Postan found indications of the sale and purchase of land among the peasants from as early as the beginning of the thirteenth century in a number of different sources - the Curia Regis Rolls from 1214 and 1219, the Winchester account rolls from 1209 and later, Bracton's notes and "the inquisition into the Bishop of Ely's estates in 1251". C. N. L. Brooke and M. M. Postan, *Carte Nativorum*, Oxford 1960, pp. xxxi-xxxii, xxxvii-xl.

114. "...A family well provided with land, but deficient in labour or stock or tools...might find their larger holding too much for them" and vice versa. Ibid., pp. xxxiv-xxxv.

115. Hyams, op. cit., pp. 19-21.

116. R. H. Hilton, *Bond Men Made Free - Medieval Peasant Movements and the Rising of 1381*, London 1973, pp. 27-28.

117. "Purquisita", not to be confused with "fines for entry, marriage, regular and irregular tax such as tithingspenny and tallage etc.". Alfred N. May, "An Index of Thirteenth-Century Peasant Impoverishment? Manor Court Fines", in *The Economic History Review*, 2nd Series, Vol. XXVI, 1973, p. 390.

118. Ibid., p. 392.

119. Ibid., p. 397.
120. Ibid., pp. 397-98.
121. Ibid., p. 399.
122. Ibid., p. 396.
123. Hilton, *Bond Men...*, op. cit., pp. 85-88.
124. Ibid., pp. 96, 109.
125. Ibid., p. 153.
126. R. H. Hilton, *The English Peasantry in the Later Middle Ages*, Oxford 1975, p. 64. See also Dyer, "A Redistribution..", op. cit., p. 33 and J. Ambrose Raftis, *Tenure and Mobility - Studies in the Social History of the Medieval English Village*, Toronto 1964, pp. 190-92.
127. See also R. B. Dobson, *The Peasant Revolt of 1381*, London 1970, p. 27.

Chapter VIII

The Historical Turning-Point
(1960-   )

VIII.1 Money and price formation before 1350

After 1960 one can hardly speak any longer of a truly "neoclassical" interpretation of the socioeconomic changes of the late middle ages. But we can still find adherents of this school who contributed to research on the history of price development, without directly linking this factor with the general developments of the period other than by referring to the previous discussion between Postan on the one side and Robinson and Schreiner on the other.

From 1969 we have D. L. Farmer's study of the movements of certain cattle prices in the thirteenth century.[1] N. J. Mayhew dealt with the connection between coinage reserves and price drops in the thirty years up to the Black Death in an article from 1974,[2] and Mavis Mate tried to demonstrate a connection between the general rise in commodity prices in the first decade of the fourteenth century and a purportedly more plentiful supply of silver.[3] In all three cases, it is attempted to demonstrate how monetary factors play a role in the price formation which in the "neoclassical" theory is a determinant of socioeconomic development in the late middle ages, and in this respect each of the three works has its own value.

Farmer points out that state money conversions four times in the course of the thirteenth century appear to have had an immediate, rather drastic braking effect on the otherwise generally rising cattle prices in the period.[4] With this he believes he has demonstrated "that thirteenth-century governments were more successful in curbing rising prices than are those today".[5] However that medieval governments implemented monetary policies with this aim in mind is not confirmed by Mate's monetary explanation of the rising prices in the initial years of the fourteenth century.

With its prohibitions against exports of silver from England and their allegedly strict enforcement, the Crown was itself, according to Mate, part of the reason for the general rise in prices at the beginning of the fourteenth century.[6] For this reason - but also because of increasing wool exports and because more silver was minted in London at the expense of the coinages in the Netherlands and France, the volume of silver and money rose considerably in England, with rising prices as the result.[7] Mate thus attempted to provide an explanation of the rise in the silver and money volume in England from around the time the crisis of the European silver mining industry began. This was an explanation which clearly differed from the one previously offered by Postan,[8] and which was later attacked from several sides - for example by Mayhew,[9] who also presented a number of statistical calculations of the money supply and coinage in England between the end of the thirteenth and the middle of the fourteenth century.

We cannot here embark upon a more detailed investigation of Mayhew's methodology and results.[10] Suffice it to say that his account of monetary developments accords well with Mate's, inasmuch as it suggests a great increase in minting and coinage volume between 1300 and 1320. After this one can note a continuous fall in the national coinage reserves up until 1349, whereas minting from London and Canterbury again seems to rise from zero around 1325 and until 1350.

Although the authors cited all maintain that monetary developments had a crucial effect on price development in the period before the Black Death, and, in more or less explicit agreement with the "neoclassical" theory, consider price development to be of prime importance for general socioeconomic development,[11] they have cast doubts on this theory, at least as far as developments in the first decade of the fourteenth century are concerned. The "neoclassical monetary" theory does not convince on the basis of these studies, and in fact their authors also assign importance to other factors - not population development, however, nor, according to Mate, any drop in agricultural productivity.[12] So despite all this, there is still after 1960 a strong opposition to the "neo-Malthusian" interpretation of late medieval developments.

It was acknowledged that the flexibility of prices for various commodities must have had a modifying effect on the otherwise regular derivation of price formation from monetary developments, and that poor harvests due to bad weather, for example, had some influence. Similarly, both Farmer and Mate thought that fiscal factors may have had some effect on price formation.[13] These are factors we shall give closer scrutiny in the following, in a more suitably developed context.

257

VIII.2 War and taxation

In a previous chapter we saw that Postan did not think that one can attribute crucial socioeconomic importance to the Hundred Years' War (see Chapter V, p. 144). He persisted in this general view in a later article from 1967, where he concluded: "To quote my earlier essay on this subject, the Hundred Years' War was at best a makeweight, not the mainspring of social change".[14] Therefore, and because we have not really seen serious attempts since the days of Denton and Cunningham to enrich the fairly narrow economic interpretations of late medieval socioeconomic developments with other aspects of importance for the process, two studies from 1975 of royal taxation in the period before the Black Death must be welcomed.

In the perpetual debate on where the dividing-line can reasonably be drawn in English history between the Middle Ages and the Late Middle Ages, both Edward Miller and J. R. Maddicott claim that it could just as easily be placed at the accession of Edward I,[15] or the beginning of the Scottish wars and the Hundred Years' War[16] as at the famine years of 1315-1317 or the inception of the alleged "Malthusian check". In fact, it could more reasonably be related to the former events, because in the middle of the 1290s Edward I involved England in a military enterprise that would not only stretch over more than a century, but would also lead to regular, recurrent taxation even in peacetime; taxation so extensive and burdensome for all groups of the population that it must have had at least a reinforcing effect on the factors that led to the social and economic changes in the period before 1350.

Maddicott, whose study focuses on the influence of taxation on the conditions of the peasants, thus thinks that if the population was reduced in the first half of the fourteenth century, it was not least due to this taxation; among other reasons, because the peasants were far more defenceless in the face of certain of the royal taxes than of the lord's rent claims.[17] Furthermore, the taxation of the peasants increased, so that it was at least far more widespread geographically in the last few decades before 1350, and perhaps was also more of a burden on the individual in these decades than before.[18]

It is hard to judge how the taxation affected different groups of the peasant population; but Maddicott is sure that the peasants as a whole were harder hit than the landlords.[19] Yet this does not mean that the latter group was unaffected by royal taxation. In the first place Maddicott

258

thinks that the taxation cannot have failed to have a negative influence on the development of the feudal rent in the period after 1300,[20] and that there are thus grounds, because of the taxation of the peasants, to speak in an indirect sense of increasing conflict between the Crown and the landlords. In the second place, part of the aim of Miller's study is precisely to show that the royal taxes also affected the landlords directly;[21] especially the religious houses and especially in the period after 1320, when price drops and increasing wages indirectly added to the direct burden of taxes.[22]

"The peasantry was most affected by three sorts of tax: by the levy on movables, by purveyance and by costs of military service".[23] This taxation had a direct influence on the reproductive conditions of the peasants, and according to Maddicott is the probable reason for the possible reduction in the population in the first half of the fourteenth century. But at any rate the taxation had such an effect on the economic conditions of the peasants that it gradually resulted in "the complex economic crisis which affected England between 1337 and 1342. The effect of low output of coin at the mint and of a run of good harvests combined with taxation on a massive scale to produce a severe shortage of money. It became extremely difficult to pay taxes and probably rents..."[24]

Against this background the landlords' tax burden was increased; in the first place because the taxation of the peasants may have affected the population figures and thus the development of wages, prices and rents; in the second, because the taxation of the peasants in itself affected the development of rents negatively, and by reducing social purchasing power also affected prices negatively. The fiscal policy of the Crown thus brought about not only the increasing desertion of peasant land and the growth in rent arrears documented by Maddicott, primarily on the basis of the *Nonarum Inquisitiones*, but also, according to Miller, to some extent "a cutting back in scale of demesne enterprise, the abandonment of some of the less fertile arable, a reduction in the rate of capital investment in repairs and improvements." This was an erosion of manorial production from the 1320s which in many places, in keeping with Maddicott's reflections on the peasants' economic situation, "was not balanced by a comparable advance of peasant economic activity."[25]

VIII.3 The turning-point

Bridbury raised a number of objections to these views in a 1977 article. He acknowledged the importance of taking the above-mentioned military and fiscal aspects into account in the debate on social and economic development in the late middle ages. He also acknowledged that the warfare that began in the 1290s should not only be seen in contrast to the fairly peaceful nature of the preceding period, but must also have been considerably more costly than previous medieval military activity. But in the first place he thought that we must consider the general price rise in the period when assessing the costs of the late medieval wars for the English economy.[26] Secondly, he objects that "we cannot hope to make sense of the costs of waging war in the reigns of Edward I and Edward II without first taking the precaution of settling those costs against the vastly increased capacity of the economy to sustain them".[27]

We note that in this critique of Miller and Maddicott, Bridbury has apparently abandoned his former view that the end of the thirteenth century and the beginning of the fourteenth were also characterized by an increasing scarcity of land due to the rising population and falling agricultural productivity (cf. Chapter VII, p. 217). Here he thinks, in keeping with the classic view of the period, that the English economy was both "various and opulent",[28] even though, or perhaps precisely because this wealth was "accompanied by the progressive impoverishment of increasing numbers of ordinary people".[29] But this increasing impoverishment did not mean, as Miller and Maddicott thought, that the wars might result in an overburdening of the English economy's overall resources and a consequent structural transformation.[30]

The boom period for the landlords continued unabated for five decades of the seven first decades of the fourteenth century. The development of rents and prices did not go against the landlords until some point in the 1370s,[31] and wage development adapted in the first three decades of the century to price development, just as it did in the period immediately after the Black Death.[32] There was plenty of labour available until long after this event. This statement should be understood in the light of Bridbury's opinion, following Russell, that the Black Death did not smite down an already reduced population. "Whatever war and famine may have done, if we may credit what the statistics of wages and prices suggest, they did not make life difficult for landlords" until after 1370.[33]

If we are to believe Bridbury, there is no basis for speaking of a turning-point for the manorial economy and population development until after the Black Death. Neither the consequences of the increasing military activity nor the famine years of the second decade of the fourteenth century had such drastic consequences. Yet Ian Kershaw had also argued just a few years before that the period between 1315 and 1322, at certain times and for certain big landlords, marked a turning-point.[34]

VIII.4 Famine and agrarian crisis, 1315-1322

Kershaw's article from 1973 is the first monograph on the social and economic consequences of these famine years in England. Before this one had to seek information in Henry S. Lucas's more general work on *The Great European Famine of 1315, 1316 and 1317* from 1930.[35] It is well known that the famine years began with a disastrous harvest in 1315, and that the failed harvest was due to heavy rainfall in the summer months of 1315. This, against the background of extensive climatological research in Europe, contributed to the view that there was a general deterioration in climate from about 1300 on.

Kershaw's chronological review of the harvest years from 1315 to 1322, based on evidence from manorial sources on climatic conditions, the research discussed above on the development of grain yield, and the statistical knowledge, also discussed above, of price developments for the most important grain types, does not appear however to confirm this view directly. In the light of price development (Rogers) the harvest year 1315-1316 seems to have been the worst, but the following year was not much better. On the other hand the 1318 harvest seems to have been a rich one, which prompted Lucas to consider this year as the end of the catastrophe - perhaps even the harvest year 1317, because "the harvest that summer must have greatly tempered the hunger".[36]

It is true that corn prices were relatively low in 1318, and also in 1319, but as early as 1320 they rose again considerably, as a foretaste of the very high prices that followed the very bad harvest of 1321, and which gave rise to a price level in the year 1321-1322 that Kershaw thinks can be compared with the prices he found for the first famine years after 1315. Only with the better harvest years and falling corn prices after 1322 does he think one can speak of the agrarian crisis being over.[37]

Kershaw justifies speaking of a real agrarian crisis, first by saying that sheep and cattle farming were also affected,[38] and secondly by saying that these losses and the years of famine, judging from his assessment of the population development,[39] not only hit the peasants hard, but also, because of the population decline, led to the dissolution of the manorial system on a number of estates; not, however, in the wealthiest and most densely populated parts of the country. And by contrast, the depopulation observable in certain regions was perhaps as much due in the longer term to the dissolution of the manorial system and emigration from these regions than to population decline itself.[40]

Although the agrarian crisis of 1315-1322 only caused long-term structural transformation regionally, Kershaw thinks that the climatic deterioration of these years can be said to have heralded a general turning-point. Yet he does not claim that the climatic factor was the only one that triggered off the social, economic and demographic changes that he thinks, along with Postan, began a few decades before the Black Death.[41]

VIII.5 Economic prosperity, 1325-1345

Yet there is little in D. G. Watts' study of the Titchfield Abbey estates to suggest such an interpretation. On the contrary, Watts thinks there are grounds to propose the following model for development between Kershaw's "agrarian crisis" and the Black Death. The years between 1325 and 1345 were economically speaking prosperous, borne up by favourable climatic developments and good harvests. The population once more overcame the temporary decimation of the great famine, and the peasants actively exploited their opportunities. More land came under cultivation, and even marginal land was productive in the good seasons. Watts thinks it unlikely one can speak of a "Malthusian" situation in the period before 1348.[42]

He arrives at this view because he sees on these estates an increasing number of land transactions among the peasants and a parcelling-out of the bigger tenant farms, while demesne production's contraction was primarily due to the transition to sheep farming, and then to leasing.[43] But he reaches this conclusion chiefly because, like Rogers, he sees generally falling corn prices followed by rising prices for animal products, which he considers "striking evidence of rising, not falling demand".[44]

VIII.6 Summing-up

This interpretation of the period between the famine years 1315-1322 and the Black Death is not surprisingly in accordance with Rogers' general assessment of the economic situation in these decades (cf. Chapter I, p. 15). Contrasted with Kershaw's article, it is yet another example of how in English research in the 1960s and 1970s there was not only great disagreement on the nature of social and economic developments in the fifteenth century, but also in accounts of those of the first half of the fourteenth century.

We have seen before how these differences were associated with various theoretical explanatory models, and further how most of these theories were related to, and as far as their internal consistency is concerned crucially dependent on, one salient empirical problem area - the demographic one. For the "neoclassicists" population factors were on the one hand a direct element in their arguments, and on the other they used demographic data as indicators to demonstrate monetary development and developments in prices, wages and rents. For both the "neo-Malthusians" and through time the "neo-Marxist" authors the thesis of overpopulation and declining population figures before 1348 was basic to the cohesion of their theories. Finally, the various epidemiological theories corresponded with different interpretations of the nature of the epidemics and estimates of population developments after the Black Death. It may therefore seem like the closing of a circle when this work must be rounded off with a discussion of the same fundamental nature as the one it began with - a review of the most recent results of demographic research and their relevance to the discussion of the socioeconomic development of the late middle ages.

In a research overview from 1965 J. C. Russell formulated the main problems of demographic research thus: the strong population growth in the thirteenth century is indisputable; but did the population density, at its peak, reach a state of overpopulation? How long before the Black Death was this peak reached? On the whole there is also agreement that the consequence of the epidemics from the Black Death on were disastrous; but did mortality vary regionally, and when did the population decline stop? Was the stabilization of the population in the fifteenth century real or did it fall until just before the quick growth set in at some point in the second half of the fifteenth century.[45]

Let us first look at some more recent studies of the population development before 1348, as an approach to the first two questions. Then we can also, in the light of the most recent research results, attempt to address the issue of the demographic consequences of the epidemics.

VIII.7 Population density at Elloe, 1250-1300

It would be unjust to call Hallam an advocate of the overpopulation theory. His study of population density in the wapentake of Elloe, a Fenland area, in the second half of the thirteenth century, does not in itself provide a basis for speaking of overpopulation, and he finds it even more difficult to prove that there was general overpopulation in England.[46] Still, his work has often been used to support this view, since Hallam found that the population density in the Lincolnshire Fenland areas in the latter half of the thirteenth century became so massive that we have to come right up to the nineteenth century before we can once more record a similar density.[47]

Yet this rather dramatic population level immediately before and after 1300 cannot necessarily be taken to mean that the area was overpopulated. As we recall from our discussion of Fenland colonization (see Chapter V, p. 165-166), the population could be reproduced here by other means than tillage, which only contributed slightly to overall production in the Fenlands. Hallam is very much aware of this,[48] and thinks therefore that "we cannot pretend to show that the Fenlanders were starving about 1300", although "it is clear that their position was not an easy one".[49]

The Fens of Lincolnshire, Cambridgeshire and Norfolk offered the population a variety of means of reproduction that we do not find in many other places. It is therefore unreasonable to see Hallam's study of population density here as documentation of overpopulation and as indicating a general demographic tendency in the area. Nor are there grounds, therefore, to go into more detail about the evidence Hallam presents for population development in the area. However, we should mention that in certain localities - Pinchbeck and Spalding - the population growth seems to have stopped between 1260 and 1287, or at least to have levelled out; and that he finds no sign of any substantial decline before 1348.[50]

VIII.8 The male population of Taunton, 1209-1330

Perhaps Titow's estimates of population growth on the Winchester manor
of Taunton are more representative. At least they are not subject to the
uncertainty that all other estimates suffer from, inasmuch as they are
constructed from material that only deals with people in tenures, and the
total population must be estimated on this basis by means of various mult-
ipliers.[51] Although it is not said directly in Titow's source that tithingspenny was also imposed on all males over twelve at Taunton, Titow
assumes that this was the case, as such an explicit age limit is found in
the annals of other Winchester manors.[52] The itemizations of tithingspenny payments thus, according to Titow, directly reflect the development
of the male population between 1209 and 1330 at Taunton.

After this period the sources no longer show any shifts in the volume
of payments, which means that after 1330 the number of payments was
probably itemized as the same every year. Titow considers that this
change in practice must be connected with the fact that the payments (cf.
his year-by-year table of payments on p. 224) reflect a turning-point in the
development of the male population at Taunton in the years after the
great famine. Similarly, he thinks that the table shows more than a doubling of the male population over twelve between 1209 and 1311.

Disregarding any remissions of tithingspenny, if we keep to the stated
payments they grow according to this table from 506 in 1209 to 1359 payments in 1311. Titow calculates this rise to represent an annual growth
rate of 0.85%.[53] Yet one may question whether this interpretation holds
water on closer scrutiny of the figures in the table, and further whether
the table can be said to support his and Postan's Malthusian views that
the famine years had their impact in a situation where a potential overpopulation became imminent.

In the first place it must be noted that between 1309 and 1310 there
was apparently a disproportionately strong rise in the number of
payments. While the figures developed fairly evenly between 1209 and
1309 - from 506 to 1167 - there is an abrupt rise between 1309 and 1310
- from 1167 to 1344. This means a rise of a good 15% between these
two years. And this level is maintained all the way up to about 1320. So
if Titow had chosen 1309 instead of 1311 as the terminus for his
calculation of population growth since 1209, it would have been rather
lower; but this would not change the fact that the table indisputably

indicates very strong growth in the male population at Taunton throughout the thirteenth century.

In the second place, and more crucially, in the light of the table one must doubt whether the famine years of the second decade of the fourteenth century really were the turning-point Titow claims. Between 1310 and 1316 the number of tithingspenny payments fluctuates between 1344 and 1334, and then, it is true, falls to an average level of 1228 in the next decade. Yet this level is still considerably higher than the number of payments for 1309 and all the previous years. The second decade of the fourteenth century is thus characterized both by a dramatic rise and a rather less dramatic fall in payments. Compared with the year when payments were most numerous - 1311 - the rise from 1309 is no less than 192, while the drop to the average level for the 1317-1328 decade is 131.

There is no disputing that the table excellently illustrates a generally acknowledged demographic consequence of the famine years. But it does not directly support the idea that we are faced here with a turning-point, nor the general existence of a Malthusian check in the second and third decade of the fourteenth century. The table rather provides grounds for speaking of a continuously rising number of males over the age of twelve at Taunton between 1209 and 1330, interrupted by great fluctuations in the first and second decade of the fourteenth century, where neither we nor Titow are able to explain the explosive growth between 1309 and 1310, although we can easily explain the rather less drastic drop after 1317 in the light of the preceding years of crop failure. The dramatic rise between 1309 and 1310 can hardly be considered to reflect the real situation, though.

VIII.9 Deaths/surviving heirs, 1250-1348

This impression of a rising male population at Taunton all the way up to the third decade of the fourteenth century is not confirmed by the constructions of so-called replacement rates drawn up by Sylvia L. Thrupp from manorial court rolls from six East Anglian manors in Lincolnshire, Norfolk, Suffolk and Cambridgeshire. For these suggest that the relationship between adult deaths (mainly unfree peasants and artisans, but also big and small free peasants) and surviving adult male heirs over the age of sixteen tended to develop negatively in the century before the Black Death - that is, towards a proportionately falling number of heirs per

266

deceased adult.[54] But, as was the case with Titow's evidence for the rising population, we are dealing here with a very small sample where we should be careful of drawing far-reaching conclusions.[55]

VIII.10 Famine and mortality, 1315-1317

By contrast, Titow's table seems to illustrate excellently J. C. Russell's view on the demographic consequences of the famine of 1315-1317. Although Russell thinks that the English population in these years experienced what was probably the worst catastrophe between the plague outbreaks of the sixth century and in 1348, he does not conclude that this catastrophe made population do any more than level out or perhaps temporarily go into a slight decline.[56] Russell's arguments start with Postan's and Titow's estimates of a 25% increase in mortality on the Winchester estates between 1310 and 1319,[57] and from his comparison of this rising mortality among the Winchester peasants with his own estimates of mortality among vassals in the same period, which is said to have seen parallel development.

Among the group of vassals Russell does not think the famine in itself can have had fatal effects. What did affect them was the subsequent dysentery epidemic which he claims to have had a particularly lethal effect on malaria and tuberculosis sufferers, of whom there were many - also in the higher social strata. Russell's estimate of mortality among vassals, which shows high excess mortality in 1316 and 1317 but a relatively low mortality in the subsequent years from 1318 to 1321, combined with these deliberations, lead him to conclude that the dysentery epidemic in the wake of the famine only hastened the incidence of death in an already moribund group.[58]

Correspondingly, he finds that infant mortality between 1320 and 1335 was no higher than at any other period between 1255 and 1346, and that this obviates the suspicion that there was an explosion in infant mortality between 1315 and 1317[59] - this despite the fact that the dysentery epidemics were supposed always to entail high infant mortality. For Russell thinks, in support of this rather dubious argument, that the same situation as he observed with vassal mortality must have existed among the rest of the population, including children. "The disease merely struck down a little earlier those who could succumb to the next serious disease", and since large sections of the population at this time had malaria or tuberculosis, "the pestilence of 1315-1317 merely hurried them along to their death, leaving the population of Europe in better health conditions than it had enjoyed before the pestilence". This helped to control the demographic consequences.[60]

VIII.11   Population development, 1300-1348

We find largely the same arguments in Russell's 1966 assessment of the population development in the period before the Black Death.[61] In this work, besides trying to arrive at a general evaluation of population development in England in the first half of the fourteenth century, he also deals with a problem of particular importance for the probability of the postulates of the overpopulation and overexploitation theory. For Russell has drawn up comparisons between peasants' and vassals' life expectancies in the period before and after 1300. He arrives at the conclusion that if there was any difference, it was to the advantage of the peasants.[62] But let us look more closely at his account of the general demographic movements between 1300 and 1348, which are certainly not without interest in this context.

His calculations of the reproductive capacity of the vassals show that the period between 1300 and 1347 was typified by growth in the numbers of this group, and that the expansion stopped temporarily in the second decade of the fourteenth century. Since he further thinks it proven that peasant life expectancy was on the whole the same as that of the vassals, he assumes that the above-mentioned developments can be generalized to the whole English population.[63] He underpins this reading in two ways, both taken from his principal work from 1948: on the one hand by comparing the number of individuals in thirty localities in surveys from the pre-1348 period with the poll tax figures from 1377; and on the other by similarly comparing Domesday and pre-1348 surveys again from 152 localities. Both methods result in a demonstration of a regularly rising population between 1290 and 1347.[64]

A more direct method of measuring population development in the first half of the fourteenth century is, as Russell remarks, comparisons of surveys from before and after 1315-1317. But unfortunately we have little chance of using this method. Another direct method is the one discussed above: using tithingspenny payments, and Russell interprets these analogously with our reading - as an illustration of the continued population growth between 1209 and 1330,[65] not as Titow would have it, as showing that 1315-1317 was a demographic turning-point.

It is striking how Russell in this work, although using demographic calculations from his earlier principal work, now seems to be arguing much more firmly for a continuation of the population growth of the high middle ages right up to the Black Death. We saw before that he was rather uncertain about general demographic development in the first half of the fourteenth century, yet still felt justified in thinking that "an S curve would illustrate the development better than a compound interest curve" as far as the period from Domesday to the Black Death is concerned. This partially revised but at least more clearly expressed view of demographic movements before 1348 resulted in his refutation of Postan's economic proof of the alleged population reduction between 1300 and 1348 and his rejection of the overpopulation theory.

VIII.12 Criticism of the overpopulation theory

Russell does not consider that the development of prices and wages in the century before 1348 supports the theory of a falling population after the second decade of the fourteenth century; he does think it supports his own view that the population rose all the way up to the Black Death.[66] As for the overpopulation theory, he thinks that it is partly due to the fact that its advocates do not take the age and gender distribution of the population into account. Thus, when Kosminsky maintains that 29% of the unfree and 47% of the free peasants had less than five acres of land, one must not forget that a large proportion of these smallholders were women and old men without much in the way of families to support. Similarly, of course, one must not overlook the fact that these smallholders to a great extent subsisted on wage labour.[67]

In fact there is not much to suggest overpopulation when one examines Kosminsky's table of "Serf and Free Peasantry" from the Hundred Rolls of 1279-1280.[68] The figures show that 13,504 tenant farmers had 187,323 acres at their disposal - that is, almost 14 acres per farm on average. And when we can further see from the table that over fifty per cent of them had more than a half virgate of land, we may conclude that there must have been a great deal of wage labour available to the good 35% who only had small plots, besides what they could get on the demesne.[69]

Russell's critique of the overpopulation theory came at the same time as Barbara Harvey took this problem and population development up in a published lecture.[70] Many of the arguments and documentation Harvey offers against the overpopulation theory agree with what has already been said - but this will not prevent us from looking more closely at them.

The lecture falls into two sections. First Harvey addresses the overpopulation theory on the basis of the development of rents, the size of peasant farms and the quality of the soil. Harvey notes what has also frequently been mentioned here - that assessments of the increasing rent burden throughout the thirteenth century are not based on fluctuations in the normal annual land rent, but on the development of entry fines. She then shows by means of a number of examples of entry fine sizes on a range of manors representing an area from the western counties to Lincolnshire and the northern part of the country, that the level at the end of the thirteenth century and the beginning of the fourteenth was in her view quite moderate.[71] And furthermore, "on many estates different fines for holdings of the same size suggest a regard at some time for the quality of the land". She also finds examples of the entry fines sometimes being fixed according to the means of the peasants.[72]

Harvey does not deny that many peasants around 1300 had plots that were too small to support a family, nor that tax payments from the areas where parcelling-out of the original farm areas had been avoided were biggest; but she discovers a connection between small farming units and the presence of available uncultivated areas. She thinks that the small farms were most prevalent where plentiful woodland and other untilled areas emancipated the peasant from complete dependence on his arable. She also maintains, again in keeping with what we have said above, that these smallholders could not only find employment as agricultural labourers, but also, as Kosminsky has pointed out, in the wool manufacturing trade. In sum, then, Harvey does not think that the relatively many smallholders in England around 1300 can be taken to indicate that the country was overpopulated.[73]

Nor does she think that the soil exhaustion theory supports the assumption that the country was overpopulated before 1350; in the first place, because she cannot accept that the colonization process around 1300 was stopped by a scarcity of potential colonization areas;[74] and secondly because she does not think there is any convincing documentation that the newly-colonized areas tended to be on marginal land that was quickly exhausted. However, land was abandoned throughout the middle ages, and villages were sometimes deserted. But there are no signs that "the pace at which settlements were totally deserted...quickened between 1300 and 1348".[75] "In the fourteenth century, as in earlier ages, certain soils may have been relatively exhausted"; but "what is lacking is evidence that this problem became more acute after 1300; we have no warrant for finding here the key to a dramatic change in the whole population trend".[76]

Having thus refuted the overpopulation theory and discussed the relationship between agrarian development and wool manufacturing with the result that she leaves the possibility of new economic growth after 1300 as an open question,[77] Harvey finally turns to her assessment of population in the first half of the fourteenth century.

She cites a number of examples of constantly rising rents right up to the fifth decade of the century. "But the general impression left by the sources is that rents were not moving much at all in the first half of the fourteenth century".[78] This could of course conceal a real drop in demand for land, and for just that reason Harvey finds it difficult to argue that the population continued rising after 1320. On the other hand she does not think that one can justifiably speak of a Malthusian check in the first half of the fourteenth century.[79]

VIII.13 The demographic significance of the epidemics

In these demographic studies from the 1960s there is not only a well-documented refutation of the "neo-Malthusian" overpopulation theory and for that matter also the "neo-Marxist" overexploitation theory, but in addition yet another indirect rehabilitation of the demographic importance of the Black Death and the subsequent epidemics. Russell's repeated assurances that the Black Death was a demographic catastrophe which in the course of between thirty and thirty-five years reduced the population by 40-50%,[80] and J. D. Chambers' account of the way the Black Death hit England like a bolt from the blue "in a situation that is only "prospectively" Malthusian",[81] is typical of this reversal. But, as mentioned before, there was disagreement over how long the population decline lasted, and how serious it was.

It will be recalled that John Saltmarsh claimed that the Black Death and the other epidemics had a cumulative effect which progressively reduced the population right up until Tudor times (see Chapter V, p. 154-155). This point of view also featured in Postan's approach to the fifteenth century, as well as in the opinion of the "neo-Marxists" - primarily Hilton - that the lot of the peasants was improved in this century. We find the view again in Chambers[82] and Thrupp, who take Russell to task for his calculations that life expectancy improved as early as the beginning of the fifteenth century, although the population figure only began to rise much later.[83]

However, Saltmarsh's most prominent opponent was J. M. W. Bean, who claimed in a 1963 article that in the wake of the epidemics of 1348/49 and 1361 there followed a stagnation in the population around 1400. After this he thought he could trace a rise up through the whole fifteenth century.[84] This evaluation was exploited by some of the scholars discussed above who wanted to revise fifteenth-century history - most directly by Du Boulay,[85] but it has been attacked since then from several quarters.[86] The stock of Saltmarsh's interpretation of the population development after the Black Death has risen, although John Hatcher has to admit that the numerical constructions he presents to confirm it "are highly speculative".[87]

Nevertheless, Bridbury does not understand why he can find high prices and relatively low wages in the period between the Black Death and the last decades of the fourteenth century, and why wages then rose out of all proportion with the now falling corn prices. Without entering into a discussion of Bridbury's evidence for these movements, we can note that the reason he can ask the last question must be related to the fact that he must still have thought in 1973 that the worst demographic shock was over after 1369.[88] As for his his view that the manorial system only broke down in earnest in the last quarter of the fourteenth century, we have seen this interpretation defended since Levett's day. Nor is his explanation of why the demographic consequences of the Black Death had no immediate impact in this respect dramatically new.[89] What is striking is that he is still asking the latter question. This gives some impression, along with the developments described in the above paragraph, of the differences of opinion in recent research as regards the demographic influence of the epidemics.

One camp - we can call it the Saltmarsh camp - claims that the epidemiological factor remained a primary determinant of mortality well up into the fifteenth century.[90] Opposing this we have a camp which claims, with Bean, that the influence of the epidemiological factor on population development was quickly weakened, at least after 1400, because the epidemic outbreaks after this were more localized and were mainly restricted to the towns.

In a recent monograph J. F. D. Shrewsbury goes so far as to claim that even the impact of the first bubonic outbreak of 1348/49 was limited to the most densely populated parts of the country. He considers it inconceivable that the disease could have spread in the fourteenth century to sparsely populated areas like Cumberland, Northumberland, Durham, Westmoreland and Cheshire, because "the epidemiology of bubonic plague renders it improbable that P. pestis could have been distributed by rat-contacts as epizootic plague in any English county in 1348-9".[91] It is therefore fanciful to reckon with an overall mortality of as much as a third of the population.

Such a high mortality may have existed in the bigger cities and densely populated East Anglia. In the rest of England he doubts that bubonic plague could have eliminated more than a twentieth of the population.[92] As if this was not enough, after having scotched the "myth" of the drastically lethal effect of the Black Death, Shrewsbury rallies to Bean in an attempt to play down the idea that the plague recurred at regular intervals in the century-and-a-half that followed 1350.[93]

Shrewsbury's opinions have been convincingly countered, not only directly in a critique by Christopher Morris,[94] but also indirectly in a later monograph on the Black Death.[95] One reason Shrewsbury can play down the mortality of the Black Death is that he exclusively identifies it with bubonic plague. For some reason he does not deal at all with pneumonic plague and the septicaemic forms that the Black Death must also have included. This and the fact that he only associates the spread of infection with the black rat - *Rattus rattus* - and its parasite *Xenopsylla cheopis*, imposes certain natural limitations on the dissemination pattern one can imagine for the plague outbreaks.

In the first place modern epidemiology distinguishes between the spread of plague by intermediaries and from human to human. Secondly, there are other bearers than *Rattus rattus* and *Xenopsylla cheopis*. Pneumonic plague, for example, is air-borne like influenza, and in this respect differs radically from insect-borne diseases like bubonic plague and malaria.[96] Pneumonic plague can thus spread directly from human to human. The same is the case when bubonic plague combined with septicaemia is spread by means of the infected human flea *Pulex irritans*.[97] It is further well known that *Xenopsylla cheopis* has other primary hosts than *Rattus*

rattus - for example *Rattus norvegicus*, the brown rat, and other rodents; but it can also be a parasite on a number of ordinary domestic animals. An exception is the horse.

These facts, and the fact that the most lethal form of the plague - pneumonic plague - is thought to have been very prevalent during the first wave of plague,[98] raise serious objections to the credibility of Shrewsbury's evaluation of plague mortality; but we need not go into the critique of his numerical and percentage calculations presented by Morris.[99]

There are thus in the most recent research not only decided disagreements on the cumulative demographic effect of the epidemics, but also a continued debate on the seriousness of the impact of the first wave of plague in the mid-fourteenth century.

VIII.14 Epidemics and the economy

But were the Black Death and the subsequent epidemics derived from socioeconomic factors? R. S. Roberts goes so far in an article as to claim that "plague is...a mere part of a "Malthusian mechanism"" and that it can be said that in the Black Death one saw an "intensification of tendencies already at work..."[100] The rising population density in particular, but also the deterioration in the reproductive conditions of large sectors of the population, paved the way in his opinion for the ravages of the plague.[101] Nor does Philip Ziegler, in his monograph on the Black Death, deny the claim, made by Hilton among others, that the ravages of the Black Death were sustained by the fact that it descended on a population that was poorly equipped to resist the disease, among other reasons because of malnutrition.[102]

Thus Roberts makes the Black Death very nearly the ultimate result of an ever more "Malthusian" situation. The epidemic catastrophe is seen as directly derived from the fact that society's social and economic reproduction process has entered into a relationship of insoluble conflict with the natural basis. This is not unlike the "neo-Marxist" variant where the social antithesis between lord and peasant is thought to have produced a social and economic situation of which the epidemics are seen as the result.

VIII. 15 Summing-up

Not without some influence from the demographic studies we have seen questioning the overpopulation and to some extent the overexploitation theory, however, in the general assessments from the 1970s of the significance of the plague epidemics, these attempts to see the outbreaks in social and socioeconomic contexts are abandoned. Once again the epidemics are considered a crucial exogenous factor in socioeconomic change. The organization of Maurice Keen's political history is a relatively clear example of this. "The changing world of the later Middle Ages" is once more directly related to the Black Death, and the chapter which introduces this part of the book therefore has the heading "Plague and the changing economy".

It is especially the transformations in the social and economic structures of agriculture that Keen sees as having been caused by the Black Death. Although some of these changes only made themselves felt in the longer term, they were all results of the plague epidemics.[103] Despite the fact that the agricultural boom was over a good while before 1350, and although it is well-known that there were years with bad harvests and famine in the second decade of the fourteenth century, Keen does not think that one can speak of an agrarian crisis before 1350. "The Pestilence marks the watershed",[104] since "when the plague struck in 1348 the whole balance of social relations and economic pressures in the countryside was thrown out by its mortality".[105]

The same views are found in Gottfried's monograph on the Black Death from 1983. In this it is striking that Gottfried, in his account of agrarian social and economic development from the thirteenth century until just before 1350, religiously follows Postan's marginal land theory and Titow's soil exhaustion theory, and that he in fact thinks that Europe around 1300 "was in the throes of a classic Malthusian subsistence crisis".[106] He also accepts Kershaw's agrarian crisis in the second decade of the 14th century,[107] but does not think that any of these led to a long-term population reduction - in the first place because of "the absence of major, exogenous, killing disease", and secondly because of "early first marriage".[108] Only when the Black Death strikes a society, precisely as an exogenous factor, is the population reduced, with immediate striking socio-psychological, social and economic effects as the result.[109]

John Hatcher confirms that despite the many generations of historians who have protested against this view, there is much to be said for seeing the advent of the plague as a historical turning-point. For him, the mortality is crucial to the population development of pre-industrial England,[110] and the population development is the crucial factor for social and economic development.[111] But since Hatcher accepts that the population decline probably began a few decades before the Black Death, he cannot defend seeing the population development as exclusively determined by the plague epidemics. We must stop explaining mortality in pre-industrial society in terms of *either* "Malthusian contradictions" *or* extrasocietal factors. "These two broad categories of mortality coexisted", thinks Hatcher.[112]

This interpretation is explicitly not an attempt to combine but to classify the two prominent bodies of theory in the debate on the social and economic transformation process of the late middle ages. Hatcher does not claim that the "Malthusian" socioeconomic circumstances at the beginning of the fourteenth century were the precondition of the later ravages of the plague; the general "Malthusian" conflict between society and nature comes to expression as a factor endogenous in society - before and while the plague itself struck the development of society as an exogenous factor. And these two factors "coexisted".

Notes

1. Farmer's material comes from southeastern and southern central England. For the period before 1270 he primarily uses the Winchester Pipe Rolls. After 1270 this source is supplemented by eighteen of the earldom of Norfolk's manors, eleven of Merton College's, eight of Westminster Abbey's, seven of Crowland Abbey's and twenty manors from various smaller manorial complexes. D. L. Farmer, "Some Livestock Price Movements in Thirteenth-Century England", in *The Economic History Review*, 2nd Series, Vol. XXII, 1969, pp. 1-6.

2. N. J. Mayhew, "Numismatic Evidence and Falling Prices in the Fourteenth Century", in *The Economic History Review*, 2nd Series, Vol. XXVII, 1974.

3. Mavis Mate, "High Prices in Early Fourteenth-Century England: Causes and consequences", in *The Economic History Review*, 2nd Series, Vol. XXVIII, 1975.

4. Fig. 3, Farmer, op. cit., p. 13.

5. Ibid., p. 16.

6. Mate, op. cit., p. 5.

7. Ibid., pp. 2-6.

8. Postan has claimed that the drop in silver production that began between 1300 and 1320 cannot have had any crucial monetary influence, as large quantities of silver must have been accumulated in the previous centuries because of the huge volume of silver compared with the limited use of coin in these centuries, M. M. Postan, "The Trade of Medieval Europe: the North", in M. M. Postan & E. E. Rich (eds.), *The Cambridge Economic History of Europe*, Vol. II, Cambridge 1952, p. 212.

9. Mayhew, op. cit., p. 2.

10. Mayhew's method is to consider various hoards' content of coin in different periods and to classify them into four demarcated periods (Table 1, p. 5), and then to estimate the total national coin reserves in the corresponding periods on this basis (Table II, p. 7 and Fig. I, p. 9). He also thinks he can give a picture of the "London and Canterbury Mint Output, 1234-1360" (Fig. 2, p. 11).

11. Thus Mate thinks that "it is possible that large-scale leasing of demesne lands, which did not gather momentum until after 1348, would have occurred earlier, if it had not been for the stimulus provided by the high prices of the early fourteenth century". Mate, op. cit., p. 16.

12. Ibid., p. 8.

13. Ibid., p. 9; Farmer, op. cit., p. 16.

14. M. M. Postan, "The Costs of the Hundred Years' War", in *Past and Present*, No. 27, 1964, p. 52.

15. Edward Miller, "War, taxation and the English Economy in the late thirteenth and early fourteenth centuries", in J. M. Winter (ed.), *War and Economic Development*, Cambridge 1975, p. 27.

16. J. R. Maddicott, "The English Peasantry and the Demands of the Crown, 1294-1341", in *Past and Present*, Supplement I, 1975, p. 75.

17. Ibid., pp. 74-75.

18. "Between 1294 and 1341 the king's taxes bit deeply into peasants' resources, at no time more heavily than during the last years of the period". At least, the number of

complaints over taxation grew, and Maddicott bases his assessment on this. But ""complaint and evasion", it has been said, indicate not so much the oppressiveness of exations as the opportunities for resistance to them in a society which was for many a very free one". Ibid., p. 67; and "only the eastern part of England (a large area) was afflicted at almost all times; not until the late 1330s were national burdens heavy and frequent enough to spread general and continuous distress to the counties of the west". Ibid., p. 68. Perhaps this view is due to the nature of the source material. The only source of national scope used by Maddicott, the *Nonarum Inquisitiones*, is, as we know, from the 1340s.

19. Ibid., p. 69.

20. Ibid., p. 71.

21. For example, "the roll for three Bedfordshire hundreds in 1297 shows that, while the peasantry paid 65 per cent of the tax, lay and ecclesiastical lords contributed 35 per cent...identical with the division of land between demesne and peasant tenures". But "even though the burden of taxation bore a rough relationship to the amount of land cultivated by different classes, peasant tenures were liable for seignorial and ecclesiastical dues which might in aggregate amount to as much as half the product of land. For that reason alone the peasant's contribution to direct taxation must obviously have been heavier, and often very much heavier, than the contribution of his lord". Miller, ibid., p. 17 (see also Table I, p. 16 and the following pages).

22. Ibid., pp. 24-25.

23. Maddicott, op. cit., p. 6.

24. Ibid., pp. 49-50.

25. Miller, p. 25.

26. A. R. Bridbury, "Before the Black Death", in *The Economic History Review*, 2nd Series, Vol. XXX, 1977, pp. 395-96.

27. Ibid., p. 397.

28. Ibid., p. 399.

29. Ibid., p. 401.

30. Ibid.

31. Ibid., p. 402.

32. Ibid.

33. Ibid., pp. 402-03.

34. Ian Kershaw, "The Great Famine and Agrarian Crisis in England 1315-1322", in *Past and Present*, No. 59, 1973. In the following Kershaw is quoted from a reprint of the article in R. H. Hilton (ed.), *Peasants, Knights and Heretics*, Cambridge 1976, p. 131.

35. Henry S. Lucas, "The Great European Famine of 1315, 1316 and 1317", in *Speculum* V, 1930, reprinted in E. M. Carus-Wilson (ed.), *Essays in Economic History*, Vol. II, London 1962.

36. Ibid., pp. 71-72.

37. Kershaw, op. cit., pp. 95-97.

38. Examples of reduced sheep farming are given from Inkpen, a Berkshire manor under Titchfield Abbey, between 1313 and 1317; from the Crown sheep farm at Clipstone in Sherwood Forest between 1316 and 1317; from Crawley (Hants.); from Teddington in Middlesex, 1315-16; from the Crown manor of Sheen in Surrey, etc. Ibid., pp. 103-04. See also Table II, "Export of Wool in Early Fourteenth Century", p. 105. "From every region of the country there are signs in manorial records of great

destruction among cattle herds between 1319 and 1321". Examples are from three Huntingdonshire manors under Ramsay Abbey, the Crown manors of Clipstone and Sheen, Teddington, three manors under Merton College, etc. Ibid., pp. 106-11.

39. The manor court rolls reviewed by Kershaw do not show a steep increase in heriots in the years 1315-22. He thinks this is because it was landless men and sub-tenants not subject to these dues who were hardest hit by the famine. On the other hand this source suggests that the famines accelerated the dissolution of the traditional tenancy conditions that was already in progress. Ibid., pp. 119-22. But he does find examples of "vacancies caused by the agrarian crisis" and a "shortage of tenants was also a frequent complaint and though, proportionate to the rental income, the vacancies were seldom serious, they could occasionally be fairly substantial". Ibid., pp. 123-24. Moreover, Kershaw's observation of a drop in entry fines in his own view suggests a drop in land value. Ibid., pp. 126-27.

40. Ibid., p. 131.

41. Ibid., p. 132.

42. D. G. Watts, "A Model for the Early Fourteenth Century", in *The Economic History Review*, 2nd Series, Vol. XX, 1967, p. 547.

43. Ibid., pp. 543-44.

44. Ibid., pp. 545-46.

45. J. C. Russell, "Recent Advances in Medieval Demography", in *Speculum*, No. 40, 1965, Cambridge, Massachusetts, p. 98.

46. H. E. Hallam, "Population Density in the Medieval Fenland", in *The Economic History Review*, 2nd Series, Vol. XIV, 1961, pp. 80-81.

47. Ibid., p. 76.

48. Ibid., pp. 76-77.

49. Ibid., p. 79.

50. Ibid., pp. 75, 78-79.

51. One of Hallam's sources, however, is more directly applicable. In three censuses of *nativi* in Spalding, Weston and Moulton the whole unfree population is counted - which prompts him, among other things, to calculate the size of individual households at an average 4.68 persons. Unfortunately the source only covers the years 1267-68. Ibid., p. 71.

52. J. Z. Titow, "Some Evidence of the Thirteenth Century Population Increase", in *The Economic History Review*, 2nd Series, Vol. XIV, 1962, p. 219.

53. Ibid., p. 220.

54. Sylvia L. Thrupp, "The Problem of Replacement-Rates in Late Medieval English Population", in *The Economic History Review*, 2nd Series, Vol. XVIII, 1965, Table I, p. 106.

55. The "replacement-rates" on the best-represented manors, Redgrave and Hinderclay in Suffolk, are calculated on the basis of a number of deaths varying between 5 and 22 in periods varying between four and twenty years. Table I, Ibid.

56. J. C. Russell, "Effects of Pestilence and Plague, 1315-1385", in *Comparative Studies in Society and History*, Vol. VIII, The Hague 1965, pp. 466-67.

57. Postan & Titow, "Heriots and Prices on Winchester Manors", op. cit.

58. Russell, "Effects of...", op. cit., p. 468.

59. See also Russell, *British Medieval Population*, op. cit., Table 10.2, pp. 240-41.

60. Russell, "Effects of...", op. cit., p. 469.

61. J. C. Russell, "The Preplague Population of England", in *The Journal of British Studies*, Vol. V, No. 2, Hartford, Connecticut 1966, pp. 8-10.

62. Russell compares the life expectancy of vassals, calculated from *inquisitiones post mortem* in Table 8.3 in *British Medieval Population*, pp. 180-81 with Hallam's observation of peasants appearing as tenants on the same farms under Spalding Priory manors in surveys from 1259-60 and 1287. H. E. Hallam, "Some Thirteenth-Century Censuses", in *The Economic History Review*, 2nd Series, Vol. X, 1958, p. 349. He further compares the life expectancy of Winchester peasants, based on Postan & Titow's "Heriots and Prices", op. cit., with the life expectancy of vassals between 1280 and 1347. From the latter comparison he concludes that there was "a close correlation of expectation of life between peasants and fiefholders....If anything, the advantage lay with the peasants. Their health may have been better: they probably ate more vegetables and less meat; their beer was not so strong as the lord's wine; their water supply was probably no worse; and they may even have lived in less crowded quarters. The royal family had a worse record for longevity than the fiefholders". Russell, "The Preplague Population...", op. cit., pp. 5-8.

63. Ibid., pp. 10-11; see also *British Medieval Population*, Table 8.5, "Generation Life Tables of Males born 1301-25", p. 182; and "Heirs Surviving" after 1301, p. 245.

64. Ibid., p. 11; see also *British Medieval Population*, Table 10.8, "Estimates of Ratio of Increase, 1086 and 1250-1348. Comparing Domesday and the Extents", p. 250; and Table 10.11, "The Ratio of Preplague Extents to Poll Tax Population", p. 258.

65. Ibid., p. 13.

66. As we recall, Postan claimed in his article "Some Economic Evidence...", op. cit., that the rising wages after 1300 indicate a population drop. "If, however, one examines the decennial averages of wages for workers in wheat (probably the commonest of wage earners) compared with the value of silver money, a different interpretation is possible. Wages varied as much as did prices of wheat from 1261 (and especially from 1281) until about 1321, because workers were often paid in wheat. Then the price of wheat jumped, even in comparison to other grains. From about 1321 wages seem to have followed the value of silver in their steadiness. If unvarying wages (in terms of wheat) accompanied an increase of population from 1261 to 1321, why should not the unvarying wages (in terms of silver) also have been accompanied by a population increase from 1321 to 1348?". Russell, "The Preplague Population...", op. cit., pp. 13-14.

67. Ibid., pp. 19-20.

68. Kosminsky, "Studies in the Agrarian History...", op. cit., p. 228.

69. Russell, "The Preplague Population...", op. cit., pp. 20-21.

70. Barbara Harvey. "The Population Trend in England between 1300 and 1348", lecture to the Royal Historical Society, February 1965, publ. in *Transactions of the Royal Historical Society*, 5th Series, Vol. 16, 1966.

71. Ibid., pp. 25-26.

72. Ibid., pp. 26-27.

73. Ibid., pp. 27-30.

74. Ibid., p. 30.

75. Ibid., p. 32.

76. Ibid., p. 36.

77. Ibid., p. 38.

78. Ibid., pp. 38-39.

79. Ibid., pp. 39, 42.

80. Russell, "Effects of Pestilence...", op. cit., p. 472.

81. J. D. Chambers, *Population, economy and society in pre-industrial England*, London 1972, p. 14.

82. Ibid.

83. S. L. Thrupp, "Plague Effects in Medieval Europe - a comment on J. C. Russell's views", in *Comparative Studies in Society and History*, Vol. VIII, The Hague 1965. Only seriously after 1470, according to Russell.

84. J. M. W. Bean, "Plague, Population and Economic Decline in England in the later Middle Ages", in *The Economic History Review*, 2nd Series, Vol. XV, 1962/63, p. 435.

85. F. R. H. Du Boulay, *An Age of Ambition*, op. cit., pp. 34, 182.

86. John Hatcher addresses the following criticism to Bean: 1. "he relies almost exclusively on chronicles for evidence of plague outbreaks"; 2. "a major part of Bean's case appears to rest on the assumption that, in order to have significant effect on the level of population, plague had to strike through national outbreaks"; 3. "Finally, another plank in Bean's case can be shown to be insecure, namely that fifteenth-century plagues were mild because they had little immediate impact on the amount of cloth exported from London". John Hatcher, *Plague, Population and the English Economy, 1348-1530*, London 1977, pp. 16-18.

87. Hatcher submits the following figures: Domesday (1.75-2.25 million); poll tax, 1377 (2.5-3 million); 1348 (4.5-6 million); 1450 (2-2.5 million); 1520 (2.25-2.75 million); ibid., pp. 68-69. We note that Hatcher thinks that Russell's figures for the population in 1348 and 1377 "are gross understatements". Ibid., p. 13.

88. A. R. Bridbury, *Economic Growth*, op. cit., p. 23.

89. It is however a new view when, with Titow's backing, he asks whether the progressive exhaustion of the soil might partly explain why prices, wages and rents rose both after the famine (1315-17) and the first wave of plague (1348-49). A. R. Bridbury, "The Black Death", in *The Economic History Review*, 2nd Series, Vol. XXVI, 1973, p. 586.

90. A clear example of an advocate from this camp is Robert S. Gottfried, who has studied various sources - poll taxes 1377-81, manor court rolls, surveys and extents etc. from East Anglia - in order to calculate the development of population in the fifteenth century here, and concludes: "East Anglia followed the biological pattern of demographic development more closely than the Malthusian pattern....Epidemic disease, including plague, was not restricted solely to urban areas in the fifteenth century.... Both fifteenth century quantitative and medical evidence indicate that plague was the primary element controlling mortality in eastern England from 1430 to 1480". Robert S. Gottfried, *Epidemic Diseases in Fifteenth Century England*, New Brunswick 1978, pp. 226, 240.

91. J. F. D. Shrewsbury, *A History of Bubonic Plague in the British Isles*, Cambridge 1970, p. 53.

92. Ibid., p. 36.

93. Ibid., pp. 126-30, 141-49.

94. Christopher Morris, "The Plague in Britain - Review of Shrewsbury's *A History of Bubonic Plague in the British Isles*", in *The Historical Journal*, Vol. XIV, 1971, pp. 205-215.

95. Robert S. Gottfried, *The Black Death*, New York 1983.

96. L. F. Hirst, *The Conquest of Plague - A Study of the Evolution of Epidemiology*, Oxford 1953, p. 220.

97. Ibid., pp. 246-47.

98. Ibid., p. 122.

99. Morris, op. cit., p. 9.

100. R. S. Roberts, "The Place of Plague in English History", in *Proceedings of the Royal Society of Medicine*, Vol. 59, 1966, p. 102.

101. Ibid.

102. Philip Ziegler, *The Black Death*, London 1969. The Dutch historian B. H. Slicher van Bath, in his *Agrarian History of Western Europe*, New York 1963, p. 84, has made the same claim. As for the question of possible malnutrition, it is curious to see how an allegedly primarily vegetable-based diet made the peasants' health better than that of the vassals around 1300 according to Russell (se note 62 above), while Thrupp maintains that a more meat-oriented diet at the end of the fifteenth century is possibly part of the explanation of the growing population around this period. Thrupp, "Plague Effects...", op. cit., p. 483.

103. Section III, "The Changing World of the later Middle Ages", Chapter 8; "Plague and the changing economy", p. 187. An example: "The effects of the plague in the long term caused a social revolution in the humbler strata of society, which brought to birth a new society of tenant farmers and labourers out of the debris of the manorial community", pp. 200-201. M. H. Keen, *England in the Later Middle Ages - A Political History*, London 1973.

104. Ibid., p. 188.

105. Ibid., p. 192.

106. Gottfried, *The Black Death*, op. cit., p. 25.

107. Ibid., p. 29.

108. Ibid., pp. 25-26.

109. In connection with these sociopsychological problems Gottfried thinks that the most important effect of the plague and the mortality it caused was "its role in the provocation of popular rebellion" - in the first place because they broke down law and order, a view we recognize from the nineteenth-century literature; in the second, because they brought about an increased class consciousness. Ibid., pp. 97-98.

110. Hatcher, *Plague, Population...*, op. cit.

111. "In the absence of significant technological advances the amount of land available per head largely determined the level of output per head and consequently the standard of living of the great rural majority. For this reason, the tendency in England, before the eighteenth century, so far as the peasantry was concerned, was for a high population level to coincide with lower living standards and a lower population level to coincide with higher standards". Miller & Hatcher, *Medieval England*, op. cit., p. xv.

112. Hatcher, op. cit., p. 72.

Conclusion and Perspective

1.

We have now analysed over a century of research on the social and economic agrarian history of England in the late medieval centuries, and have seen how the period has been described as one of upheavals, and how in the course of the century the period was identified in various ways as one of crisis: as "the crisis of the landlords", "the crisis of the manorial system", "the crisis of feudalism", and as an ordinary agrarian crisis, partly in the sense that other social groups than the landlords were hit by the crisis, partly because it was claimed that national agricultural production fell after 1350. We have seen how all but one of these crisis definitions, the "political crisis" which is peripheral to our field, have been toppled or at least must be considered doubtful in the light of the overall results of research as they have been compared and analysed here.

There is little point in speaking of "the crisis of feudalism", because there is no basis for identifying the late medieval transformations in the structure of agrarian production with a formal rupture in the feudal organization of production. Seebohm and Rogers laid the foundations of a controversy over this which persisted for many years, since they spoke respectively of a dissolution of the feudal relations of production through the dispossession of the primary producers,[1] and a change in form accomplished by the abolition of villeinage, commutation and the establishment of the stock and land lease.[2] Research on the chronology and forms of the enclosure movement firmly scotched Seebohm's radical view,[3] while Rogers' position increasingly gained ground, although such that the focus was on commutation and leasing in general. How these two forms developed quantitatively in relation to each other remained long undetermined, just as a more specific identification of the nature of the leasing relationships in particular was still lacking around 1930, although in the preceding period it had been regarded more or less as a capitalist form.[4]

But just a few years after the debate between Seebohm and Rogers, Hyndman introduced the view that the welfare of the fifteenth-century peasantry was ensured by a quasi-petit-bourgeois organization of production, where on the one hand they had the right to free disposal and exchange of their own labour and on the other had retained their rights to land and the means of production.[5] It is in principle a development of this view that we get from Hilton almost seventy years later, whereas Kosminsky went nowhere near so far in his assessment of the forms of the relations of production in the fifteenth century.[6] Kosminsky concluded that neither commutation nor the spread of leasing involved an absolute break with the feudal forms of landholding or the introduction of capitalist property rights with purely financial relations regulated by a free market.

As we have mentioned several times in this work, it is indisputable that commutation in itself did not entail a break with the feudal relations of production; it was only a change in form,[7] while it is open to discussion whether the advance of leasing heralded such a break. But it should be emphasized that in the unfortunately few cases where the issue has been taken up for assessment in recent years, it emerges that personal rather than market conditions seem to have determined the decision to lease and the fixing of the rent.[8] In this light leasing in the fifteenth century can at most be conceived, as Kosminsky puts it, as "a bridge to the capitalist economy".

The thesis that a general "agrarian crisis" set in with a drop in national agricultural production in the fifteenth century[9] was refuted about twenty years after it had been postulated at the end of the 1930s, since it could be shown that any demonstrable decline in the gross agricultural and national product was anyway explicable in terms of the declining population - this had incidentally been pointed out long before the thesis was presented.[10] A further body of evidence was presented to show that the per capita product must have developed positively in the fifteenth century. Kosminsky saw the manorial system as an obstacle in several respects to the development of agricultural productivity, and thought that the productivity increase related to its decline, in terms of labour value theory, was reflected in falling grain prices.[11] In keeping with this, Bridbury justified his revision of the historical view of the fifteenth century by saying that the population decline, because it eased the pressure on the peasants and the land, contributed to a rise in productivity in agriculture which spread to urban industries, from which he presented evidence for his claim.[12] The very few direct studies of agricultural productivity that we have do not favour either of the conflicting views, so cannot be used as documentation of a crisis in agricultural productivity. The scanty documentation of agricultural productivity development in the fifteenth century suggests a stabilization, perhaps a slight rise, in the final years of the century.[13]

The idea of a general "agrarian crisis" - in the sense that more segments of the population than the landlords and perhaps the bigger peasants got into difficulties after the Black Death, was doubtful even when this crisis definition was launched around 1880.[14] It will be recalled that it was proposed in opposition to the prevailing opinion that the period after 1350 was an age of prosperity for the rural labourers and peasants.[15] But later research does not seem to justify this negative assessment of the situation of the agricultural labourers and peasants in the latter half of the fourteenth century and in the fifteenth century.[16]

The most durable crisis definition followed from Thorold Rogers' early assessment of the *situation of the landlords* in connection with the erosion of the manorial system after the Black Death. The shortage of labour and rising wages led to "distress" for the landlords although corn prices rose in the period immediately afterwards.[17] This view was supported by Seebohm's claim for a drop in land rents,[18] and later by the demonstration - for example by Tout and Postan - that the general level of prices gradually stagnated or directly dropped.[19]

However, various objections were raised to this assumption at an early stage. Charles H. Pearson expressed the view that the great landlords were able to counter the labour scarcity that had arisen, partly by forcible retention of their own villeins, partly by attracting other villeins, free yeomen and agricultural labourers with high wages and low rents; and that the labour scarcity problems were thus passed on to the smaller landlords.[20] Similarly, Jessopp thought that the high mortality among tenants and landlords alike must inevitably have meant a high concentration of landholdings, favouring not only the surviving tenants but also their lords,[21] a view that later won ground among modern historians;[22] and A. H. Johnson maintained that the labour shortage must have had a different impact on different manors.[23] This view was confirmed by Levett's and Ballard's studies of the Winchester estates in the years after the Black Death.[24]

A series of studies of the manorial economy and evaluations of the landlords' incomes after 1350 from the fifties and sixties,[25] as well as Levett's from 1916, make it difficult to speak of the "crisis of the landlords" in general. As has been demonstrated, there is no firm general evidence of financial problems among the landlords after the restructuring of agricultural production had gathered speed. The great landlords were said to be the big losers, but there is ample documentation to counter this classic interpretation.[26] It should also be remembered that the main premiss of the revision of fifteenth-century history at the beginning of the sixties was that the restructuring of agrarian production in fact led to a rise in productivity, and that this revision seems to have developed as a result of or during the writing of these studies of the manorial economy and the incomes of the landlords.[27]

The *breakdown of the manorial system* was at the core of the agrarian reorganization, but was this a manifestation of crisis? Can it be described as the crisis of the manorial system, and did it have social causes and consequences that allow us to speak of late medieval crisis? Can one speak of a general crisis of society in late medieval England?

In the longer historical perspective the manorial system in England appears to have been a way of organizing production that came and went. It was a salient, but probably not the dominant, mode of organization in the thirteenth century.[28] But perhaps its prime, as Du Boulay writes of the Canterbury estates, was "hardly more than a substantial interlude in the age-long system of lease-hold farming".[29] That means, as we have already documented, that the transformation of the late medieval centuries did not lead to a rupture with the feudal mode of production, since manorial organization gave way to other existing feudal forms through commutation and leasing.

Even if the feudal manorial system never again appeared on the arena of history in England, in this light it seems hard to speak of the crisis of the feudal social system, or for that matter of crisis in general. If the progressive transformation of one mode of organizing feudal production into another can be called a crisis, then one could with equal justification speak of a crisis in the period around 1200 when the manorial system won new ground again at the expense of the leasing, money and kind-based forms of production; and no one has ventured so far to do this.

The long-term nature of the process moreover suggests that we cannot describe it as the crisis of the manorial system in the normal sense of the term. In the absence of a much-needed general survey of the breakdown of manorialism in England our research overview seems to point in the direction of a gradual process,[30] accelerating in the latter half of the fourteenth century.[31] Examples are given of the manorial system ceasing, with consequent commutation, the abolition of villeinage and the development of leasing as far back as the thirteenth century,[32] but there is much to suggest that the reorganization only speeded up a few decades after the Black Death, and that it was most intense at the end of the fourteenth and the beginning of the fifteenth century.[33] But this did not lead to the total disappearance of the manorial system in England.[34] As we have discovered, this interpretation had already been verified in the first two decades of the twentieth century,[35] and to some extent it confirms Rogers' early assumption that the breakdown of the manorial system was hastened by the Black Death.[36]

It was in the nature of the transformation that it cannot have affected the whole agrarian reproduction process. Since demesne cultivation primarily seems to have been prevalent on the big religious manorial complexes,[37] and since it seems to have been particularly associated with the eastern, southeastern, and to a lesser extent the southwestern regions of the country,[38] the structural transformation of agriculture must have affected England in different ways regionally. These facts preclude any talk, prompted by the dissolution of the manorial system, of a socioeconomic crisis in the normal sense, since many manors and parts of the country cannot have been directly affected by this structural change. On the other hand we are not thereby justified in considering this process of reorganization to have had no national consequences, since the manors and areas that were directly affected must be regarded as very important to the national economy.

2.

In accordance with the crisis definition given in the introduction as a guideline for this study, one cannot speak on the basis of what has been said so far of a socioeconomic crisis in late medieval England. The gradual breakdown of the manorial system cannot be described as the *reproductive crisis of the feudal social system*, since there is nothing to suggest that the feudal agrarian relations of production in the late middle ages had as yet lost their reproductive capacity. The formal change in the relations of production that took place with the breakdown of the manorial system did not involve a break with their *feudal* form, nor is it likely to have extended to the majority of the feudal relations of production, not even in those parts of the country where the manorial system held its greatest sway.[39] It must further be stressed that the change to a great extent seems to have taken place as a result of factors exogenous to society - this despite the fact that more than a century's research on the late medieval period has been typified by constant efforts to propose theories and present empirical proof that the agrarian relations of production generated crisis and/or change.

The first fifty or sixty years of this research thus concentrated increasingly strenuous efforts on rejecting the significance of the Black Death for the late medieval transformation process. These efforts produced an "epidemiological",[40] a "neoclassical"[41] and a "political"[42] theory as well as certain theoretical elements where it was thought that population growth in the thirteenth century could explain the incipient breakdown of the manorial system before 1350.[43] These theories and elements of theories originated on the one hand in attempts to place this socially exogenous factor, the Black Death, under the influence of social factors, and on the other in the effort to explain the late medieval developments exclusively in terms of factors endogenous in society in an awareness of the factual chronology of the agrarian transformation process. These efforts continued in the subsequent decades.

The 1940s saw the emergence of a "neo-Marxist" theory,[44] and in 1950 the "neo-Malthusian" theory was formulated.[45] Now the five theories of development in the late medieval epoch existed. The period between 1930 and 1960 was, as we have seen, characterized by the continued development of the theories. This process, between 1950 and 1960, resulted in strongly marked, mutually incompatible bodies of theory, but was soon afterwards resolved, in a progress away from monocausal explanatory models, in an incipient fusion after 1960 between the two theories - "neo-Marxist" and "neo-Malthusian" - that had been especially prevalent in the decade.[46]

Yet a substantial element in the common ground for this fusion, the alleged population decline around 1300 or at least in the second decade of the fourteenth century, had never been given convincing empirical support. The discovery that the Black Death did not immediately result in the desertion of farms and difficulties for the landlords had at a fairly early stage led to the conjecture that a large population surplus before 1348 must have cancelled out or modified the effects of the first plague outbreak in this respect.[47] The man behind the first major demographic study of the English middle ages, J. C. Russell, could partly explain this by saying that a high excess mortality among the old in the period after 1350 paved the way for an acceleration in the succession of generations in agriculture;[48] but he got into serious difficulties with his assessment of population development between 1300 and 1348, although he hesitantly took the stand that population growth seems to have levelled out just before the Black Death.[49]

In the 1950s Postan tried, at first alone, then with Titow, to prove that the population was falling in the first half of the fourteenth century, as an element in the development of the "neo-Malthusian" theory. But we found none of these attempted proofs convincing.[50] Nor could we accept Titow's later study of "tithingspenny" payments on the Winchester estate of Taunton as proof of the alleged population development before the Black Death, since it indicated above all that the male population at Taunton rose up until the third decade of the fourteenth century, where the survey stopped.[51] Like Hallam, we cannot see that his calculations of population density in the wapentake of Elloe around 1300 meant that the area was overpopulated, although it was in fact very densely populated because of the particular natural and occupational character of the area.[52]

Although we found a single study from 1965 which, on the basis of a particularly narrow range of empirical material, might suggest a drop in the male population on the six East Anglian manors studied,[53] our scepticism as regards the theory of overpopulation and a consequent drop in the population in the first half of the fourteenth century has been confirmed by the most recent demographic research. Russell, formerly uncertain about his assessment of population development in this half-century, arrived in a couple of studies published in the mid-sixties at a clearer view and rejected both the overpopulation theory and the claim for a declining population before the Black Death.[54] Barbara Harvey rejected the overpopulation theory at the same time as Russell, but concluded that her material made it hard to argue that the population continued to rise after 1320.[55] It is also characteristic that in the most recent general accounts of the social and economic history of the late middle ages one consequently sees that the Black Death and later epidemics have regained much of their former explanatory power.[56]

Research on medieval desertions of farms and villages has further shown that any drop in population before 1350 did not at least manifest itself as a measurable rise in the number of abandoned settlements between 1300 and 1350. The research, which is mostly concentrated in the 1950s, indicates that it was only after 1400 - perhaps even only in the latter half of the fifteenth century - that an increase in the desertions that had taken place throughout the middle ages began.[57]

The original "neo-Malthusian" justification for the thesis of a falling population before 1350 was that a "Malthusian check" made its impact as "after a time the marginal character of marginal land was bound to assert itself".[58] But this theory could not be verified by the intensified research on the colonization process done in the 1950s,[59] and on the whole there is very little and very tenuous evidence for it.[60] After 1960 Postan also claimed the opposite: that newly-colonized land was most valuable, and that there was general exhaustion of the soil due to the increasing imbalance between animal farming and tillage.[61]

Here Postan was invoking an old theory,[62] and a debate that was almost as hoary over the general exhaustion of the soil in the late medieval centuries - a theory that had otherwise seemed to receive its death blow as early as the twenties and thirties;[63] but which Titow now resuscitated with his study of the "Winchester Yields". Titow's documentation for the increasing exhaustion of both marginal and normally fertile soil is, as has been demonstrated, open to attack.[64] And the theory was also later refuted by Barbara Harvey.[65]

The documentation for "neo-Marxist" claims of overexploitation and a consequent deterioration in the reproductive conditions of the peasants towards 1300 is equally doubtful. We have direct evidence that entry fines rose on some Winchester estates, and that "week work" was intensified, among other places on the manor of Wilburton and other estates of the Bishop of Ely in the course of the thirteenth century.[66] Beyond this indirect evidence has been presented that labour and money rents rose, insamuch as it has been pointed out that there was an increasing number of complaints over rent rises in the same century,[67] and as there are indications that labour services were intensified when the expansion of demesne cultivation was not followed up by a corresponding expansion in peasant land.[68]

Vinogradoff had already noted complaints from the peasantry but did not exclude the possibility that they might be exaggerated;[69] and even Hilton had to admit that the rise in the observable number of complaints might be related to the fact that there are more plentiful sources as we move into the thirteenth century.[70] The claim that the increase in demesne land took place without a corresponding increase in peasant land was disproved in several places even before it was made.[71]

If we accept, despite these objections, that the level of rents was rising, we must raise further objections to the assessment of the consequences of this rise for the reproductive capacity of the peasants. In the first place the development of any money dues must be evaluated in relation to price developments in the period, which in the case in question must have compensated for or at least modified nominal rises.[72] This objection is all the more important when we know that money dues were dominant all over England, even if it cannot be disputed that labour services must have won ground with the advance of the manorial system in the thirteenth century.[73] In the second place, the burden of labour services too must be judged in this light - i.e. in the context of the static or falling money dues,[74] and in connection with a couple of other important points. We have seen examples of how the so-called "boon-days" were partly compensated by payments in kind. Such payments, along with the rising population and the consequently rising number of household members, must be supposed to have counterbalanced the burden of the labour services incumbent on the individual farm.[75]

It has further been demonstrated that the conditions of the peasants in the thirteenth century, even within a single manorial complex, depended greatly on the type of manor they were subject to.[76] Moreover, studies of social stratification among the peasants, indicated by, among other things, the existence and development of a land market among them, seem to contradict the view that there can have been general exploitation of the English peasantry towards 1300.[77] The impression that land transactions among the peasants increased in the course of the thirteenth century seems to suggest that there was increasing social stratification among the peasants. For this reason among others it cannot be denied that, in contrast to the group of well-situated peasants around 1300, there were many smallholders and more or less landless agricultural labourers.

In the face of these facts it must be remarked that the many smallholders are not in themselves proof of increasing exploitation and a deterioration in reproductive conditions. A number of factors like age and sex distribution on the individual farm, opportunities for supplementary employment and access to uncultivated areas must also be taken into account, since they have direct influence on the assessment of a farm's reproductive potential.[78] Nor can one assume that the conditions of rural labourers necessarily deteriorated with the price rises and the drop in wages. For these were not wage-earners proper; they were labourers whose wages, often paid in kind, were related to their disposal over small or larger plots of land.[79] Finally it should be mentioned that it is doubtful whether the rising royal taxes from the beginning of the fourteenth century had any great influence on the reproductive conditions of the peasantry. The question is whether they did not rather have a negative influence on the development of rents - that is, helped to keep them in check.[80]

It can be established that there cannot have been general overexploitation of the primary producers, and that any deterioration in their reproductive conditions cannot have been universal either. So one cannot speak of a general deterioration in reproductive conditions for the peasants at the beginning of the century, any more than one can speak of a general revenue crisis for the landlords after 1350.

Some of the first scholars to question the thesis of the falling population before 1350 were representatives of the "neoclassical" school.[81] They sought the explanation of the incipient agrarian restructuring in the falling price of grain and saw the drop in prices as something caused by the similarly falling silver and money volume from about 1300.[82] We demonstrated how the theory was in conflict with the empirical facts of the period between 1369 and 1400,[83] and saw doubt cast on it by Postan, who did not think that price developments could be considered the reason for either a crisis or fundamental changes in a society dominated by primary relations of production.[84] More doubt has been cast on the monetary theory, inasmuch as recent estimates from the seventies of the development of the English money supply between 1250 and 1350 do not suggest a consistent fall between 1300 and 1350 but a fairly rapid rise and a consequent rise in prices in the first decade and an increase in minting from the third to the fifth decade of the fourteenth century.[85] The "neoclassical monetary" view of the reasons for the early erosion of the manorial system is thus no longer convincing, and various fiscal considerations, proposed most recently in the mid-seventies,[86] nor can alone stand as an explanation of early agrarian reorganization.

It has been demonstrated how difficult it has been to explain the social and economic changes of the late middle ages in terms of theories of socioeconomic development before 1348. In the course of this demonstration we saw that, although it is still claimed that the reproductive conditions of the peasantry in the thirteenth century gradually approached subsistence level,[87] it is at least a debatable claim. There is no documentation for the existence of a "Malthusian" situation in the first half of the fourteenth century, and just as little proof that a deterioration in reproductive conditions due to overexploitation and/or general "Malthusian" circumstances in fact turned into a general population decline before 1348. These facts, given the crisis definition stated in the introduction, provide no grounds for describing the decades before the Black Death as typified by the *reproductive crisis of feudal society* - society's incapacity to reproduce its members. Apart from the famine years in the second decade of the fourteenth century, when society really did have difficulty reproducing its members for reasons exogenous to society itself - climatic reasons[88] - it is hard on the given basis to speak of reproductive problems in society before 1348.

Shortly after this traditional historical turning-point it is also difficult. The manorial system seems to have resisted the effects of the first, most lethal plague epidemics, because society appears to have had a surplus capacity for reproducing itself before this. And even if the epidemics entailed, in both the short and the long term, a rise in mortality - whose extent was itself debatable - these terrible events can only justify us in speaking of crisis in the sense that an exogenous factor struck at and disturbed society's otherwise excellent ability to reproduce its members. There is no demonstrable connection between the conditions of social reproduction in the first half of the fourteenth century and the epidemics, and therefore no basis either for seeing the Black Death as "the consequence of oppression and poverty", as Hilton maintained in 1950, or to connect the epidemics in any other way with socially created and generated circumstances, as others have done. The effects of the Black Death and the later epidemics can probably on the whole be regarded as a demographic crisis, but it was at least not a social one in the sense that it can be said to have developed from causes endogenous in society.

3.

In the dominant explanatory models the economic fluctuations of the manorial system are associated with general socioeconomic fluctuations - primarily price, wage, rent and population development. Falling population, prices and rents and rising wages are thought to explain the breakdown of the manorial system and vice versa. We have shown that the empirical basis of these theories does not exist; but even if it did, or could be procured in future, it may be doubted how adequate this type of explanation is.

By means of a comparative study of the development of the manorial system between the twelfth and thirteenth centuries the historian Edward

Miller has convincingly brought out the problems inherent in this type of explanation. Miller concludes that two features were characteristic of the situation of the manorial system in the twelfth century, and that these, in keeping with the general opinion, make it differ fundamentally from the situation of the manorial system in the thirteenth century: first, the reductions in demesnes and the waiving of peasant labour services; and secondly the leasing out of demesne against fixed rents, that is, a development that recalls the general tendency in the period after the Black Death, although this development in the twelfth century cannot have been influenced by socioeconomic fluctuations like those which took place after 1350. "On the contrary, like the thirteenth century [the twelfth] bears many of the marks of an age of intensified economic activity".[89]

The existing disagreement on the general course of economic development in the last decades before the Black Death,[90] and to some extent the discussion of the corresponding course of events in the fifteenth century, accentuate the degree of caution one must exercise in trying to explain the development of the manorial system in the light of the "general economic environment". Nor does our knowledge of the history of the manorial system in the thirteenth century in the broader western European perspective appear to offer any confirmation of this type of explanation. Although the history of the manorial system is rather more obscure on the Continent than in England, there is little to suggest that there was the same reflorescence of this way of organizing production anywhere else in the thirteenth century than in England, and in the eastern and southeastern areas in particular. On the contrary it seems that the manorial system in Flanders, Italy, and to a lesser degree in France continued to be phased out in this century, despite the fact that we find indications of the same price, wage, rent and population fluctuations as in England.[91]

The unravelling of the reasons for the apparent revitalization of the manorial system in its old age, and its final decline in England, is still an unsolved historical problem - despite more than a century of research and despite the fact that ever since the beginning of this research process, rudimentary versions of other explanatory models than those which primarily focus on the breakdown of the system in the context of the above-mentioned socioeconomic fluctuations have been proposed. These models have remained at the rudimentary stage because they have never been seriously tested or even developed. The signs of counter-productivity that Thorold Rogers claimed were typical of the manorial system[92] have never

294

been taken up for serious consideration and systematic development in thorough studies of the structures of the manorial system and its productivity in the context of the other forms of the feudal relations of production - leasing, money and kind dues, and their varying economic scope and conditions in differing socioeconomic situations.[93]

Studies like this would bring us closer to an understanding of the history of the manorial system, provided they were done with due consideration for the influence of natural geographical and other regional factors - and not least for the political, ideological and religious associations of this form of production. It has been shown that medieval and late medieval economic cycles must be seen to a great extent in their regional context - cf. for example N. S. B. Grass's distinction between local and central markets. As has also been established in this work, there is a clear dividing-line between the development of church and lay estates, and the distribution of the various feudal ways of organizing production was related to differences in the regions of the country.

This demonstration of the difficulties inherent in describing the late medieval centuries as a period of crisis, and our rejection of the idea of a general social and economic crisis in late medieval England, breaks with a deep-rooted view of the epoch and provides grounds for doubting an interpretation of the social and economic development of the middle ages that is just as traditional. The familiar view that it can be described as one long boom period from about 1000 AD followed by a general crisis lasting about two centuries, beginning either in the early or mid-fourteenth century, seems, given the present results, doubtful; or at best is only a greatly simplified description of economic fluctuations.

Our demonstration of the relativity of the concept of crisis, the prevailing disagreement over the economic cycles of the fourteenth and fifteenth centuries, and especially our rejection of the concept of a general late medieval social and economic crisis, lead us to propose the hypothesis that general social and economic development in northwestern Europe in the medieval centuries should rather be depicted as several S-curves than as a generally rising trend in the so-called high medieval epoch from about 1000 to 1300 or 1350, and that after this a general economic downturn set in.

This hypothesis could conceivably be verified in several ways. Studies of the history of social and economic phenomena in the medieval centuries from 1000 to 1500 - for example, the development of the manorial system and/or comparative studies of the development of identical phenomena in different northwestern European countries over long or short periods of years - are some of these ways. Another is the investigation of the development of conditions of tenure in the relatively short period between 1300 and 1348 with a view to demonstrating how one or more of the S-curves mentioned could not only illustrate the development of conditions of tenure, but could also reveal fluctuations in other social and economic phenomena and perhaps even indicate what could be described as the general economic trend of these decades.

Covering both church and lay holdings, such a study of the decades before the Black Death could address itself to the development of the feudal forms of rent - that is, the relationships among labour, product and money rent, and the quantitative fluctuations in rents; a clarification of the peasants' land rights - that is, a study of the duration and inheritability of tenancies; the scope and nature of land transactions among peasants; the peasants' opportunities for subletting etc. Such a registration and definition of agrarian conditions of tenure would create a basis for evaluating the conditions necessary for the survival of the manorial system, the incomes of the landlords and the peasants' level of reproduction in the period just before the Black Death, and would thus outline fluctuations in the most important social and economic aspects of certain agrarian sectors.

Such a study would contribute to a much-needed refinement of our concepts of medieval crises and economic cycles by providing a canvas for the depiction of a relatively brief and strongly differentiated course of economic events. Done on the basis of English source material, more of which is extant from the period than is normally supposed, it could further be expected to contribute to the clarification of a still very obscure period, and would this create the preconditions for a qualified effort in what is now a century-long debate on the significance of the Black Death and the subsequent epidemics for the dissolution of the manorial system and of villeinage in England

Notes

1. Chap. I, pp. 6-7.
2. Chap. I, pp. 12-14.
3. Chap. III, pp. 89-96.
4. Chap. IV. pp. 117-118.
5. Chap. I, pp. 25-25.
6. Chap. VI, pp. 206-208.
7. Chap. II, p. 46; Chap. III, p. 69, Note 5, pp. 87-88.
8. Chap. VII, p. 222.
9. Chap. V, pp. 139-141.
10. Chap. II, p. 76.
11. Chap. VI, pp. 208-209.
12. Chap. VII, pp. 217-218.
13. Chap. VII, p. 226.
14. Chap. I, p. 22; Chap. II, pp. 47, 52, 55.
15. Chap. I, pp. 5, 10, 22-25.
16. Chap. IV, p. 106; Chap. V, pp. 139-140; Chap. VI, p. 206-207; Chap. VII pp. 217-218
17. Chap. I, pp. 12-13 .
18. Chap. I, p. 6.
19. Chap. IV, p. 106; Chap. V, p. 142.
20. Chap. I, pp. 22-23.
21. Chap. III, p. 76.
22. Chap. VII, pp. 219-220.
23. Chap. IV, pp. 107-108.
24. Chap. IV. pp. 111-112.
25. Chap. VII, pp. 218-223.
26. Chap. VII, pp. 219-221.
27. Chap. VII, p. 218.
28. Chap. III, pp. 84-86; Chap. V, pp. 146-149; Chap. VI, pp. 203-204.
29. F. R. H. Du Boulay, *The Lordship of Canterbury: An Essay on Medieval Society*, London 1966.
30. Chap. I, pp. 12-14; Chap. III, p. 81; Chap. IV, pp. 114-116; Chap. V, pp.142-143.
31. Chap. IV, pp. 106-109, 116.
32. Chap. I, pp. 11-13, 21; Chap. III, p. 82.
33. Chap. III, pp. 78-79; Chap. IV, pp. 107-110, 113.
34. Chap. VII, pp. 220, 224.
35. Chap. IV, pp. 114-115.
36. Chap. I, p. 14.
37. Chap. V, pp. 146-147.
38. Chap. III, p. 86; Chap. V, pp. 142, 147; Chap. VI, p. 204.
39. Chap. VI, p. 204.
40. Chap. III, pp. 70-71, 73.
41. Chap. I, pp. 8-9, 15-19.

42. Chap. II, pp. 43-44, 54-55, 61.
43. Chap. III, pp. 82-83.
44. Chap. V, pp. 152-154.
45. Chap. VI, pp. 179-181.
46. Chap. VII, pp. 244-247.
47. Chap. IV, p. 112; Chap. V, p. 156; Chap. VI, pp. 188-189.
48. Chap. V, p. 160.
49. Chap. V, pp. 159.
50. Chap. VI, pp. 181-185.
51. Chap. VIII, pp. 265-266.
52. Chap. VIII. p. 264.
53. Chap. VIII, pp. 266-267.
54. Chap. VIII, pp. 267-269.
55. Chap. VIII, pp. 270-271.
56. Chap. VIII, pp. 274-276.
57. Chap. VI, p. 188.
58. Chap. VI, p. 180.
59. Chap. VI, pp. 188-189.
60. Chap. V, p. 156.
61. Chap. VII, pp. 229-230.
62. Chap. I, pp. 26-24
63. Chap. II, p. 56-58; Chap. IV, pp. 118-129; Chap. V, pp. 151-152, 165-166.
64. Chap. VII, pp. 232-235.
65. Chap. VIII, pp. 270-271.
66. Chap. V, pp. 149-150 Chap. VI, p. 201.
67. Chap. V, p. 149; Chap. VI, pp. 202-203.
68. Chap. V, p. 149-150.
69. Chap. VI, p. 202.
70. Chap. VI, p. 202.
71. Chap. V, pp. 163-164.
72. Chap. IV, pp. 110, 113.
73. Chap. VI, pp. 203-204.
74. Chap. V, p. 149.
75. Chap. III, p. 83; Chap. IV, p. 113; Chap. VI, p. 203.
76. Chap. VII, p. 241.
77. Chap. VII, pp. 241-242.
78. Chap. VII, p. 242; Chap. VIII, pp. 269-271.
79. Chap. VI, pp. 204-205.
80. Chap. VIII, pp. 258-259.
81. Chap. VI, p. 191.
82. Chap. VI, pp. 191. 193.
83. Chap. VI, pp. 192, 194.
84. Chap. VI, p. 195.
85. Chap. VIII, pp. 256-257.
86. Chap. VIII, pp. 258-261.
87. J. L. Bolton, *The Medieval Economy, 1150-1500*, London 1980, p. 345.
88. Chap. VIII, pp. 261-262.

89. Edward Miller, "England in the twelfth and thirteenth Centuries: An Economic Contrast?" in *The Economic History Review*, 2nd Series, Vol. XXIV, 1971, pp. 4, 7.

90. Chap. I, p. 15; Chap. VII, pp. 259-264, 270-271.

91. Perry Anderson, *Passages from Antiquity to Feudalism*, London 1974, pp. 189-91; Georges Duby, *Rural Economy and Country Life in the Medieval West*, Columbia 1976, pp. 237-38, 264-68. 273-79.

92. Chap. I, pp. 14-15.

93. But see Chap. III, pp. 82-83; Chap. IV, pp. 110-111, 113; Chap. V, pp. 152-153; Chap. VI, pp. 197-198.

Sammenfatning

Der er en lang og rodfæstet tradition for at sætte Nordvesteuropas historie i de senmiddelalderlige århundreder i forbindelse med begrebet krise. Af den grund, og fordi der vitterligt i disse århundreder indtraf en række vidtgående sociale og økonomiske forandringer i denne del af verden, er perioden udvalgt som historisk grundlag for den belysning af det samfundsøkonomiske fænomen krise, som er sigtet med denne afhandling.

En af hovedhjørnestenene i den europæiske kriseforskning har været engelsk historieskrivning. Forestillingerne om en senmiddelalderlig krise har her været meget fremtrædende. Da det af praktiske grunde var hensigtsmæssigt at indskrænke undersøgelsen til et velafgrænset område i Nordvesteuropa, faldt valget blandt andet derfor på England. Undersøgelsen er endvidere indsnævret til hovedsageligt at omfatte landbruget af den simple grund, at det feudale samfund uomtvisteligt fundamentalt set var et agrarsamfund. Enhver efterforskning af almindelige samfundsmæssige kriser i de feudale århundreder må derfor tage udgangspunkt i konjunkturbeskrivelser af denne sektor.

Afhandlingens problemområde er følgelig det senfeudale agrare Englands mulige sociale og økonomiske reproduktionskriser. Indenfor dette område er rejst en række strukturerende hovedspørgsmål for udforskningen af, hvor vidt man kan tale om én eller flere socioøkonomiske samfundsmæssige kriser i engelsk senmiddelalder og i givet fald i hvilken forstand og med hvilken geografisk udbredelse. Først og fremmest ønskes det undersøgt om den agrare produktionsorganisering var af en sådan art, at den i sig selv var krisegenererende, eller om det feudale agrare samfunds/samfundssystems eventuelle reproduktionsvanskeligheder blev affødt af i forhold til samfundets socioøkonomiske strukturer eksogene faktorer.

Undersøgelsens generaliserende og begrebspræciserende sigte har nødvendiggjort en metode, hvori metabegrebslige, metateoretiske og empiriske studier er kombineret. Besvarelsen af de rejste spørgsmål udvikles gennem sammenstillingen og kritikken af bestående begreber og teorier ud fra deres empiriske og teoretiske forudsætninger.

Metoden er derfor historisk i dobbelt forstand. Problemområdet bliver ikke kun belyst ud fra den primære historiske baggrund - de senmiddelalderlige århundreder - men også de sidste mere end hundrede års historieteoretiske og historiografiske udvikling i England. Fremstillingsmæssigt udtrykker denne metode sig strukturelt i, at forskningsudviklingen bliver blotlagt kronologisk retvendt fra kapitel til kapitel, mens de enkelte kapitler og afsnit er struktureret indholdsmæssigt i forhold til problemfeltets mangesidede aspekter og deres indbyrdes forbindelser.

Undersøgelsen er gennemført i et forsøg på så vidt muligt at inddrage al den tilgængelige litteratur indenfor problemkredsen i perioden fra 1860'erne til omkring 1980. Denne tidsmæssige afgrænsning er faldet naturlig, fordi vi i midten af det 19. århundrede i England på den ene side kan se en blomstrende interesse for social og økonomisk historie, samtidigt med at den empirisk orienterede økonomiskhistoriske videnskab opstod gennem kritikken af den mere spekulative eller metafysiske tradition. På den anden side finder vi netop i 1860'erne en diskussion om "den sorte døds" sociale og økonomiske konsekvenser i England mellem to af pionererne i det omtalte videnskabsteoretiske opbrud - en diskussion som skulle vise sig at få stor betydning for de følgende års debat om de senmiddelalderlige århundreders sociale og økonomiske historie. Med denne diskussion påbegyndes en sejlivet strid om "den sorte døds" historiske betydning.

Frederic Seebohm argumenterede for, at "den sorte død" førte til ophævelse af de agrare feudale produktionsforhold, mens Thorold Rogers hovedsageligt så pesten som en stimulerende faktor i en allerede igangværende proces, hvorunder produktionsforholdene blot forandredes gennem livegenskabets og godsdriftens ophør. Rogers' syn på denne transformationsproces omfattede både et evolutionært og et revolutionært perspektiv. Sidstnævnte knyttede han til pestens demografiske følger. Førstnævnte til en "neoklassisk" opfattelse af, at en fri løn- og prisdannelse var gældende i senmiddelalderen. På trods af hans opgør med de politiske økonomer introducerede han en "neoklassisk" teori om godsbesiddernes og godsdriftens krise.

Fra denne banebrydende diskussion og efter dokumentation af, hvordan den havde et langt efterspil i de følgende årtier frem mod slutningen af 1880'erne, vender vi os i andet kapitel mod nogle generelle historiske fremstillinger fra omkring 1890. Her finder vi et teoretisk og metodisk opgør med Rogers, som kastede et andet lys over periodens historie. W. J. Ashley negligerede krisen, W. Cunningham gav den et andet indhold og W. Denton tilbagedaterede dens gennemslag til begyndelsen af det 14. århundrede.

Cunningham og Denton opprioriterede politiske og konstitutionelle aspekter og introducerede en "politisk" kriseteori. I Denton's fortolkning blev politisk labilitet ikke alene årsag til godsdriftens krise, men til en almindelig agrar krise, fordi andre sociale grupper end godsbesidderne mentes berørt af krisen, og fordi godsdriftens opbrud mentes at have medført en stigende udpining af agerland. Cunningham definerede derimod udelukkende krisen som en politisk og konstitutionel krise.

I kapitel III bliver det demonstreret, hvordan der i debatten om de senmiddelalderlige århundreders sociale og økonomiske historie blandt historikerne i pionerfasen blev udsondret nogle centrale emneområder, og hvordan de blev genstand for en intensiv forskningsindsats i perioden mellem 1880'erne og 1915. For det første blev de senmiddelalderlige epidemiers karakter og samfundsmæssige følger undersøgt i en udstrækning, som aldrig før var set i engelsk historieskrivning. For det andet blev overgangen fra arbejdsydelser til pengeafgifter og indhegningsbevægelsens årsager og kronologi efterforsket grundigt med henblik på en nøjere datering af godsdriftens ophør og bestemmelsen af dens sociale og agrartekniske følger.

Selv om Charles Creighton fremsatte en egentlig "epidemiologisk" teori og den ikke uvæsentlige betragtning, at pesten havde en betydelig langtidseffekt, berigede udforskningen af epidemiernes historie ikke med afgørende nyt vedrørende sammenhængen mellem epidemierne og de socioøkonomiske og teknologiske ændringer i senmiddelalderens agrare verden.

Der blev derimod med den forøgede udforskning af overgangen fra arbejdsydelser til pengeydelser i landbruget opnået resultater af afgørende betydning. Det blev dokumenteret, at denne proces strakte sig over flere århundreder - fra det 13. til langt ind i det 15. århundrede - og at der i vurderingen af dette spørgsmål gør sig markante geografiske forskelle gældende. Tilsvarende bidrog udforskningen af indhegningsbevægelsens historie med viden om, at de senmiddelalderlige former for indhegninger rettede sig mod udviklingen af de middelalderlige marksystemer i visse regioner. Forskningen fastslog desuden, at den form for indhegninger som forbindes med overgangen fra agerbrug til fårehold, og dermed altså bøndernes fordrivelse fra jorden, tidligst kan dateres til de allersidste årtier af det 15. århundrede og de følgende århundreder. Hermed var Seebohm's opfattelse af "den sorte døds" umiddelbare konsekvenser blevet afvist.

Udforskningen af de anførte emneområder indebar også rudimentære forstadier til eftertidens udlægning af epidemiernes samfundsmæssige betydning, og i forbindelse med dokumenteringen af hoveriafløsningen før 1350 kan der spores elementer til en demografisk teori i forklaringer af

godsdriftens tidlige opbrud set i lyset af højmiddelalderens befolkningsvækst. Men forskningen mellem 1880'erne og 1915 bragte altså først og
fremmest dokumentation for, at godsdriftens fald var en evolutionær proces, som om ikke umiddelbart så dog på længere sigt blev fremskyndet
af "den sorte død", og som heller ikke umiddelbart førte til sprængning af
de agrare feudale produktionsrelationer.

Selv om der i tiden omkring 1900 ikke var absolut enighed herom, ser
vi i kapitel IV, at den fremlagte dokumentation i det mindste havde en
sådan styrke, at man i de generelle fremstillinger fra perioden mellem
1905 og 1915 var blevet overbevist om ikke at tillægge "den sorte død" en
for overvældende historisk betydning. I dette kapitel påvises det endvidere,
hvordan undersøgelser af Winchestergodserne mellem 1916 og 1929 da
heller ikke tydede på, at der her skete nogle alvorlige strukturforand-
ringer, som kan tilskrives denne begivenhed. Forskningsudviklingen mellem
1900 og 1930 må i det hele taget siges at bekræfte de evolutionære
aspekter, som dominerede Rogers' syn på senmiddelalderens socioøkonomiske forandringer. Spørgsmålet om årsagerne til godsdriftens begyndende
afvikling før 1350 trængte sig derfor stadigt stærkere på.

Foruden de tidligere omtalte "politiske", "epidemiologiske" og "neoklassiske" teorier samt de demografiske teorielementer, som kan spores i forbindelse med udforskningen af hoveriafløsningsprocessen i andet årti af
det 20. århundrede, blev der i samme årti udviklet en "jordudpiningsteori",
som mentes at kunne forklare godsdriftens tidlige opbrud. Denne teori
kan betragtes som en teknisk variant af de nævnte demografiske
betragtninger, hvori højmiddelalderens befolkningsvækst blev knyttet til
problemer i den agrare produktionsorganisering. "Jordudpiningsteoriens"
fortalere hævdede, at det stigende befolkningstal medførte en voksende
ubalance mellem kvæghold og agerbrug, og at dette misforhold
efterhånden opbrød et afgørende middelalderligt agrarteknisk system - det
åbne marksystem - som ellers var indrettet på at kunne tilgodese en vis
befolkningstilvækst. "Jordudpiningsteorien" blev imidlertid allerede falsifice-
ret i 1920'erne, og den fik, selv om den senere blev genoptaget, aldrig
alvorlig betydning i sin oprindelige form.

Tiden mellem 1930 og 1950 var for det første præget af endnu et forsøg
på udvidelse af begrebet om den senmiddelalderlige krise. For det andet
af empirisk forskning indenfor områder af betydning for udviklingen af de
senest opstillede kriseteorier. I kapitel V ser vi, hvordan M. M. Postan
argumenterede for tilstedeværelsen af en almindelig samfundskrise i det
15. århundrede. En "landbrugskrise" med rødder tilbage til 1350 - måske

endog tidligere - og med forbindelse til en tilsvarende krise i de urbane erhverv, hvilke kriser tilsammen udtrykte sig i et faldende nationalt produkt. Vi ser endvidere, hvordan Postan og E. A. Kosminsky, gennem udforskningen af indfæstningsafgifternes udvikling i det 13. og begyndelsen af det 14. århundrede, tog tråden op efter en indsats, der var blevet påbegyndt i andet årti af det 20. århundrede. Gennem dette arbejde skabte de et afgørende empirisk støttepunkt for Maurice Dobb's lancering af en "neomarxistisk" kriseteori. Tilsvarende skabte J.C. Russell's omfattende demografiske undersøgelser, en række studier af den middelalderlige koloniseringsproces, samt ikke mindst Postan's kritik af den "neoklassiske" teori forudsætninger for den snart efterfølgende opstilling af en "neomalthusiansk" kriseteori. Endelig blev den "epidemiologiske" teori rekapituleret og styrket af John Saltmarsh.

Med Postan's opstilling af den "neomalthusianske" kriseteori i 1950 forelå i engelsk historieskrivning de fem eksisterende teorier om senmiddelalderkrise. Kapitel VI handler herom og om "neomalthusianernes" forsøg på forankring af teorien gennem fremlæggelse af mere entydig demografisk bevisførelse for den. Kapitlet behandler endvidere udforskningen af den senmiddelalderlige ødelæggelsesproces, og hvordan denne forsknings resultater synes at tale imod de "neomalthusianske" betragtninger på væsentlige områder. Desuden inddrages marxistiske forfatteres - R. H. Hilton's og Kosminsky's - teoretiske og empiriske indsats mellem 1950 og 1960. De definerede krisen som "feudalismens" krise, det vil sige det feudale samfundssystems krise, og især Hilton så den afledt af stigende vanskeligheder med at reproducere befolkningen fra slutningen af det 13. århundrede.

Denne udlægning af senmiddelalderkrisen samt kritik rettet mod forudsætningerne for Postan's krisebestemmelse lagde op til den revision, som fulgte efter 1960. Revisionen er et af kapitel VII's hovedtemaer. Mens man tidligere i langt de fleste tilfælde havde dateret krisegennemslaget til efter eller tidligst umiddelbart før "den sorte død", men i hvert fald betragtede århundredet efter 1350 som kriseperioden "par excellence", hævdede Anthony Bridbury og andre nu, at det 15. århundrede var en generel ekspansionsperiode. I opposition til Postan, der havde talt om en agrarkrise i det 15. århundrede, hævdede Bridbury, at den agrare produktivitet steg stærkt i dette århundrede, hvorimod den havde udviklet sig negativt i det 13. århundrede. Revisionen hentede støtte hos Hilton, og også empiriske undersøgelser af godsøkonomien efter hovedgårdsdriftens afvikling samt i nogen grad den sparsomme eksakte viden om den agrare produktivitetsudvikling før og efter 1350 synes at tale for den.

Det andet hovedtema i kapitel VII er, at der i samme periode - tiden efter 1960 - kan bemærkes et begyndende sammenfald mellem den "neomarxistiske" og den "neomalthusianske" kriseteori. Befolkningsvæksten i det 13. århundrede blev i stigende grad afgørende for den "neomarxistiske" overudbytningsteori, mens traditionelle marxistiske argumenter som godsbesiddernes uproduktive forbrug og manglende investeringer i samme århundrede blev stadigt vigtigere forudsætninger for den "neomalthusianske" overbefolkningsteori.

I det afsluttende kapitel VIII er de seneste årtiers kritik af den "neomarxistiske" og "neomalthusianske" kriseteori og krisebestemmelse opsamlet og undersøgt. Påstandene om den faldende agrare produktivitets og befolkningsudviklingens betydning for den agrare omlægning er blevet angrebet af "neoklassiske" fortalere, samtidigt med at fiskale, politiske og klimatiske forhold med fornyet kraft er blevet draget ind i debatten fra anden side. Selve påstanden om overbefolkning i de første årtier af det 14. århundrede og udviklingen af en "malthusiansk" situation med følgende befolkningsnedgang fra omkring tredie årti af århundredet er endvidere blevet imødegået i forskellige undersøgelser af tiden mellem 1300 og 1348.

I disse undersøgelser forefindes ikke alene ret veldokumenterede tilbagevisninger af den "neomalthusianske" overbefolkningsteori og for så vidt også den "neomarxistiske" overudbytningsteori, men også indirekte en rehabilitering af "den sorte døds" og de efterfølgende epidemiers demografiske betydning og deres samfundsmæssige følger. Et forhold der afspejles i de seneste årtiers mere generelle litteratur om de senmiddelalderlige århundreders sociale og økonomiske udvikling.

Man kan sondre mellem denne afhandlings historiografiske og dens historieteoretiske resultater. Under de historiografiske resultater hører demonstreringen af, at man kan finde såvel krisebestemmelser som mere eller mindre veludviklede kriseteorier allerede fra 1860'erne, uanset den hidtil almindeligt gældende opfattelse af at kriseproplemet først blev påtrængende i historieskrivningen om de senmiddelalderlige århundreder omkring og efter 1930. Det er ligeledes et historiografiske resultat, når det bliver dokumenteret, at et hovedproblem og den dominerende anstrengelse blandt pionererne lang tid før 1930 var tilbagevisningen af "den sorte døds" betydning som faktor i det historiske forløb. Det er og har ellers været en almindeligt udbredt forestilling blandt historikere både før og efter 1930, at "tidligere tiders historikere tillagde denne begivenhed en overvældende betydning".

Under de historiografiske resultater hører endvidere påvisningen af, hvordan meget tidligt udviklede teorier og teorielementer senere er blevet taget op og er indgået i udviklingen af de moderne teorier, samt udredningen af hvordan to af teorierne - den "neomalthusianske" og den "neomarxistiske" - først kontrasteredes for senere at tendere mod fusion. Selve udsondringen af de eksisterende bestemmelser af den senmiddelalderlige krise er i sig selv et historiografisk resultat, mens udfaldet af sammenstillingen og kritikken af den foreliggende forskning gennem mere end ét århundrede med henblik på besvarelsen af undersøgelsens hovedspørgsmål er at betragte som historieteoretiske resultater.

Det er historieteoretiske resultater, når det kan dokumenteres, at ingen af de fundne krisebestemmelser kan anses for at være historisk holdbare, og at ingen af de eksisterende teorier om den senmiddelalderlige krise eller transformationsproces kan stå sig, overfor den empiriske og teoretiske prøvelse de bliver udsat for. Disse historieteoretiske resultater samt afhandlingens hovedpointe - afvisningen af, at en almindelig samfundsøkonomisk krise udspillede sig i det senmiddelalderlige England - giver anledning til en revurdering af vort syn på middelalderens sociale og økonomiske konjunkturforløb og til anderledes tænkning, for så vidt angår vor forståelse af årsagerne til de produktionsstrukturelle forandringer vi med sikkerhed kan konstatere i landbruget i de senmiddelalderlige århundreder.

Bibliography

W. Abel,

Agrarkrisen und Agrarkonjunktur in Mitteleuropa vom 13. bis zum 19. Jahrhundert, Berlin 1935

W. Abel,

Die Wüstungen des Ausgehenden Mittelalters, 1934

W. Abel,

Strukturen und Krisen der Spätmittelalterlichen Wirtschaft, Stuttgart 1980

K. J. Allison,

The Lost Villages of Norfolk, 1952

Perry Anderson,

Passages from Antiquity to Feudalism, London 1974

W. J. Ashley,

An Introduction to English Economic History and Theory, Vol. I, Book 1, 1888, Book 2, 1893

A. R. H. Baker,

"Evidence in the Nonarum Inquisitions of Contracting Arable Lands in England During the Early Fourteenth Century", in *The Economic History Review*, 2nd Series, Vol. XIX, 1966

A. R. H. Baker & R. A. Butlin, (eds.),

Studies of Field Systems in the British Isles, Cambridge 1973

A. R. H. Baker & J. B. Harvey (eds.),

Man made the Land, 1973

A. Ballard,

"The Manors of Witney, Brightwell and Downton", in *Oxford Studies in Social and Legal History*, Oxford 1916

E. Barger,

"The Present Position of Studies in English Field Systems", in *The English Historical Review*, CCXI, Vol. LIII, 1938

F. H. Baring,

"The Making of the New Forest", in *The English Historical Review*, Vol. XVI, 1901

M. Ley Bazely,

"The Extent of the English Forest in the Thirteenth Century", in *Transactions of the Royal Historical Society*, 4th Series, Vol. IV, 1921

J. M. W. Bean,

"Plague, Population and Economic Decline in England in the Later Middle Ages", in *The Economic History Review*, 2nd Series, Vol. XV, 1962/63

H. S. Bennett,
Life of the English Manor, Cambridge 1937

M. K. Bennett,
"British Wheat Yield per Acre for Seven Centuries", in *Economic History*, Vol. III, 1935

M. Beresford,
The Lost Villages of England, London 1954

M. Beresford & J. G. Hurst (eds.),
Deserted Medieval Villages, London 1971

W. Beveridge,
"The Yield and Price of Corn in the Middle Ages", in *Economic History*, 1927

W. Beveridge,
"Wages in the Winchester Manors", in *The Economic History Review*, Vol. VII, 1936

T. A. M. Bishop,
"Assarting and the Growth of the Open Fields", in *The Economic History Review*, Vol. VI, 1935

J. L. Bolton,
The Medieval Economy, 1150-1500, London 1980

H. Bradley,
The Enclosures in England - An Economic Reconstruction, New York 1918

P. F. Brandon,
"Cereal Yields on the Sussex Estates of Battle Abbey during the Later Middle Ages", in *The Economic History Review*, 2nd Series, Vol. XXV, 1972

A. R. Bridbury,
Economic Growth - England in the Later Middle Ages, London 1962

A. R. Bridbury,
"The Black Death", in *The Economic History Review*, 2nd Series, Vol. XXVI, 1973

A. R. Bridbury,
"Before the Black Death", in *The Economic History Review*, 2nd Series, Vol.XXX,1977

R. H. Britnell,
"Agricultural Technology and the Margin of Cultivation in the Fourteenth Century", in *The Economic History Review*, 2nd Series, Vol. XXX, 1977

C. E. Britton,
A Meteorological Chronology to 1450, London 1937

C. N. L. Brooke & M. M. Postan,
Carte Nativorum, Oxford 1960

E. M. Carus-Wilson (ed.),
Essays in Economic History, Vol. II, London 1962

G. Chalmers,
An Estimate of Comparative Strength of Britain during the Present and Four Preceding Reigns, London 1772

J. D. Chambers,
Population, Economy and Society in Pre-Industrial England, London 1972

E. P. Cheyney,
"The Disappearance of English Serfdom", in *English Historical Review*, Vol. XV, 1900

J. H. Clapham (ed.),
The Cambridge Economic History of Europe, Vol. I, Cambridge 1942

J. H. Clapham,
A Concise Economic History of Britain, Cambridge 1949

A. M. Cooke,
"The Settlement of the Cistercians in England", in *English Historical Review*, No. 625, 1893

G. B. Coulton,
The Medieval Village, 1926

C. Creighton,
History of Epidemics in Britain - From A.D. 664 to the Extinction of the Plague, Vol. I, Cambridge 1891

C. Creighton,
History of Epidemics in Britain - From the Extinction of the Plague to the Present Time, Vol. II, 1894

C. Creighton,
A History of Epidemics in Britain, Vol. 1, 1965

W. Cunningham,
The Growth of English Industry and Commerce, 1890

F. Curschman,
Hungersnöte im Mittelalter, Leipzig 1900

W. H. R. Curtler,
A Short History of English Agriculture, 1909

H. C. Darby,
A Historical Geography of England before A.D. 1800, Cambridge 1936

H. C. Darby,
The Medieval Fenland, Cambridge 1940

F. G. Davenport,
"The Decay of Villeinage in East Anglia", in *Transactions of the Royal Historical Society*, New Series, Vol. XIV, 1900

W. Denton,
  England in the Fifteenth Century, 1888
A. Dimock,
  "The Great Pestilence", in *Gentleman's Magazine*, Vol. 283, 1897
F. R. H. Du Boulay,
  "A Rentier Economy in the Later Middle Ages: the Archbishopric of Canterbury", in
  The Economic History Review, 2nd Series, Vol. XVI, 1963-64
F. R. H. Du Boulay,
  "Who were farming the English Demesnes at the End of the Middle Ages?", in *The Economic History Review*, 2nd Series, Vol. XVII, 1964-65
F. R. H. Du Boulay,
  The Lordship of Canterbury: An Essay on Medieval Society, London 1966
F. R. H. Du Boulay,
  An Age of Ambition - English Society in the Later Middle Ages, London 1970
Georges Duby,
  Rural Economy and Country Life in the Medieval West, Columbia 1976
M. Dobb,
  Studies in the Development of Capitalism, London 1947
R. B. Dobson,
  The Peasant Revolt of 1381, London 1970
C. Dyer,
  "A Redistribution of Incomes in Fifteenth-Century England?", in *Past and Present*, No. 39, 1968
C. Dyer,
  Lords and Peasants in a Changing Society - The Estates of the Bishopric of Worcester, 680 - 1540, Cambridge 1980
R. Ernle,
  Untitled article in *Quarterly Review*, April 1885
Lord Ernle,
  English Farming Past and Present, 1912
Lord Ernle,
  "The Enclosures of Open-Field Farms", in *Journal of the Ministry of Agriculture*, Dec. 1920 (I), Jan. 1921 (II)
R. J. Faith,
  "Peasant Families and Inheritance Customs in England", in *The Agricultural History Review*, Vol. XIV, 1966

D. L. Farmer,
"Some Livestock Price Movements in Thirteenth-Century England", in *The Economic History Review*, 2nd Series, Vol. XXII, 1969

D. L. Farmer,
"Grain Yields on the Winchester Manors in the Later Middle Ages", in *The Economic History Review*, 2nd Series, Vol. XXX, 1977

K. G. Feilings,
"An Essex Manor in the Fourteenth Century", in *English History Review*, Vol. XXVI, 1911

C. W. Foster & T. Longley,
"Lincolnshire Domesday", in *Lincolnshire Record Society*, XIX, 1924

G. E. Fussel,
"Social Change but Static Technology - Rural England in the Fourteenth Century", in *History Studies*, Vol. 2, 1968

F. A. Gasquet,
The Great Pestilence, 1893

F. A. Gasquet,
The Black Death, 1908

E. F. Gay,
"Inquisitions of Depopulation in 1517", in *Transactions of the Royal Historical Society*, Vol. XIV, 1900

E. F. Gay,
"Inclosures in England", in *Quarterly Journal of Economics*, Vol. XVII, 1903

E. F. Gay,
"The Midland Revolt and the Inquisition of Depopulation of 1607", in *Transactions of the Royal Historical Society*, Vol. XVIII, 1904

Gibbins,
Industry in England, New York 1897

R. Gneist,
Englische Verfassungsgeschichte, Berlin 1882

E. C. K. Gonner,
Common Land and Inclosure, 1912

Jack Goode et al. (eds.),
Family and Inheritance - Rural Society in Western Europe, 1200 - 1800, Cambridge 1976

R. S. Gottfried,
Epidemic Diseases in Fifteenth Century England, New Brunswick 1978

R. S. Gottfried,
The Black Death, New York 1983

N. S. B. Gras,
The Evolution of the English Corn Market, New York 1915

N. S. B. & E. C. Gras,
The Economic and Social History of an English Village, 1930

J. Graunt,
Natural and Political Observations upon the Billes of Mortality, London 1662

F. Graus,
"Die Erste Krise des Feudalismus", in *Zeitschrift für Geschichtswissenschaft*, Vol. III, 1955

F. Graus,
"Das Spätmittelalter als Krisenzeit, in *Mediævalia Bohemica*, Supplementum 1, 1969

H. L. Gray,
"The Commutation of Villein Service in England before the Black Death", in *The English Historical Review*, Vol. XXIX, 1914

H. L. Gray,
A Study of English Field Systems, Cambridge, Massachusetts 1915

J. R. Green,
A Short History of the English People, 1876

H. J. Habakkuk,
"The Economic History of Modern Britain", in *Journal of Economic History*, Vol. XVIII, 1958

M. Hale,
The Primitive Origination of Mankind, London 1677

A. D. Hall,
The Book of the Rothamsted Experiments, London 1905

H. Hall (ed.),
Pipe Roll of Bishopric of Winchester, 1208-9, 1903

H. E. Hallam,
"Some Thirteenth-Century Censuses", in *The Economic History Review*, 2nd Series, Vol. X, 1958

H. E. Hallam,
"Population Density in the Medieval Fenland", in *The Economic History Review*, 2nd Series, Vol. XIV, 1961

A. Hamilton,
"The Registers of John Gynewell, Bishop of Lincoln, for the Years 1349 - 50", in *The Archaeological Journal*, Vol. 68, 1911

A. Hamilton,

"Pestilences of the Fourteenth Century in the Diocese of York", in *The Archaeological Journal*, Vol. 71, 1914

J. L. & Barbara Hammond,

The Village Labourer, 1911

B. Harvey,

"The Population Trend in England Between 1300 and 1348", in *Transactions of the Royal Historical Society*, 5th Series, Vol. 16, 1966

B. Harvey,

"The Leasing of the Abbot of Winchester's Demesnes in the Later Middle Ages", in *The Economic History Review*, 2nd Series, Vol. XXII, 1969

B. Harvey,

Westminster Abbey and its Estates in the Middle Ages, Oxford 1977

W. Hasbach,

A History of the English Agricultural Labourer, 1908

J. Hatcher,

"A Diversified Economy: Late Medieval Cornwall", in *The Economic History Review*, 2nd Series, Vol. XXII, 1969

J. Hatcher,

Plague, Population and the English Economy, 1348 - 1530, London 1977

J. F. C. Hecker,

The Epidemics of the Middle Ages, publ. by the Sydenham Society, 1894

R. H. Hilton,

The Economic Development of Some Leicester Estates in the 14th and 15th Centuries, Oxford 1947

R. H. Hilton,

"Peasant Movements in England before 1381", in *The Economic History Review*, 2nd Series, II, 1949

R. H. Hilton & H. Fagan,

The English Rising of 1381, London 1950

R. H. Hilton,

"Y Eut-Il une Crise Générale De La Féodalité?", in *Annales - E.S.C*, No. 1, 1951

R. H. Hilton,

Medieval Agrarian History, V. C. H., Leicestershire, II, London 1954

R. H. Hilton,

"A Study in the Pre-History of English Enclosures in the Fifteenth Century", in *Studi in Onore Armado Sapori*, Vol. I, Milan 1955

R. H. Hilton,
"The Content and Sources of English Agrarian History before 1500", in *The Agricultural History Review*, Vol. III, Part I, 1955

R. H. Hilton,
"L'Angleterre Économique et Sociale des XIVe et XVe Siècles - Théories et Monographies", in *Annales - E. S. C.*, Vol. 13, 1958

R. H. Hilton,
"Rent and Capital Formation in Feudal Society", in *Second International Conference of Economic History*, Vol. II, Paris 1962

R. H. Hilton,
"Freedom and Villeinage in England", in *Past and Present*, No. 31, 1965

R. H. Hilton,
A Medieval Society - The West Midlands at the End of the Thirteenth Century, London 1966

R. H. Hilton,
The Decline of Serfdom in Medieval England, London 1969

R. H. Hilton,
Bond Man Made Free - Medieval Peasant Movements and the Rising of 1381, London 1973

R. H. Hilton,
The English Peasantry in the Later Middle Ages, Oxford 1975

R. H. Hilton (ed.),
Peasant, Knights and Heretics, Cambridge 1976

M. A. C. Hinton,
Rats and Mice as Enemies of Mankind, British Museum of Natural History Economic Series, No. 9, 3rd Edition, London 1931

L. F. Hirst,
The Conquest of Plague - A Study of the Evolution of Epidemiology, Oxford 1953

G. A. Holmes,
The Estates of the High Nobility in Fourteenth-Century England, Cambridge 1957

G. C. Homans,
English Villagers of the Thirteenth Century, Cambridge, Massachusetts, 1942

G. C. Homans,
"The Explanation of the English Regional Differences", in *Past and Present*, No. 42, 1969

W. G. Hoskins,
"The Deserted Villages of Leicestershire", in *Transactions of Leicestershire Archaeological Society for 1944 - 45*, Vol. XXII, 1946

W. G. Hoskins,
The Making of the English Landscape, 1955

P. R. Hyams,
"The Origin of a Peasant Land Market in England", in *The Economic History Review*, 2nd Series, Vol. XXIII, 1970

H. M. Hyndman,
The Historical Basis of Socialism in England, 1883

A. H. Inmann,
Domesday and Feudal Statistics, London 1900

E. F. Jacob,
"The Fifteenth Century, 1399 - 1485", Vol. 6 of *The Oxford History of England*, Oxford 1961

A. Jessopp,
"The Black Death in East Anglia", in *Nineteenth Century*, April 1885 and December 1884

A. H. Johnson,
The Disappearance of the Small Landowner, 1909

M. H. Keen,
England in the Later Middle Ages - A Political History, London 1973

Ian Kershaw,
"The Great Famine and Agrarian Crises in England 1315 - 1322", in *Past and Present*, No. 59, 1973

E. A. Kosminsky,
"The Hundred Rolls of 1279-80 as Source for English Agrarian History", in *The Economic History Review*, 1931

E. A. Kosminsky,
Studies in the History of the English Village, Moscow 1935

E. A. Kosminsky,
"Services and Money Rents in the Thirteenth Century", in *The Economic History Review*, Vol. V, 1935

E. A. Kosminsky,
"The Evolution of the Feudal Rent in England from the XIth to the XVth Centuries", in *Past and Present*, No. 7, 1955

E. A. Kosminsky,
Studies in the Agrarian History of England in the Thirteenth Century, Oxford 1956

J. Krause,
"The Medieval Household: Large or Small?", in *The Economic History Review*, 2nd Series, Vol. IX, 1957

Ingela Kyrre,
  "Teorier om Førkapitalistisk Samfundsdynamik", in *Kritiske Historikere*, no date or number
H. H. Lamb,
  "Britain's Changing Climate", in *Geographical Journal*, Vol. 133, Part 4, 1967
E. Lamond (ed.),
  Walter of Henley's Husbandry, London 1890
J. R. Lander,
  Conflict and Stability in Fifteenth-Century England, London 1969
R. Lennard,
  "The Alleged Exhaustion of the Soil in Medieval England", in *The Economic Journal*, Vol. XXXII, 1922
R. Lennard,
  "Statistics of Corn Yields in Medieval England", in *Economic History*, Vol. III, 1936
E. M. Leonard,
  "The Inclosure of Common Fields in the Seventeenth Century", in *Transactions of the Royal Historical Society*, New Series, 1905
A. E. Levett,
  "The Black Death on the Estates of the See of Winchester", in *Oxford Studies in Social and Legal History*, Oxford 1916
A. E. Levett,
  Studies in Manorial History, Oxford 1938
H. Levy,
  Entstehung und Rückgang des Landwirtschaftlichen Grossbetriebes in England, Berlin 1904
H. Levy,
  Large and Small Holdings, 1911
A. R. Lewis,
  "The Closing of the Medieval Frontier, 1250 - 1350", in *Speculum*, Vol. XXXIII, 1958
H. Lindkvist,
  Middle English Place-Names of Scandinavian Origin, Upsala 1912
E. Lipson,
  An Introduction to the Economic History of England, Vol. I, "The Middle Ages", London 1915
A. G. Little,
  "The Black Death in Lancashire", in *The English Historical Review*, Vol. V, 1890 and Vol. VI, 1891

H. S. Lucas,
The Great European Famine of 1315, 1316 and 1317, 1930

D. MacPherson,
Annals of Commerce, Vol. I, London 1805

J. R. Maddicott,
"The English Peasantry and the Demands of the Crown, 1294 - 1341", in *Past and Present*, Supplement I, 1975

F. W. Maitland,
"The History of a Cambridgeshire Manor", in *The English Historical Review*, Vol. XXXV, 1894

F. W. Maitland,
"The Material for English Legal History", in *Collected Papers*, Vol. II, 1911

K. Marx,
Das Kapital, MEW Vols. 23 and 25, Berlin 1977

M. Mate,
"The Indebtedness of Canterbury Cathedral Priory 1215 - 95", in *The Economic History Review*, 2nd Series, Vol. XXVI, 1973

M. Mate,
"High Prices in Early Fourteenth-Century England: Causes and Consequences", in *The Economic History Review*, 2nd Series, Vol. XXVIII, 1975

J. M. Mathews (ed.),
Records of Cardiff, Vol. I, 1898

A. N. May,
"An Index of Thirteenth-Century Peasant Impoverishment? Manor Court Fines", in *The Economic History Review*, 2nd Series, Vol. XXVI, 1973

N. J. Mayhew,
"Numismatic Evidence and Falling Prices in the Fourteenth Century", in *The Economic History Review*, 2nd Series, Vol. XXVII, 1974

M. McKisack,
"The Fourteenth Century 1307 - 1399", Vol. 5 of *The Oxford History of England*, Oxford 1959

E. Miller,
"The English Economy in the Thirteenth Century: Implications of Recent Research", in *Past and Present*, No. 28, 1964

E. Miller,
"England in the Twelfth and Thirteenth Centuries: An Economic Contrast?", in *The Economic History Review*, 2nd Series, Vol. XIV, 1971

E. Miller & J. Hatcher,
Medieval England - Rural Society and Economic Change 1086 - 1348, London 1978

P. G. Mode,
The Influence of the Black Death upon English Monasteries, Chicago 1916

J. R. H. Moorman,
Church Life in England in the Thirteenth Century, Cambridge 1955

C. Morris,
"The Plague in Britain - Review of Shrewsbury's "A History of Bubonic Plague in the British Isles"", in *The Historical Journal*, Vol. XIV, 1971

E. Nasse,
Über die Mittelalterliche Feldgemeinschaft und die Einhegungen des Sechszehnten Jahrhunderts in England, 1869

E. Nasse,
The Land Community of the Middle Ages, 1871

N. Neilson,
Economic Conditions on the Manor of Ramsay Abbey, 1898

W. v. Ochenkowski,
Englands wirtschaftliche Entwicklung im Ausgang des Mittelalters, 1879

C. Oman,
The Political History of England, Vol. IV, 1906

C. S. & C. S. Orwin,
The Open Fields, Oxford 1938

F. M. Page,
The Estates of Crowland Abbey, 1934

T. W. Page,
Die Umwandlung der Frohdienste in Geldrenten, Baltimore 1897

T. W. Page,
"The End of Villeinage in England", in *American Economic Association*, 3rd Series, Vol. I, New York 1900

C. H. Pearson,
English History in the Fourteenth Century, 1876

H. Pirenne,
Economic and Social History of Medieval Europe, London 1936

D. S. Pitkin,
"Partible Inheritance and the Open Fields", in *Agricultural History*, Vol. 35, 1961

A. F. Pollard,
Factors in Modern History, 1910

318

A. J. Pollard,
"Estate Management in the Later Middle Ages: The Talbot and Whitchurch, 1383 - 1525", in *The Economic History Review*, 2nd Series, Vol. XXV, 1972

J. J. Pontanus,
Rerum Danicarum Historia, 1631

A. L. Poole (ed.),
Medieval England, Vol. I, 1958

M. M. Postan,
"The Chronology of Labour Services", in *Transactions of the Royal Historical Society*, 4th Series, Vol. XX, 1937

M. M. Postan,
"Revisions in Economic History - The Fifteenth Century", in *The Economic History Review*, Vol. IX, 1938-39

M. M. Postan,
"Some Social Consequences of the Hundred Years' War", in *The Economic History Review*, Vol. XII, 1942

M. M. Postan,
"The Rise of a Money Economy", in *The Economic History Review*, Vol. XIV, 1944

M. M. Postan,
"Moyen Age", in *Rapports du IXe Congrès International Des Sciences Historiques*, Paris 1950

M. M. Postan,
"Some Economic Evidence of Declining Population in the Later Middle Ages", in *The Economic History Review*, 2nd Series, Vol. II, 1950

M. M. Postan & E. E. Rich (eds.),
The Cambridge Economic History of Europe, Vol. II, Cambridge 1952

M. M. Postan,
"The Famulus, the Estate Labourer in the XIIth and XIIIth Centuries", in *The Economic History Review*, Supplement 2, London 1954

M. M. Postan & J. Titow,
"Heriots and Prices on Winchester Manors", in *The Economic History Review*, 2nd Series, Vol. XI, 1959

M. M. Postan, "Village Livestock in the Thirteenth Century",
in *The Economic History Review*, 2nd Series, Vol. XV, 1962

M. M. Postan,
"The Costs of the Hundred Years' War", in *Past and Present*, No. 27, 1964

M. M. Postan (ed.),
The Cambridge Economic History of Europe, Vol. 1, 2nd Edition, Cambridge 1966

319

M. M. Postan,
"Investment in Medieval Agriculture", in *The Journal of Economic History*, Vol. 27, 1967

E. Powell,
The Rising in East Anglia, 1896

E. E. Power,
"The Effects of the Black Death on Rural Organization in England", in *History*, Vol. III, 1918

E. Power & M. M. Postan (eds.),
Studies in English Trade in the Fifteenth Century, London 1941

E. Power,
The Wool Trade in English Medieval History, London 1941

M. Powicke,
"The Thirteenth Century, 1216 - 1307", Vol. 4 of *The Oxford History of England*, Oxford 1953

B. H. Putnam,
The Enforcement of the Statutes of Labourers, New York 1908

J. A. Raftis,
The Estates of Ramsay Abbey - A Study of Economic Growth and Organisation, Toronto 1957

J. A. Raftis,
Tenure and Mobility - Studies in the Social History of the Medieval English Village, Toronto 1964

W. Rees,
"The Black Death in England and Wales, as exhibited in Manorial Documents", in *Proceedings of the Royal Society of Medicine*, Vol. 16, 1923

D. Ricardo,
Principper for den politiske ¢konomi og Beskatningen, Copenhagen 1978

H. Robbins,
"A Comparison of the Effects of the Black Death on the Economic Organization of France and England", in *Journal of Political Economy*, Vol. XXXVI, Chicago 1928

B. K. Roberts,
"A Study of Medieval Colonization in the Forest of Arden, Warwickshire", in *The Agricultural History Review*, Vol. 16, 1968

R. S. Roberts,
"The Place of Plague in English History", in *Proceedings of the Royal Society of Medicine*, Vol. 59, 1966

W. C. Robinson,

"Money, Population and Economic Change in Late Medieval Europe", in *The Economic History Review*, 2nd Series, Vol. XII, 1959

E. Robo,

"The Black Death in the Hundred of Farnham", in *The English Historical Review*, Vol. XLIV, 1929

D. Roden,

"Demesne Farming in the Chiltern Hills", in *The Agricultural History Review*, Vol. 17, 1969

J. E. T. Rogers,

A History of Agriculture and Prices in England, Vol. I - IV, 1866 - 1881

J. E. T. Rogers,

"England Before and After the Black Death", in *Fortnightly Review*, 1866

J. E. T. Rogers,

Six Centuries of Work and Wages, Vol. I, 1884

J. E. T. Rogers,

The Economic Interpretation of History, 1888

E. J. Russell,

The Fertility of the Soil, Cambridge 1913

J. C. Russell,

"Clerical Population of Medieval England", in *Tradition*, No. 2, 1944

J. C. Russell,

British Medieval Population, Albuquerque, New Mexico 1948

J. C. Russell,

Late Ancient and Medieval Population, Philadelphia 1958

J. C. Russell,

"Recent Advances in Medieval Demography", in *Speculum*, No. 40, Cambridge, Massachusetts 1965

J. C. Russell,

"Effects of Pestilence and Plague, 1315 - 1385", in *Comparative Studies in Society and History*, Vol. VIII, The Hague 1965

J. C. Russell,

"The Preplague Population of England", in *The Journal of British Studies*, Vol. V, No. 2, Hartford, Connecticut 1966

J. Saltmarsh,

"Plague and Economic Decline in England in the Later Middle Ages", in *Cambridge Historical Journal*, Vol. VII, 1941

H. W. Saunders,
  Introduction to the Obedientary and Manor Rolls of Norwich Cathedral, 1930
R. S. Schofield,
  "The Geographical Distribution of Wealth in England, 1334 - 1649", in *The Economic History Review*, 2nd Series, Vol. XVIII, 1965
J. Schreiner,
  Pest og Prisfall i Senmiddelalderen, Oslo 1948
J. Schreiner,
  "Wages and Prices in England in the Later Middle Ages", in *Scandinavian Economic History Review*, Vol. II, 1954
T. E. Scrutton,
  Commons and Common Fields, or the History and Policy of the Laws Relating to Commons and Enclosures in England, Cambridge 1885
E. Searle,
  Lordship and Community - Battle Abbey and its Banlieu, 1066 - 1538, Toronto 1974
F. Seebohm,
  "The Black Death and its Place in English History", I and II, in *The Fortnightly Review*, 1865
F. Seebohm,
  "The Population of England Before the Black Death", in *The Fortnightly Review*, 1866
F. Seebohm,
  "The Land Question - Part II - Feudal Tenures in England", in *The Fortnightly Review*, 1870
F. Seebohm,
  "The Rise in the Value of Sliver between 1300 and 1500", in *The Archaeological Review*, 1889
F. Seebohm,
  The English Village Community, 1926
M. E. Seebohm,
  The Evolution of the English Farm, London 1927
J. F. D. Shrewsbury,
  A History of Bubonic Plague in the British Isles, Cambridge 1970
V. G. Simkhovitch,
  "Hay and History", in *Political Science Quarterly*, Vol. XXVII, 1913
S. B. J. Skertchly,
  The Geology of the Fenland, 1877
G. Slater,
  The English Peasantry and the Enclosure of Common Fields, London 1907

B. H. Slicher van Bath,

The Agrarian History of Western Europe, New York 1963

R. A. L. Smith,

Canterbury Cathedral Priory, Cambridge 1943

A. Steensberg,

"Archaeological Dating of the Climatic Changes in North Europe About A.D. 1300",
in *Natura*, Vol. 168, 1951

J. Steenstrup,

Historieskrivningen i det nittende Aarhundrede, Copenhagen 1921

G. F. Steffen,

Studien zur Geschichte der Englischen Lohnarbeiter, Vol. I, Stuttgart 1901

F. M. Stenton,

Facsimiles of Early Charters from Northamptonshire Collections, 1930

F. M. Stenton,

Documents Illustrative of the Social and Economic History of the Danelaw, 1920

W. Stubbs,

The Constitutional History of England, Vol. II, 1875

R. H. Tawney,

The Agrarian Problem in the Sixteenth Century, 1912

J. Thirsk,

"The Common Fields", in *Past and Present*, No. 29, 1964

S. L. Thrupp,

"Plague Effects in Medieval Europe - a comment on J. C. Russell's views", in
Comparative Studies in Society and History, Vol. VIII, The Hague, 1965

S. L. Thrupp,

"The Problem of Replacement-Rates in Late Medieval English Population", in *The
Economic History Review*, 2nd Series, Vol. XVIII, 1965

J. Z. Titow,

"Some Evidence of the Thirteenth Century Population Increase", in *The Economic
History Review*, 2nd Series, Vol. 14, 1961-62

J. Z. Titow,

"Some Differences between Manors and their Effects on the Condition of the Peasant
in the Thirteenth Century", in *The Agricultural History Review*, Vol. X, 1962

J. Z. Titow,

"Medieval England and the Open-Field System", in *Past and Present*, No. 32, 1965

J. Z. Titow,

English Rural Society, 1200 - 1350, London 1969

J. Z. Titow,
   Winchester Yields. A Study in Medieval Agricultural Productivity, Cambridge 1972
T. F. Tout, *The Political History of England*, Vol. III, 1905
R. Trow-Smith,
   A History of British Husbandry to 1700, London 1957
P. Vinogradoff,
   Villainage in England, 1892
P. Vinogradoff,
   "Agricultural Services", in *Economic Journal*, Vol. X, 1900
P. Vinogradoff,
   "Review of "The End of Villainage in England" by T. W. Page", in *The English Historical Review*, Vol. XV, 1900
P. Vinogradoff,
   The Growth of the Manor, 1905
D. G. Watts,
   "A Model for the Early Fourteenth Century", in *The Economic History Review*, 2nd Series, Vol. XX, 1967
J. M. Winter (ed.),
   War and Economic Development, Cambridge 1975
Philip Ziegler,
   The Black Death, London 1969

Index

account rolls *11, 17, 240*
Abel, W *138, 180*
afforestation *162, 167*
agrarian: productivity *190, 195, 196*; revolution *45, 46, 53*; structural reorganisation *111, 178, 191*
agricultural: boom *275*; distress *12, 17, 18*; labour *205, 208*; labourers *5, 17, 46, 52, 56, 110, 117, 141, 142, 198, 270*; surplus production *4*; techniques *48-50, 54, 58, 223, 226*; technology *4, 26-28 , 54, 56, 57, 88, 95, 112, 114, 119-121, 162, 165, 167, 197, 208, 224, 225*; wage labour *114, 115*; workers *56*; yield *118*
animal farming *186, 228, 231, 233*
arable areas *163, 166, 224, 227*; farming *30, 48, 54, 58, 76, 88, 90, 126, 166, 189*; land *89, 120, 140, 152, 231*
Arden, Forest of *229*
Ashley, W.J. *42, 45-49, 52, 56, 58-60, 75, 87-89*
assarts *164, 166, 229*
Avesbury, chronicler *74*

Baker, R.H. Alan *231*
Ballard, A *111, 113, 122, 123, 155*
barter economy *27, 28, 79, 111*
Battle Abbey, Sussex *224, 225*
Bazeley. Margaret Ley *167*
Bean, J.M.W. *272, 273*
Bec, monastry *242*
Bedfordshire *231*
Bennett, M.K. *152, 196, 230, 231*
Beresford, Maurice *187-189*
Berkeley *240*
Berkshire *111*
Beveridge, Lord *128, 129, 152, 181, 196, 224, 225, 230*
birth rate *160*

Bishop, T.A.M. *162-164, 166*
Black Death *1-5, 7-14, 16, 19, 20, 22, 23, 29-31, 41, 43, 45, 47, 49-55, 68, 70-77, 79-86, 105, 106, 109, 111, 112, 114-118, 122, 123, 129, 140, 141, 151, 155, 158, 160, 161, 164, 178, 179, 183-185, 188, 189, 191-194, 196, 198, 200, 218, 219, 224, 227, 244, 256-258, 261-263, 268, 269, 272-274*
black rat *273*
Bloch, Marc *138, 180*
boom period *42, 245, 260*
boon-days *83, 84, 203*
boon-services *109*
Bradley, Harriet *121-124, 126, 230*
Brandon, P.F. *224, 225*
Bridbury, Anthony *217, 218, 223-227, 238, 245, 260, 261, 272, 275*
Brightwell, Winchester manor *111, 113*
Bristol, city of *3*
Britnell, R.H. *223, 224, 226, 231, 234*
brown rat *274*
bubo-plague *70*
bobunic plague *155, 273*
Buckinghamshire *85, 187, 231*

Calais *79*
Cambridgeshire *79, 85, 162, 264*
Canterbury *257*
Canterbury, Archbishop of *219, 221, 222, 236, 237*
capitalist agriculture *143, 190* capitalist demesne production *196*
cash nexus *50, 56, 108*
Cattle: breeding *165, 186*; farming *162, 166, 189, 230, 262*
Celtic field system *93-95*

325

21, 29, 43, 57, 81-83, 109, 150, 151, 156, 158, 159, 161, 164, 165, 197, 240, 244, 245; growth rate 158, 159, 265; increase 119, 153, 154, 164, 167, 192, 194, 229; losses 1, 9, 14, 30, 41, 158, 189, 192, 217; pressure 235, 246; reduction 95, 269, 275; rise 158, 193, 201, 217, 244, 260, 267-269; surplus 82, 83, 117, 189

Postan, M.M. 139-146, 149, 153, 154, 155, 163, 179-185, 187, 188, 190-194, 196, 197, 201, 202, 205, 207, 208, 218, 219, 221, 222, 224, 226, 227, 229-231, 237, 238, 241, 242, 245-247, 256-258, 265, 267, 271, 275

poverty 123, 124, 231

Power, E.E 114.115

Power, Eileen 140

predial services 80

previous accummulation 7, 9, 48

price, in general 24, 30, 51, 52, 144, 181, 183, 190, 195, 259, 260, 263, 256, 261, 269

price: of cattle 256; of corn 12, 89, 106, 191-194, 208, 261, 272; decline 145; drop 256, 259; fall 106, 144, 191; formation 256, 257; of labour 17, 81; of lease 222; of manufactured goods 106; rise 12, 76, 83, 109, 115, 117, 144, 203, 204, 228, 235, 242, 256, 260; of silver 191-193; stagnating 142, 144; of wheat 122; of wool 89, 121, 122

product rent 46, 204

productivity 10, 15, 123, 152, 208, 209, 217, 218, 223, 224, 226-229, 232, 234, 237-239, 246, 257, 260

proletariat 69, 208

quit-rents 18

Ranke, L. von 138

ratio: animal/arable 235; animal farming/area sown 225; land/manpower 4, 30; land/population 157, 218, 241, 244; tillage/stock breeding 26

reciprocal appropriation 17

reclamation 163-165

Ree, William 116

rent, in general 5-7, 18, 21, 45, 50, 54, 69, 77, 93, 110, 142, 145, 181, 195, 199, 200, 259, 263, 270, 271

rent: arbitrary 11, 108; battle of 244; declining 220, 221; falling 76, 106, 107, 116, 140, 142, 182, 206; fixed 26, 84, 182; increase 270; rising 149, 150, 182, 200-202, 228, 235, 236, 241, 244; unpaid 182

resistance, of peasants 202

revolution 4, 5, 10, 12, 15, 27, 71, 82, 89, 107, 108, 115, 121, 138, 144

revolutionary, effect of the plague 8, 10, 12, 13

revolutionary, significance of the plague 105

Ricardo, David 16

Richard II 50

Richard III 50

Robbins, Helen 115, 116

Roberts, B.K. 229

Roberts, R.S. 274

Robinson, W.C. 192-194, 256

Robo, E 111, 112, 155

Roden, David 228

Rogers, James E. Thorold 9-24, 26-28, 30, 41, 42, 46, 47, 50-52, 55, 58, 59, 68, 74, 76, 79, 80, 82, 86, 89, 105, 110, 111, 114-117, 122, 138, 142, 148, 152, 181, 200, 227, 246, 263

rotation system 224, 225

Rothamsted, Hertfordshire 126, 127, 231-233

runrig 92

rural labourers 69, 74, 76, 164, 205

Russell, J.C. 157-161, 166, 178, 184, 192, 263, 267-272

Saltmarsh, John 154-156, 161, 187, 229, 271, 272

scarcity: of grassland 228; of labour 56; of land 188, 217, 228, 260; of seed 231

Schreiner, J. 180, 191-194, 256

Schofield, R.S. 221

Scottish War 43

Medieval England

1. Durham
2. Northumberland
3. Redesdale
4. Tynedale
5. Hexhamshire
6. Cumberland
7. Sadberge
8. Lordship of Furness
9. Westmorland
10. County Palatine of Lancaster
11. Yorkshire
12. County Palatine of Flint
13. County Palatine of Chester
14. Derby
15. Nottingham
16. Lincoln
17. Shropshire
18. Stafford
19. Leicester
20. Rutland
21. Norfolk
22. Worcester
23. Warwick
24. Northampton
25. Huntingdon
26. Isle of Ely
27. Hereford
28. Bedford
29. Cambridge
30. Suffolk
31. Gloucester
32. Oxford
33. Buckingham
34. Hertford
35. Essex
36. Somerset
37. Wiltshire
38. Berkshire
39. Surrey
40. Cornwall
41. Devon
42. Dorset
43. Hampshire
44. Sussex
45. Kent